Logging Railroads of the Adirondacks

LOGGING RAILROADS OF THE ADIRONDACKS

BILL GOVE

SYRACUSE UNIVERSITY PRESS

First Edition 2006
19 20 21 22 23   7 6 5 4 3

Publication of this volume is made possible in part by a gift made in memory of Dr. Wilmer H. Randel, Jr.

Frontispiece: Spruce and white pine trees frame the newly laid railroad track twisting through the timber land of the Rich Lumber Company. Circa 1904.

All photographs courtesy of the author unless otherwise indicated.

∞ The paper used in this publication meets the minimum requirements of the American National Standard for Information Sciences—Permanence of Paper for Printed Library Materials, ANSI Z39.48-1992.

For a listing of books published and distributed by Syracuse University Press, visit https://press.syr.edu.

ISBN: 978-0-8156-0794-6 (hardcover)

Library of Congress Cataloging-in-Publication Data

Gove, Bill.
Logging railroads of the Adirondacks / Bill Gove.— 1st ed.
p. cm.
Includes bibliographical references and index.
ISBN 0–8156–0794–6 (alk. paper)
1. Logging railroads—New York (State)—Adirondack Mountains—History.
2. Steam locomotives—New York (State)—Adirondack Mountains—History.
I. Title.
TF25.A34G68 2005
385.5'4097475—dc22   2005028867

Manufactured in the United States of America

Bill Gove is a retired forester with a deep interest in railroading and lumbering history. He is the author of numerous magazine articles and has authored or co-authored seven books, including *Sky Route to the Quarries, Log Drives on the Connecticut River, J. E. Henry's Logging Railroads*, and *Logging Railroads of the Saco River Valley.*

Contents

Illustrations vii

Tables xi

Maps xiii

Introduction xv

Part One: Railroads in the Woods

1. Steam Enters the Timber 3
2. Early Rails in the Adirondacks 7
3. Gandy Dancers and Light Rail: The Construction of Logging Railroads 10
4. Logging Locomotives and Rolling Stock 16
5. Rails Through the Timber: Operations and Mishaps 25
6. Company Villages and Common Carriers 30
7. Logging the Old-Growth Timber 37
8. Forest Fires and the Railroads 54

Part Two: Individual Logging Railroads

SECTION ONE: *Logging Railroads That Connected with the New York Central and Hudson River Railroad, Adirondack Division* 61

9. Adirondack and St. Lawrence Railroad, 1890–1982; Mohawk and Malone Railroad, 1892–1893; New York Central and Hudson River Railroad, Adirondack Division, 1893–1972 65
10. Moose River Lumber Company Railroad 72
11. John A. Dix Railroad Logging Operation and the Raquette Lake Railway 76
12. Woods Lake Railroad 80
13. Mac-A-Mac Railroad 87
14. Whitney Industries Railroad 96
15. Partlow Lake Railroad 99
16. Horse Shoe Forestry Company Railroad 101
17. The Emporium Forestry Company and the Grasse River Railroad 106
18. Paul Smith's Electric Railway 138
19. Kinsley Lumber Company Railroad 142

SECTION TWO: *Logging Railroads That Connected with the Northern Adirondack Railroad* 145

20. Northern Adirondack Railroad 147
21. St. Regis Falls and Everton Railroad 162
22. Watson Page Lumber Company Railroad 165
23. Brooklyn Cooperage Company Railroad 169
24. Bay Pond Railroad 182
25. Oval Wood Dish Corporation Railroad 189
26. Sisson-White Company Railroad 202

SECTION THREE: *Logging Railroads That Connected with the New York Central and Hudson River Railroad, Carthage and Adirondack Branch* 203

27. Mecca Lumber Company Railroad 205

28. Newton Falls Paper Company Railroad 207

29. Post and Henderson Logging Railroad 212

30. Newton Falls and Northern Railroad 215

31. Rich Lumber Company Railroad and Cranberry Lake Railroad 220

32. Jerseyfield Lumber Company Railroad 237

Epilogue: The Slow Bell 242

Glossary 249

Bibliography 253

Index 255

Illustrations

1. Log drive on an Adirondack river 3
2. An old-growth white pine tree 4
3. Harvesting old-growth spruce 4
4. Sled loaded with spruce logs 5
5. Newly laid railroad track 6
6. Horse-powered pole railroad 8
7. Old-growth hard maple tree 9
8. Hardwood logs piled along track 9
9. Removing earth on roadbed construction 10
10. Railroad roadbed construction 11
11. Railroad ties laid out in new construction 11
12. Crew laying new rail 12
13. Construction crew with transit 12
14. Heisler locomotive laying new track 12
15. Construction crew displaying tools 12
16. Sawmill waste used in fill 13
17. Section crew tamping and leveling rail 13
18. Long trestle before being filled in 14
19. Railroad trestle bent design 14
20. Steam shovel in a burrow pit 15
21. Shay No. 39, Emporium Forestry Company 16
22. Baldwin No. 33, Emporium Forestry Company 17
23. Mogul No. 68, Emporium Forestry Company 18
24. Shay No. 4, Rich Lumber Company 19
25. Diagram of Shay trucks and gears 19
26. Climax No. 5, Emporium Forestry Company 20
27. Diagram of Climax truck 20
28. Heisler, Mac-A-Mac Corporation 21
29. Locomotive ads from *American Lumberman* 22
30. Skeleton log cars 22
31. Pulpwood rack cars 22
32. Frank Cutting's letterhead 23
33. Barnhart log loader 23
34. American Hoist and Derrick Company log loader 23
35. Interior of Barnhart loader 24
36. Log loader at work 24
37. Raymond gasoline log loader 24
38. Loading logs by hand 24
39. Logs stacked railside in winter 25
40. Jackworks in operation 26–27
41. Unusual log-loading deck 28
42. Explosion of boiler on a Shay 28
43. Wreck of a Heisler 28
44. General store at Wanakena 31
45. Brandreth village 32
46. Interior of store at Brandreth 32
47. Sash gang saw 32
48. Display of saw types in filing room 33
49. Interior of band sawmill 33
50. Locomotive taking water from brook 34
51. Depot at Conifer 35
52. Rail bus at Childwold depot 35
53. Wanakena depot, passengers arriving 36
54. Logging with crosscut saw 37
55. Large sled load of pine logs 38
56. Logging camp 39
57. Older logging camp 39
58. Interior, logging camp bunk room 40
59. Interior, logging camp mess hall 42
60. Loggers with tools for felling trees 42
61. Skid horse at a skidway 43
62. Scaling a deck of logs 43
63. Harvesting hemlock bark 43
64. Bark piles along railroad 44

65. Loading bark in a boxcar 44
66. Logs and crew on skidways 45
67. Loaded sleds on sled road 46
68. Sprinkler sled for icing sled roads 46
69. Hay spread in sled ruts on hill 46
70. Barienger brake for use on steep hills 47
71. Frame for loading tall sled loads 47
72. Loaded sled with spring-pole binding 48
73. Record sled load of spruce 49
74. Huge sled load of pulpwood 49
75. Log slide 49
76. Log slide patent drawing 50
77. Log slide 50
78. Water-trough slide for pulpwood 50
79. Crib work for high chute 51
80. Snubbing drum for steep slopes 51
81. Linn tractor 52
82. Holt tractor with sleds 53
83. Map of 1903 forest fires 55
84. Fire destruction at Sabattis 56
85. New York Central fire train 57
86. New York Central testing of fire equipment 57
87. New York Central firefighting crew 57
88. Nehasane Park fire train 57
89. New York Central letterhead 65
90. William Seward Webb 66
91. Adirondack and St. Lawrence locomotive 67
92. Mohawk and Malone locomotive 67
93. Adirondack and St. Lawrence Line 1893 timetable 68
94. Adirondack and St. Lawrence train loading logs 69
95. Mountain View station 69
96. Childwold station 70
97. Sawmill at Beaver Falls 72
98. Moose River Lumber Company sawmill 73
99. Hardwood logs piled trackside 75
100. Loading log train 75
101. General store at Carter 76
102. Raquette Lake Railway locomotive 77
103. Depot at Raquette Lake 79
104. International Paper Company letterhead 80
105. Champlain Realty Company letterhead 82
106. Large sled load of spruce 82
107. Holt tractor with trainload of sleds 83
108. Loading tractor sleds 84
109. Woods Lake station 85
110. Flag stop at Brandreth 87
111. First depot at Brandreth 87
112. Carlson's logging camp 88–89
113. Skidways filled with logs 89
114. Dimelow's log landing 89
115. Carlson's Landing on North Pond 92–93
116. Jackworks 93
117. Heisler with loaded cars 93
118. Heisler and loaded cars at Thayer Lake 94–95
119. New York Central locomotive with log train at Brandreth 94
120. Baldwin No. 33 at Whitney Park 97
121. River drivers at Bog River Falls 102
122. Locomotive on A. A. Low's dam 102
123. Locomotive Washington gathering sap 104
124. A. A. Low riding on flat car 104
125. Railroad track to sugar house 105
126. Mill at Conifer when first built 110–11
127. Diagram of machinery layout in sawmill 110
128. Conifer sawmill, empty log cars 111
129. Sawmill log carriage 111
130. Emporium store and office 112
131. Emporium Forestry Company letterhead 113
132. Emporium Forestry Company stock certificate 113
133. Emporium founders and officials 113
134. Locomotive No. 43 with log train 113
135. Shay No. 51 114
136. Pine Curve on the North Tram 116
137. Log train on cold day 118
138. William L. Sykes and train loading logs 118
139. Skeleton log cars 118
140. Climax No. 35 in yard 118
141. Shay with Barnhart loading train 119
142. Mogul No. 68 119
143. Baldwin No. 33 119
144. Carload of logs at mill pond dump 120
145. Cleaning switch points in Conifer yard 120

146. Grasse River Railroad dispatcher 120
147. Grasse River No. 2 in Conifer yard 120
148. Grasse River Railroad timetable 121
149. Grasse River Railroad timetable 122
150. Grasse River Railroad ticket 123
151. Grasse River No. 68 in yard 124
152. Grasse River Railroad promotion 125
153. Motor car No. 11 at Childwold 126
154. Loading mail at Childwold 127
155. Cranberry Lake sawmill 127
156. Aerial view of Cranberry Lake village 127
157. Interior of locomotive shop 129
158. Shay No. 40 129
159. Three railroad presidents 130
160. Railroad passes 130
161. Railcar No. 11 130
162. Railcar No. 12 130
163. Engine house fire remains 131
164. Portable log slide 131
165. Horses working on log slide 131
166. Proulx's logging camp 131
167. Barnhart loading a raft 133
168. Barnhart loader working at Cranberry Lake 133
169. Linn tractor with eight sleds 133
170. Emporium Forestry Company letter 134
171. Lumber scaler with wide maple boards 134
172. Conifer mill in 1955 136
173. Grasse River Railroad diesel locomotive 136
174. Diesel locomotive with log cars 136
175. Caboose, Perry's Pride 137
176. Cover of Grasse River Railroad annual report 137
177. Lake Clear Junction 140
178. Paul Smith's electric combine 140
179. Logs trackside at Paul Smith's 141
180. Electric combine with log cars 141
181. St. Regis Falls depot 148
182. Second St. Regis Falls depot 148
183. Track-laying crew 150
184. Locomotive No. 7, Northern Adirondack Railroad 150
185. Engine house at Santa Clara 151
186. Santa Clara depot 151
187. Interior of Santa Clara Lumber Company sawmill 152
188. Settlement of Derrick 152
189. Tupper Lake depot 156
190. Tupper Lake Junction 156
191. Northern Adirondack timetable 157
192. Charcoal kilns 157
193. Tupper Lake's Big Mill 158
194. New York and Ottawa Railroad letterhead 160
195. New York and Ottawa Railroad timetable 160
196. New York Central timetable 161
197. Lumbering settlement 161
198. St. Regis Falls and Everton Railroad locomotive 163
199. Advertised lumber prices 165
200. Watson Page electric locomotive 167
201. Load of lumber on horse-powered track 167
202. Watson Page electric locomotive 167
203. Brooklyn Cooperage Company letterheads 170
204. Pile of barrel staves 170
205. Finishing and bundling staves 170
206. Brooklyn Cooperage Company plant, St. Regis Falls 171
207. Workers from cooperage mill 172
208. Climax locomotive and Barnhart loader 172
209. Shay at woods landing 174
210. Climax No. 2 with train being loaded 175
211. Frank A. Cutting letterhead 175
212. Pulpwood debarking plant 177
213. Bernhard E. Fernow 177
214. Brooklyn Cooperage Company plant at Tupper Lake 178
215. Loading logs at Cross Clearing 179
216. Fernow's plantation today 180
217. Baldwin No. 33 185
218. Mogul No. 2 186
219. Loaded cars at log landing 186
220. Mogul with trainload 187
221. Oval Wood Dish Company letterhead 190
222. Oval Wood Dish Company factory, Tupper Lake 190
223. Kildare 191
224. Kildare school 192
225. Cutting blocks of ice at Kildare 192
226. Hermit at Kildare 192

227. Heisler at log landing 193
228. Heisler in woods 193
229. Loading logs on railroad 194
230. Raymond derrick crane 194
231. Men on speeder 196
232. Oval Wood Dish Camp 7 198
233. Cornell forestry students at Camp 7 198
234. Heisler with load of machinery 200
235. Two-pole log slide in use 200
236. Logging officials 201
237. Woods superintendent Roy LaVoy 201
238. Linn tractor 201
239. Unloading logs at Mecca sawmill 206
240. Newton Falls Paper Company letterhead 207
241. William N. Kellogg and family 208
242. Newton Falls Paper Company sawmill crew 208
243. Locomotive No. 1 at Aldrich 209
244. Loading spruce logs on railroad 209
245. Loading crew ride log 211
246. Benson Mines depot 212
247. Post and Henderson locomotive 214
248. Logging camp at Spruce Pond 216
249. Dining room at logging camp 216
250. Climax No. 1 216
251. Climax No. 1 217
252. Rail-mounted log skidder 217
253. Rich Lumber Company letterhead 221
254. Loading train on the Plains 222
255. Log train in reverse 224
256. Barnhart loader at work 224
257. Shays at the engine house at Wanakena 225
258. Panorama of mill complex at Wanakena 226–27
259. Sawmill at Wanakena 226
260. Rich Lumber Company officials 227
261. Lath-manufacturing crew 228
262. Piles of bundled laths 228
263. Whip-butt mill 228
264. Heading mill 228
265. Shoe-last mill 229
266. View of mill complex 229
267. Veneer mill crew 229
268. Cranberry Lake Railroad letterhead 229
269. Log train at Big Fill 230
270. Cranberry Lake Railroad passenger train 231
271. Speeder at Benson Mines depot 231
272. Railcar 231
273. Wanakena depot 232
274. Bissell's lumber loading yard 232
275. Claude Carlson and family 233
276. Log sleds and horses with frosty beards 233
277. Planting shrubs in front of general store 234
278. Boats at lake dock 234
279. Original buildings of Ranger School 235
280. R. M. Whitney Company letterhead 238
281. Wagon hubs 238
282. Brooklyn Cooperage Company plant, Salisbury Center 238
283. Marion log loader 239
284. Shay No. 1 239
285. Shay No. 6 239
286. Brooklyn Cooperage Company barrels 239
287. Merriman's railroad doodle bug 241
288. Row of logs and young growth along railroad 242
289. Shay on display, once owned by Horse Shoe Forestry Company 243
290. Former Mac-A-Mac roadbed 243
291. Large pine stump on the Plains 244
292. Present location of former Brandreth Station 244
293. Former flow ground for logs 244
294. New growth of timber 245

Tables

1. Timber Values in 1911 38
2. Logging Costs in 1911 47
3. Moose River Lumber Company Railroad 72
4. John A. Dix Railroad 76
5. Raquette Lake Railroad Common Carrier 77
6. Woods Lake Railroad 80
7. Mac-A-Mac Railroad 88
8. Whitney Industries Railroad 96
9. Partlow Lake Railroad 99
10. Horse Shoe Forestry Company Railroad 101
11. Emporium Forestry Company and Grasse River Railroad 107–9
12. Hourly Wages of Emporium Sawmill Employees in 1915 112
13. Approximate Cost of a New Locomotive in 1916 119
14. Emporium Forestry Company Costs 132
15. Paul Smith's Electric Railway 138
16. Kinsley Lumber Company Railroad 142
17. Northern Adirondack Railroad 147
18. Northern Adirondack Railroad Company Operations, 1889 157
19. St. Regis Falls and Everton Railroad 162
20. Watson Page Lumber Company Railroad 165
21. Brooklyn Cooperage Company Railroads 169
22. Bay Pond Railroad 182
23. Oval Wood Dish Corporation 189
24. Cost of Operations for a Log Train, 1919 196
25. Crew and Wages for a Twenty-Man Camp 199
26. Railroad Equipment and Construction Costs, 1919 199
27. Total Cost for Logs Delivered to Tupper Lake Mill, 1919 200
28. Sisson-White Company Railroad 202
29. Mecca Lumber Company Railroad 205
30. Newton Falls Paper Company Railroad 207
31. Post and Henderson Logging Railroad 212
32. Newton Falls and Northern Railroad 215
33. Rich Lumber Company Railroad and Cranberry Lake Railroad 220
34. Cranberry Lake Railroad 232
35. Jerseyfield Lumber Company Railroad 237

Maps

1. Main Line Railroads in Adirondacks 60
2. New York Central and Hudson River Railroad 62
3. New York Central and Hudson River Railroad 63
4. Moose River Lumber Company Railroad and McKeever Village 74
5. Raquette Lake Railway 78
6. Woods Lake Railroad and Woods Lake Depot 81
7. Mac-A-Mac Corporation Logging Railroad 90
8. Brandreth Station 91
9. Partlow Lake Railroad 100
10. A. A. Low's Railroad 103
11. Conifer Village 112
12. Childwold Station 114
13. Grasse River Railroad and Logging Branches 115
14. "North Tram" Railroad Line 117
15. Cranberry Lake Sawmill 128
16. Paul Smith's Electric Railway 139
17. Kinsley Lumber Company Railroad 143
18. Northern Adirondack Railroad 149
19. St. Regis Falls Village 150
20. Village of Santa Clara 153
21. Northern Adirondack Railroad 155
22. Tupper Lake track plan 159
23. St. Regis Falls and Everton Railroad 163
24. Watson Page Lumber Company Electric Railroad 166
25. St. Regis Falls Village 171
26. Brooklyn Cooperage Company Railroad, Meno and Lake Ozonia Branches 173
27. Brooklyn Cooperage Company Railroad, Tupper Lake Branch 179
28. Bay Pond Inc. Logging Railroad 184
29. Oval Wood Dish Company Kildare Railroad Camp 190
30. Oval Wood Dish Company Logging Railroad 195
31. Camp No. 7, Kildare 197
32. New York Central and Hudson Railroad, Carthage and Adirondack Branch 203
33. Mecca Lumber Company Railroad 206
34. Newton Falls Paper Company Railroad 210
35. Post and Henderson Company Logging Railroad 213
36. Newton Falls and Northern Railroad 218
37. Rich Lumber Company Railroads 223
38. Wanakena Village 225
39. Jerseyfield Lumber Company Railroad 240

Introduction

The era of steam power was synonymous with the robust era of the most intense timber harvest ever seen in the Adirondack forest. The harvest of timber accessible for river drives had been going on for decades, with always more big trees around the next bend of the river. But there was an end to it, after all; accessible timber became largely thinned out by the mid–nineteenth century. Vast areas were still beyond arm's reach; mechanization had yet to come to the woods.

Then came the railroads. From the viewpoint of the wood industries, the railroads appeared at just the appropriate time. Many thousands of acres of old-growth timber became available for an industry hungry for raw material. Changes were occurring in the industry's wood usage; the railroads were able to adapt and provide the services needed in the woods of the Adirondack Mountains.

For sixty years, the whistle of the logging locomotives echoed through the tall spruce and lofty maple of the North Country. To some it was a glorious era of a sweaty work ethic that one seldom encounters now; to others it was an era of devastation. Long before steam died into obscurity, the old-growth timber was practically gone. The Adirondack forest had changed character, and most of the large steam-powered industrial sites had been scrapped.

Steam offers romantic charm, as does logging. Together they present a historic portrait of grime and sweat and the power of a steam piston. The glory days of steam are many generations behind us now, but during its prime the logging railroads of the Adirondacks made a large impact on the timber resource.

PART ONE

RAILROADS IN THE WOODS

C H A P T E R O N E

Steam Enters the Timber

STEAM ENTERED THE ADIRONDACK TIMBER at a significant point in the history of the North Country. Changes had begun to occur in the political scene and on the industrial scene that were forcing some adjustments for the Adirondack wood industry. The lumber industry, along with tourism, was the backbone of the Adirondack economy, but the raw material supply had become a matter of concern. The introduction of logging railroads into the Adirondacks during this period of change produced a fascinating chapter in Adirondack history. The railroads played a role in shaping the forest that we have today.

Early efforts to harvest Adirondack timber had been confined, for the most part, to the waterways, the lakes, and the rivers. Timber cutting in the earliest years of the nineteenth century had consisted of small-scale incursions into the edges of the North Country wilderness. White pine was the principal species sought; as with other areas in the northern United States, the ponderous and abundant pine trees provided the preferred wood for construction. Then, in mid-century, the large sawmills became established on the main rivers draining the Adirondack region.

Log drives on the rivers were the only way at that time to move logs any distance of consequence. By 1850 lumberman had penetrated well into the Adirondacks' interior, following the courses of major rivers such as the Hudson, the Moose, and the Raquette. As early as 1806, the state had adopted a law declaring a river a "public highway" when the Salmon River was declared available for log rafts. Other rivers gradually came under legal obligations to be kept open and free from restriction for the movement of logs downriver to the sawmills.

Logs were boomed together on lakes and "warped" across to the outlet by means of a man-powered windlass mounted on a raft. By 1850 the loggers had uncontested use of the Adirondack water system. And the bulk of the logging was done within a reasonable sled-haul distance from water suitable for river driving.

The river driving of logs had begun in the Adirondacks on the upper Hudson River back in 1813 and had soon developed into a large yearly event. Large mills were built in concentration at such locations as Glens Falls on the Hudson River and Forestport on the Black River. The first gang mill was built in the 1840s at Fort Edward on the Hudson. Gang mills had a combination of many saws arranged to process a log rapidly; there might be as many as thirty-two upright saws held within a single sash or frame.

Timber also flowed northward. French Canadian loggers came down to harvest timber along the Raquette River and up as far as the mouth of the Raquette and Oswegatchie

1. Rivers were once the principal avenue of transportation for pine and spruce logs. River drivers are shown at work releasing a large log jam on one of the Adirondack's major rivers.

2. An old-growth white pine along the shore of Cranberry Lake, typical of the timber once found in the region. Circa 1928. Photograph by Ted DeGroff.

Rivers. Rafts of logs were sent down into the St. Lawrence River in the mid-1800s.

Only coniferous or softwood logs have sufficient floatation for river drives, of course, but hardwood had a very limited market at that time. All of the large sawmills around the edges of the Adirondacks were softwood consumers, producing construction and industrial-grade lumber from spruce, pine, and hemlock.

By the mid-1800s the preferred old-growth white pine was about exhausted in the accessible timber areas. Lumbermen then turned to an ever more abundant species, red spruce. Pine had been quite abundant in the eastern portion of the Adirondacks but was a minor component within the interior forest. Spruce frequently occupied from 30 to 50 percent of the stand volume in the interior; if there was only a scattering of spruce, lumbermen classed the stand as unacceptable.

Spruce has always proven to be quite versatile in site requirements. On moist flats, even swamplands, where most tree species could not survive, spruce often became the dominant tree. Hardwoods thrive best on the low ridges and moderate slopes with deeper soils, but spruce again dominated, along with balsam fir appearing on the upper slopes with shallow soil. The best spruce was found on the side hills. In old-growth stands, spruce almost attained the impressive size of the white pine, though a hundred-foot-tall spruce with a butt diameter of thirty-six inches was an exceptional tree. The favorable spruce sites had trees 350 years of age and sometimes older.

3. Harvesting an old-growth spruce forest.

The nineteenth-century Adirondack loggers cut only the larger spruce trees for floating down to the big mills. There were plenty of spruce trees to harvest, and the large sawmills with their multiple gang saws worked most efficiently with large logs. Waste in the milling process was excessive, but as far as the lumbermen were concerned there were plenty more spruce trees further up the river. Cutting only the larger spruce trees meant that a residual stand was left that continued to be well stocked with hardwood species and younger spruce. The loggers were cutting in an old-growth forest that was often multiaged, that is, with trees of all stages of maturity, though there were some even-aged stands resulting from fires or other catastrophies. The Adirondack forest was not cut heavily prior to about 1890.

By the year 1885, only 15 to 30 percent of the forest cover had been cut from slightly over one-third of the original Adirondack Park area. But the situation was not that rosy for the spruce lumber industry. The volume of spruce timber accessible to drivable streams was fast dwindling. The thirty thousand workers employed by the Adirondack lumber industry were threatened with layoff.

By the 1890s lumber companies in the vicinity of Tupper Lake village had to begin cutting second-growth timber. Accessible spruce timber had not kept the industry supplied for as many years as the loggers had envisioned. Changes had to come, and they did.

CHANGES COME TO THE ADIRONDACK FOREST INDUSTRY

Two series of events occurred in the late 1800s that served to change the character of the Adirondacks' principle industry, lumbering.

The first series of events was a combination of political movements. Because of the rapid advance of the lumberman through the North Country wilderness, the cry of "save the Adirondacks" was gaining in volume. The voices emanated not from the North Country residents, but from residents elsewhere who looked upon the Adirondacks as a vacation retreat of unsurpassed natural grandeur. The period 1875–1910 made up the gilded years of the Adirondacks; nineteenth-century tourism was at its height of

4. A fine sled load of spruce logs was chained and ready to move behind a single team. It may take a few sledge blows on the runners to free them from the ice in the ruts. Source: N.Y. Assembly Doc.

popularity and fashion, with the region's famed hotels, private camps, hunting lodges, and salty guides.

At the same time, state land ownership was rapidly increasing, from 17,000 to 500,000 acres in the period 1870–85, primarily through tax sales. But problems arose in obtaining clear titles and attempting to limit the excessive timber trespass that was occurring. Thus, in 1885 the state established the Forest Preserve, which maintained that the land would forever remain in state ownership. A May 20, 1892, legislative decree designated the North Country the Adirondack State Park, delineated on the maps by a blue line. The Park initially encompassed about 3.5 million acres, of which one quarter was state owned.

The final political act of the nineteenth century, which was to alter the character of the Adirondacks, occurred shortly afterward, in 1894. Recognizing that there was a hole in the Park legislation that could still allow the sale of standing timber, the citizens of New York State adopted a new provision in the state constitution. From 1894 onward, no trees were to be cut on any state land within the Blue Line for any purpose.

The second series of events to bring changes in the forest of the North Country was a combination of industrial events. Beginning in 1883, logging railroads began to penetrate into timber tracts heretofore considered inaccessible. Shortly afterward, the newly arising pulp and paper industry began to demand more and more spruce wood for the new pulping process. Small-diameter wood was acceptable, and loggers returned to the land previously logged for sawlogs and stripped the spruce. Then another development in the 1890s, a market for hardwood logs, opened up, creating more freight for the logging rails. Also, the toil of the lumberjack became shortened somewhat about that time with the introduction of the crosscut saw to the woods.

Thus, with the state's acquisition of more land into the Park and the establishment about this time of large private land preserves and with the growing industrial market for more and more wood of almost all species, loggers were cutting more timber from less land.

This harvesting of the Adirondack old-growth forests had its most significant impact during the period approximately 1890–1910. During this brief span there were about fifteen logging railroads active in moving logs or pulpwood to market. The spruce sawlog cut had peaked in 1882, but the growth of the spruce pulpwood market more than filled the void. The spruce pulpwood market had a 567 percent growth from 1890 to 1905. By the end of the century there were sixty-seven pulp mills within the Adirondacks or along the edge, obtaining much of their wood from the spruce forests of the Adirondacks.

5. Spruce and white pine trees frame the newly laid railroad track twisting through the timber land of the Rich Lumber Company. Circa 1904.

Thus, steam entered the Adirondack woods at a decisive time. There was a turning point in the character and treatment of the forest and a turning point in the decline and development of various wood markets. For the tourists, hunters, and vacationers from the urban areas seeking rest among primeval beauty, steam in the timber meant another threat for the Adirondacks. For the lumbermen and the residents of the region, it meant a boost to the industry and the local economy.

CHAPTER TWO

Early Rails in the Adirondacks

THE ADIRONDACK REGION became an interesting arena for the staging of a host of grandiose and hotly contested railroad schemes. Few of the proposed railroads progressed any further than wishful thinking, however; many a desk cubbyhole became stuffed with worthless railroad stock. If all of the railroads that were proposed to cross the North Country had been built, the area would have been a cobweb of rail lines.

Charters to lay rail across the Adirondacks came early in railroad development. Not long after the first railroad was chartered in New York State in 1826, a plan was proposed whereby the settlement of Little Falls would be connected by rail with the headwaters of the Raquette River. That was in 1834. And like other dreams and schemes to follow over the next few years, the idea was to link this new concept of railroads with existing water routes and newly constructed canals in order to reach the markets of Canada or the population of the Great Lakes. Chartered as the Mohawk and St. Lawrence Rail Road and Navigation, the scheme never got off the paper.

Other proposals came forth over the next half century; charters were even granted to lay rail completely across the vast wilderness forest expanse. But the Adirondacks remained impenetrable by rail—except, that is, for a short incursion by Thomas C. Durant in 1871 with his Adirondack Railroad, which extended along the Hudson River only as far north as North Creek. And there were also the extensions that the Delaware and Hudson Railroad built into Ausable Forks in 1874 and into Saranac Lake in 1887. However, there were still no rails into the heart of the Adirondack forests.

By the late 1880s, the attraction of the timber resources could no longer be ignored. Connecticut industrialist John Hurd entered the Adirondacks from the north with his Northern Adirondack Railroad in 1883. Reaching Tupper Lake in 1889, he built the largest sawmill ever seen in this part of the northeast and was responsible for the creation of Tupper Lake village. John Hurd's railroad was the first to cross over the then existing Blue Line, but it was his final achievement before financial collapse.

As Hurd breathed his last breath of success, a remarkable project was underway that was to lay down the first line of rail to traverse completely the wilderness of the Adirondack Mountains. It was a stupendous bit of construction, but in 1892 William Seward Webb completed his Mohawk and Malone Railroad, soon to become the Adirondack Division of the New York Central and Hudson River Railroad.

Although Webb completed his dream of a through route from the Mohawk valley to Montreal for passenger revenue, it was Adirondack timber that provided a large share of the income in the early years. These two major Adirondack rail lines built by Hurd and Webb were the principal roads opening up the Adirondack timber resources for the expanded timber harvest yet to occur. And from these two roads most of the Adirondack logging railroads were to branch out over the next forty-five years of railroad logging.

The first genuine logging railroad built in New York State was actually built quite a fair distance from the Adirondacks. In 1860 the firm of Fox, Weston and Bronson built a road using wooden rails in the Town of Lindley in Steuben County, next to the Pennsylvania state line. According to forest historian William F. Fox, the railroad used a steam locomotive and small platform cars to haul logs to the bank of the Tioga River, where the logs were rolled in for the river drive to the large sawmills at Painted Post.

Wooden Rails in the Timber

Wooden rails began to penetrate the edges of the Adirondacks at about the same time. History records a number of small wooden rail lines built to service small sawmills in remote locations along the edge of the Adirondacks. But they weren't actually logging railroads; they hauled sawn lumber down to a location where it could be reloaded onto canal barges or onto another railroad.

In the mid-1800s Henry Devereaux had built a mill on the west branch of the Sacandaga River in the Town of Arietta, in Hamilton County. A tram railroad about ten miles long was built to haul out the lumber, but the project reportedly failed.

The village of Forestport, just outside the western edge of the Adirondacks, was a large lumber manufacturing center in the mid–nineteenth century. Located at the head of navigation on the Black River feeder canal, the area hosted sundry sawmills sending lumber down the Black River on barges. In 1868 the Black River and Woodhull Railroad was incorporated to extend out to sawmills east of Forestport at the edge of the great Adirondack forest. The charter authorized use of horses, mules, or steam for motive power and rails of either wood or iron not exceeding a weight of twenty-eight pounds per yard. Steam engines on wooden rails would not be allowed to exceed a speed of eight miles per hour. When built, the thirteen-mile railroad was laid with wooden rails; horse power pulled the cars of lumber from the gangsaw mills on Woodhull Creek to the wharves on the feeder canal. But the sawmill industry at Forestport went into a decline after the tragedy of April 21, 1869, when the dam at the North Lake Reservoir on the Black River broke open. Floodwater cascaded through Forestport, destroying many of the sawmill installations.

Narrow-gauge wooden-rail tram roads became popular in other locations along the western edge of the Adirondacks in the 1870s and 1880s. Albert Eaton built a tram road in 1871 from his mill on Otter Creek to a banking ground on the Black River above Glenfield; Lewis, Crawford and Company built a five-mile tram to the river from their tanning extract mill and two sawmills on Chase Lake in 1871; Henry Abbey built a sawmill on Independence River in 1885 that included a three-quarter-mile tram to the Black River. In all cases these crude railroads were used to haul sawn lumber to a point where it could be loaded on canal barges or other conveyances.

In the nearby Tug Hill plateau region, Page, Fairchild and Company purchased a fifty-thousand-board-foot-per-day band mill in the Town of West Turin in 1892, built a second mill on Alder Creek shortly afterward, and then

6. A horse-powered pole railroad was used in 1915 by the Emporium Forestry Company to haul logs from Wheeler Mountain to railside. Note the wide flanged wheels used to run on pole rails and the chain binder poles made from beech saplings, bent over to tighten the chain.

constructed a wooden-rail tram to haul the lumber down to the wagon roads in Martinsburg. From there it was handled again to load it on wagons for the trip to the canal at East Martinsburg, where it was handled once more into barges. In the 1890s the Black River canal was essentially a carrier of lumber and other forest products, reaching a peak of 143,561 tons in 1893.

As mentioned previously, logging railroads first entered the Adirondack woods at that point in history when the Adirondack forest was to receive the greatest assault on its old-growth timber. Logging took on a different flavor; timber once out of reach became available, hardwoods had a means of transport, and new markets opened up for species of trees once passed over. Lumbermen were quick to realize that railroads would relieve them of a dependence on seasonal river driving and would have the large advantage of being available most months of the year. In time it became evident that railroad logging was more economical than was river driving.

Adirondack logging by rail had its actual beginnings in the mid-1880s and early 1890s, when short rail lines were built into softwood timber to supplement river driving. Spruce and pine logs were floated downstream to a point where they could be conveniently jacked out and loaded onto railroad cars. These early rail lines were built off of either Hurd's Northern Adirondack Railroad or Webb's Mohawk and Malone Railroad.

Then as hardwood markets opened up for the lumbermen, new rail lines were built into the hardwood stands. Adirondack railroads began to haul hardwood logs about the year 1900, and eventually almost half of them were hardwood transporters. In fact, almost all of the larger logging railroads were hardwood movers. By far the largest of all was the Emporium Forestry Company, with over eighty miles of rail laid during its thirty-two years of operation.

Eventually there were twenty-two logging railroad operations established in the Adirondacks, plus the early logging activity on Hurd's Northern Adirondack and Webb's Mohawk and Malone Railroads. Five of them became chartered as common carriers and also hauled passengers, sawn lumber, and other freight items.

The first of the logging railroads became established about 1884, and the last of them shut down in 1945, a span exceeding sixty years. Each of these logging railroads is documented in the second part of this book.

7. A large hard maple tree in an old-growth timber stand being harvested by the Robert W. Higbie Company at his Newton Falls and Northern Railroad. Courtesy of the New York State Museum.

8. Rails of the Emporium Forestry Company pass alongside Blue Stream in the northern town of Colton, flanked by the winter's cut of hardwood sawlogs and pulpwood. Circa 1925. Photograph by William L. Distin.

CHAPTER THREE

Gandy Dancers and Light Rail

The Construction of Logging Railroads

The function of a logging railroad was simple: it had to provide a means to carry logs to a mill or to a transfer point on a mainline railroad—an out-and-back proposition. When all the merchantable timber was removed, the rail was pulled up and the roadbed abandoned. The result, therefore, was a railroad lacking high standards of construction that avoided as many of the higher costs as was possible. Logging economy prescribed the shortest and cheapest route.

Routes of railroad construction were dictated in part by the terrain encountered and the ground condition. The Adirondacks region is not characterized by an abundance of steep mountain terrain, thus excessive grades were seldom a barrier to overcome. Of greater concern to the surveyors were the numerous lakes, the boggy ground, and the swamps. The bends and moderate sweeps as the logging pike made its way between the rolling Adirondack hills were sometimes looked upon as advantageous. The curves tended to slow down a heavy log train powered by a light locomotive and at times equipped only with hand brakes. The northern portion of the Adirondacks has large sections of gentle terrain and sandy soils, easing the task of the builder.

Railroad logging in mountainous areas of the Adirondacks often meant a problem of slopes or grade. Mainline railroads would lengthen the line to avoid grades that could restrict load limits; logging railroads could not avoid the hills. Grades on the main logging trunk were kept to a maximum of 3 percent, if possible; those on the spurs usually did not exceed 5 or 6 percent. More often than not, the limitations imposed by grade restricted the number of empty cars the locomotive could take in, not the number of loaded cars going out.

A principal method of overcoming steep grade, if it was deemed essential to do so, was the use of a switchback, a series of doglegs climbing up the slope. Unlike the region of the Adirondack High Peaks, with its numerous steep slopes, the areas traversed by the logging railroads had grades that were relatively mild, and thus there was only one case where a switchback system was employed. The Oval Wood Dish Company used a switchback to surmount the north side of Blue Mountain (Mt. Matumbla) in Piercefield.

Although logging railroad construction was usually completed only to a temporary-use standard, the operation still required a heavy dose of manpower. In the Adirondacks there was insufficient labor available for the arduous task of

9. A work crew of Italian immigrants grading a new rail location along the edge of Silver Lake for the Emporium Forestry Company in 1913. Note the narrow rail in the background for the small dirt car.

10. Digging a cut using wheelbarrows during the construction of the Newton Falls Paper Company Railroad in 1915. Courtesy of Bill Kellogg.

11. Newly hewn birch and maple railroad ties, made on location, are laid out for new roadbed construction on the Mac-A-Mac logging railroad at Brandreth, 1912. Photograph by H. M. Beach. Courtesy of The Adirondack Museum.

grubbing out railroad beds. Thus the majority of construction labor was brought in from outside the Adirondack region. And one of the best sources of available strong-backed labor was the recent flood of immigrants to the country.

The railroad construction camps, therefore, were often filled with Italian and Irish workers, plus occasional labor crews from Spain, Mexico, and Quebec. The Emporium Forestry Company used Italian crews on their large railroad construction. John Hurd used Italians also on his Northern Adirondack Railroad construction, paying laborers $1.25 per day and charging $6.00 or $7.00 per month for room and board. The Rich Lumber Company brought along their loyal camp of hard-working Swedes when they moved operations from Pennsylvania to the Adirondacks.

Railroad building was by nature very labor intensive. Construction crews were large, but none as large as the army of men employed by W. Seward Webb when constructing his Mohawk and Malone Railroad across the heart of the Adirondacks in 1891. He had a total of four thousand laborers rapidly grubbing a hundred-foot-wide swath through the virgin forest.

Railroad construction required the movement of large amounts of dirt and rock, even with those logging grades built for short duration. The first chore after the clearing of a swath through the timber was the removal of the stumps by horse power, followed by the loosening of the soil and rocks. Ground surface was broken up by picks, mattock, and horse-drawn brush-breaking plows if needed. The plow was designed to break up the earth into manageable sections rather than to turn the soil over as an agricultural plow would do. Ledges and large rocks were disposed of with blasting powder.

Ground material was laboriously removed from the cuts and dumped into fills or into a "spoil bank" if not needed for fill. If a deep gully required more fill than the cuts could provide, a "burrow pit" was created nearby for additional material. Large crews of men would be working side by side shoveling dirt into wheelbarrows or into horse-drawn dump carts. A common piece of equipment was the drag or slip scraper, drawn by one or two horses. To operate, the teamster would lift the handles slightly while in motion, forcing the scraper's cutting edge into the dirt and filling the scraper bowl. After dragging the contents to a dump point, the handles were lifted even higher in order to dig the cutting edge

12. The rail-laying crew paused for photographer H. M. Beach in 1912 during construction of Mac-A-Mac Corporation's new logging line at Brandreth. All of these two-faced ties are yellow and white birch. In the background note the two men hewing ties as well as the crew spiking the rails. Courtesy of The Adirondack Museum.

14. Mac-A-Mac Corporation's Heisler was at work laying new track. A load of rails is being backed down to the construction crew in the Brandreth woods. Circa 1912. Photograph by H. M. Beach.

13. The construction crew included a transit man to maintain acceptable grades on the Mac-A-Mac Railroad at Brandreth, 1912.

15. The railroad construction crew on the Rich Lumber Company Railroad displays the tools used for laying rail, 1904. Left to right: spiking hammer, shovel, tamping bar, rail jack to raise ties for tamping, grub hoe, and peavey. Standing on the right with the peavey is construction contractor Claude Carlson and behind him is the rail gauge. Section boss Charles Abramson is second from right. The others were Italian immigrants.

firmly into the ground and dump the scraper upside down. With the use of a scraper, it frequently was not necessary to first loosen up the ground surface.

With the cuts and fills completed, the roadbed was leveled with a drag or possibly with some type of grader, and it was then ready for the ties and rails. The new bed might be allowed to settle a while, but the temporary nature of logging railroads often precluded such a wait.

Railroad ties were fashioned from available local hardwood trees, usually beech, birch, or maple plus hemlock and cedar, if found in the area. Oval Wood Dish Company, like many others, would cut ties from young trees found along the newly cleared right-of-way. They were hand-hewn and sometimes flattened only where the light rail was secured with a couple of spikes. Other builders used crudely fashioned ties that had been sawn only on the two opposite faces. A long life was not expected from the ties. The Emporium Forestry Company, however, anticipated a longer life on their main lines and would give the ties a cold-dip creosote treatment.

Rail was light in weight, understandably lighter than that used on the common carrier railroads. Many of the logging railroad builders used sixty-pound rail (sixty pounds per yard) but some used even lighter rail. Webb's Montreal and Malone Railroad, built to be a common carrier, used seventy-five-pound rail. And occasionally, early logging lines even used logs or poles as rails, the power to haul the cars being either horses or a lightweight locomotive with concave wheels. The Post and Henderson logging railroad south of Benson was originally laid with crude pole rails for its small locomotive. Even the sophisticated Emporium Forestry Company had a pole railroad for horse-drawn log cars in a remote location.

17. The Rich Lumber Company section crew was tamping under the ties and driving blocks where necessary to spike the rail on every tie. Note the round birch ties, slightly slabbed on top for seating of the rail.

A common practice of the "gandy dancers" (the railroad builders) was to hastily lay the ties on uneven ground surfaces, fasten the rails, and then dump some gravel fill on top. The workers would tamp down the gravel, jacking up the track where necessary and even placing a wooden block underneath to patch up where a rail wouldn't seat properly.

On logging spurs of short duration, the builders were known to fill depressions and gullies with whole logs or trees of little value, such as hemlock. It wasn't unusual to bring out a flat car loaded with slabwood from the sawmill to use as fill on temporary lines. Rail and ties were laid right over the top of the fill.

When W. Seward Webb accomplished the monumental

16. Slabwood and other sawmill waste was used to fill the gullies and wooden blocks were used to shim the crude ties during new railroad construction on the Emporium Forestry Company's north tram, 1913. Courtesy of the Sykes family.

task of building the 160-mile Mohawk and Malone Railroad through the Adirondack wilderness in the short span of eighteen months, his contractors had the construction workers divided into five groups. The first gang followed the line marked out by the surveyors and felled all standing trees in a hundred-foot-wide path. A second gang worked close behind to cut up the trees and pile them all in the center of the path, where they were burned. Even before the embers turned cold, another gang was busy with stump pullers, block and tackle, and grubhooks to remove the stumps. Right behind them were the pick-and-shovel men moving dirt to throw up a grade. Tight on their heels were the crews laying down the ties and then spiking down the rails. Each gang followed closely behind the other.

Extensive trestle building or large fills were avoided as much as possible on the temporary logging lines. Bridges were considered only as a cost-governed substitute for solid fill; they were expensive and were fair game for forest fires. When long fills were made, it was on the main logging tram or on the common carrier rail lines that some lumber companies had to build to provide access from the mill town to the major railroad in the area. The Rich Lumber Company had an impressive four-hundred-foot trestle built with round timbers on their common-carrier Cranberry Lake Railroad. As was customarily done in railroad construction, it was eventually filled in by means of dump cars on the tracks above, burying the trestle and creating what became known as the "big fill."

To create more permanent stable roadbeds, gravel ballast

18. *The long trestle on the Cranberry Lake Railroad had just been completed in 1901 and soon the entire stretch would be filled in with iron ore sand and become known as the Big Fill.*

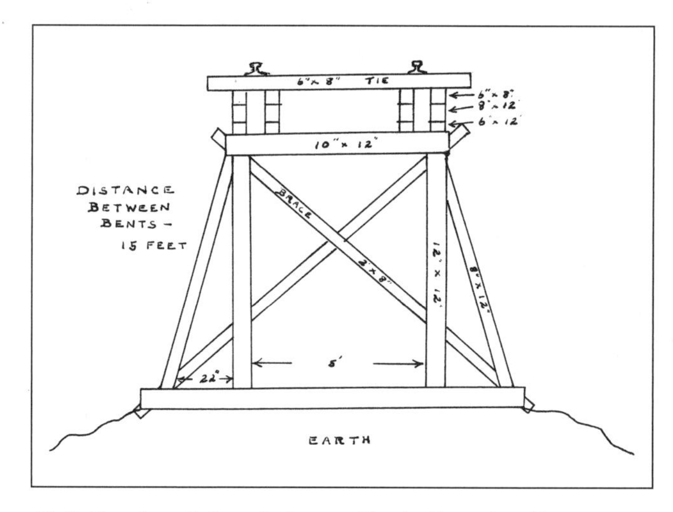

19. *Railroad trestle bent design used by the Emporium Forestry Company.*

was hauled in with a work train. When John Hurd was building the Northern Adirondack Railroad, originally a logging line, he had in his hire a construction supervisor who devised an innovative method of unloading the gravel cars. Although gravel cars were usually loaded by hand, Hurd brought in a large $7,000 steam shovel capable of hefting three tons of sand per bucketful. They called the ponderous machine Jumbo. It could load a nine-gravel-car train in eighteen to twenty minutes, fourteen to fifteen yards of fill in each car.

Unloading the gravel involved a unique system. The brakes would be set on the gravel cars, and the engine would uncouple and then pull a V-shaped plow along the car floors by means of a long cable. Starting from the rear car, the plow ran on rollers and followed a beam in the center of each car, spreading the dirt out each side as it advanced. It took about five minutes to unload the train.

The majority of logging lines were built at standard gauge (4'8$^1/_2$" between rails), despite the substantial cost that would be saved by constructing and outfitting a narrow-gauge operation. The reasons for standard gauge were obvious. The majority of logging pikes branched off of mainline railroads and sometimes delivered loaded log cars to sidings for pickup by the mainline freights. Another major factor dictating standard gauge was the use of secondhand locomotives and rolling stock obtained from the major railroads. Seldom did a logging operation purchase a new locomotive.

There was also one more motivation behind the choice of standard gauge in the case of a few of the operators. There

20. A Marion steam shovel was used on the Grasse River Railroad to provide fill from a convenient burrow pit. Circa 1913.

was the occasional builder who envisioned his logging line as eventually extending through to a terminus at an industrial site or a village, creating an important reason for the railroad to be upgraded and to remain in place. An improved permanent roadbed condition might provide a new source of profit with busy freight traffic and passenger revenue.

For example, when the Moose River Lumber Company was building its short logging railroad easterly from McKeever, it was hoped that the line eventually would be extended across the southern edge of the Adirondacks through other properties controlled by the same owners, and would make connections with a paper mill on the Hudson River. But the railroad never got beyond six miles in length.

John Hurd had dreams of extending the Northern Adirondack Railroad into a network across the Adirondacks. Rail extensions were proposed northward to Ottawa and from the Tupper Lake terminus east to Saranac Lake and southwest to Utica. Bankruptcy killed that dream.

The Delaware and Hudson Railroad took an option on the Kinsley Lumber Company's railroad to DeBar Mountain with the vision that it might become a link to railroad construction westward from Loon Lake to St. Regis Falls, thus opening up new markets. That dream also failed.

Although the tendency was to build short-life logging roads as cheaply as possible, the expense was still relatively high. The somewhat fixed cost of surveying, grading, bridge construction, and tie production had to be written off in a short period of time. Equipment life was reduced by hard usage. Logging railroad construction costs were estimated to be between $5,000 and $10,000 per mile in the early 1900s. When Webb built his Mohawk and Malone, there was no expense spared to push the new rails through the Adirondacks in record speed in 1892. His estimated cost topped $40,000 per mile.

C H A P T E R   F O U R

Logging Locomotives and Rolling Stock

LOGGING LOCOMOTIVES seldom mirrored the image of the powerful mainline engines. The rugged nature of their work and the flimsy roadbed conditions required a special design of locomotive that was compatible with the conditions imposed by the loggers. The heavy engines with large-diameter driving wheels that were common in mainline service were quite unsuitable for loggers.

As tired and worn as it might be, the locomotive was still the most valuable possession owned or borrowed by the railroad logger. Rarely did it present a handsome appearance, the exception being the occasional wealthy Adirondack landowner who was able to build his own little railroad. When William G. Rockefeller purchased a large timber tract near Loon Lake in 1899, he acquired with it an existing logging road and a well-used locomotive. He promptly sent the engine to the shops and had it repaired and painted a lime green with red running gear. The name William Rockefeller was lettered on the side.

Regardless of whether there was a name, the logger's lokie was always a "she." As explained by Kramer Adams in his book *Logging Railroads of the West,* the engine had to be a she because it wore an "apron, binder, bonnet, collar, hood, hose, jacket, muffler, pumps, sash, wrapper and ate twice what it was worth." And also the age was not discussed beyond twenty years.

Adirondack loggers were about equally divided in their

21. Shay No. 39 of the Emporium Forestry Company has the appearance of a hard-working logger, battered and loaded with accessories. This sixty-five-ton machine was built in 1905 and scrapped by Emporium in 1934. Pictured at the company's operation in Galeton, Pa., before being shipped to Conifer, N.Y., in 1917. From collection of Thomas T. Taber III.

preferences for rod (direct-drive) and geared locomotives. On the twenty-one logging railroads in the region, there were at least fifty-three different locomotives used over the six decades of railroading in the woods. Twenty-four of the locos, almost half, were rod locomotives of various makes. Twenty-five were geared locomotives (Shay, Climax, and Heisler), plus there were three electric engines and one diesel.

By far the largest roster of locomotives was that found in the yard of William L. Sykes's Emporium Forestry Company at Conifer. Emporium, and its successor, the Heywood-Wakefield Company, had a total of twenty-two of those fifty-three locomotives over their thirty or more years of operation in the Adirondacks. The Sykes family collected locomotives whether they needed them or not.

The record for longevity of any logging locomotive in the Adirondacks was held by Shay No. 40 on the Emporium roster. Built in 1902, it was a powerful ninety-two-ton, three-truck brute that was not only the company's largest locomotive but also their best, according to the railroad crews. Its final days were spent on the Elk River Coal and Lumber Company line in West Virginia, where it was scrapped in 1962, culminating sixty years of tough logging.

Rod Engines

Rod locomotives were frequently used by Adirondack loggers because of the gentle terrain; it was seldom steep enough to require the slower and more expensive geared locomotives. Rod engines were mechanically simpler than geared engines because power was applied directly from the cylinders to the drive wheels by connecting rods. Although rod engines were the type always used on the mainline railroads, the logger lokies showed little in comparison. On the woods lines the locomotives were small, light in weight, and often with a saddletank draped over the boiler to carry the water supply.

Saddletanks were popular with loggers because they not only carried the water tank over the drive wheels, adding tractive weight, they also did not require a tender. Tenders were a nuisance in the woods when a locomotive often had to run backward going in (there were no turntables) and when the engineer didn't want his field of vision impeded during close-up maneuvers. Fuel (wood or coal) was carried in a bunker behind the cab. There might be either four or six drive wheels and generally small lead and tail trucks. The

22. Emporium Forestry Company's little thirty-eight-ton Baldwin saddletank No. 33 saw service on at least four logging lines during its sixty years of life. Emporium used it on their Grasse River Railroad, shown above arriving at Cranberry Lake village with the daily passenger run, pulling a single combine and a few empty log cars. Circa 1920. Courtesy of the Sykes family.

23. Emporium locomotive No. 68 was a Schenectady-built mogul used on the Grasse River Railroad for almost thirty years for such duties as hauling pulpwood cars from the loading dock at Cranberry Lake to the Childwold Station. While being rebuilt in the Emporium shop, the tender was equipped with a larger, five-thousand-gallon tank. Photographed at Childwold in September 1949, soon after purchase of the railroad by Heywood-Wakefield Company, who operated the engine for two years before scrapping it. Photograph by Philip R. Hastings.

manufacturers Baldwin and Porter dominated the field with saddletanks.

Loggers were prone to pick up cast-off locomotives and would use whatever might be available if it was light in weight and still serviceable. Moguls (2-6-0; the numbers refer to the wheel arrangement on the locomotive) were a popular type in the sixty- to seventy-ton range. It was not unusual for a logger to acquire a locomotive that had already seen twenty to thirty years of service elsewhere, and then nurse a few more years of hard work out of it before passing it on to another operator. Being light in weight, these logging locomotives were obviously limited in pulling power. But that was seldom a shortcoming on these small operations. The expense of building a railroad that was able to support heavy engines was not justified.

Geared Engines

Geared railroad locomotives were originally designed and developed specifically for the needs of the loggers. Power was applied to all of the driving wheels, increasing traction without greatly increasing the weight of the engine. The entire weight of the locomotive was available for adhesion, even the fuel and water, and was distributed over all of the axles.

Drive wheels were small, usually in the range of twenty-eight to thirty-six inches in diameter. No counterweights were needed with a geared drive, which of course eliminated the pounding on the rail. Thus lighter rail could be used. The various systems of gearing the wheels allowed for the needed flexibility on the crooked and uneven track of the logger. Speed was sacrificed for power. Whereas a rod locomotive with loaded cars would be confined to a maximum grade of 1.5 to 2 percent, a geared locomotive could pull a grade of 6 or 7 percent. Switching maneuvers have been accomplished on inclines as steep as 10 percent. Braking power was also improved with the smaller-diameter wheels.

There were three distinct basic designs used in the manufacture of geared locomotives. All three of them saw use in the Adirondacks.

Shay

By far the most popular of the geared locos around the nation was the Shay, a workhorse actually developed about 1880 by a Michigan logger named Ephraim Shay. Shays were the favorite in the Adirondacks also; there were fifteen of them used there, half of those being on the Emporium roster. The Shay was produced by the Lima Locomotive Works of Lima, Ohio, in sizes ranging from 10 to 150 tons. From 1880 until 1945, there were 2,761 Shay engines manufactured, so varied in design and customer preferences that seldom were there two identical locomotives.

The Shay's unique design employed three vertical cylinders mounted on the right side of the boiler (two cylinders in the earliest model). The cylinders drove a crankshaft mounted on a horizontal drive shaft that went the length of the entire locomotive on the outside of the right-hand wheels. The shaft powered the right wheel on each wheel axle by means of bevel gears on the shaft and on the wheel face. The drive shaft was rendered quite flexible by means of slip joints and universal joints. The driving gears were easily adjusted by shims in the axle journals. Thus all wheels remained powered with full freedom of movement under the engine frame when traveling on sharp curves or poor track conditions. A Shay would have two trucks unless it exceeded sixty tons in weight, when three could be installed.

24. *Shay locomotives were a favorite with the Rich Lumber Company. No. 4, a three-truck engine, had recently arrived at Wanakena in 1904 from the company's Pennsylvania operation when the photo was taken and still retained the lettering of Rich Lumber Company's South Branch Railroad. This was a fifty-ton locomotive with a wagon-top style of boiler, built in 1898. Conductor Bob "Chub" Matteson (center) always wore a derby hat.*

25. *Shay locomotive trucks, showing the method of power transmission from the horizontal drive shaft to the wheels.*

With all of the moving parts along the right side, they were quite accessible for repair. There were disadvantages to this arrangement, too. The Emporium shop crew complained that the snow would pack into the exposed bevel gears on the face of the wheels, soak up the grease, and then cause the gears to clatter loudly. With the weight of the engines and gears on the right side, the boiler had to be moved over to the left of the center line to compensate for the excess weight on the right side. This gave the Shay a very distinctive lopsided appearance.

This remarkably rugged locomotive had a distinctive sound all its own. As the engineer adjusted the notched throttle quadrant and began to surge steam into the three small cylinders, the gears would slowly begin to turn and soon build up to a fast churning. The rapid rhythm of the exhaust from the three cylinders is an unforgettable sound, especially when laboring up a grade with a heavy load, buried in exhaust steam and smoke.

Climax

The second most popular geared locomotive in the Adirondacks was the Climax, made in Corry, Pennsylvania. The

26. Climax No. 5 of the Emporium Forestry Company was built in 1905 and saw forty years of logging service by the company before being sold and eventually scrapped in 1950. The fifty-ton engine had a straight-top boiler containing 174 two-inch tubes nine feet long, working under a pressure of 180 lbs. per sq. inch. The steam brake applied on all eight wheels. Photograph taken in Hammersly, Pa. where it was in use by Emporium until shipped to Conifer in 1912. From collection of Thomas T. Taber III.

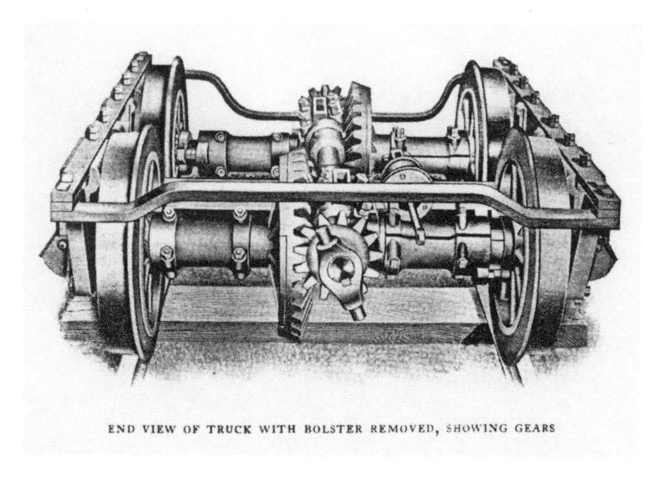

27. Climax locomotive truck, showing the arrangement of the power transmitting gears between the central drive shaft and the wheel axles.

Climax employed a long-stroke piston on each side of the smoke box, mounted at a 45-degree angle. The cylinders powered a cross-crank shaft located just in front of the cab. This shaft was connected by a master gear to a longitudinal center drive shaft, which powered each of the four axles by means of level gears. Although the cylinders were not large, the Climax did not lack for power and was preferred over the Shay by some locomotive engineers.

But other train crews found the Climax less to their liking because of the rough riding vibration set up in the machine by the flying main rods. It was also the slowest of the three styles of geared locomotives. However, nobody denied that the Climax could stand up with the best in pulling power.

Heisler

The Heisler geared locomotive, made in Erie, Pennsylvania, never found acceptance on any New England logging jobs, but there were four used on Adirondack logging lines.

The Heisler also employed two cylinders, on opposite sides of the boiler and underneath it. However, they were inclined at an angle of 45 degrees from the vertical and set in a V fashion. They were positioned just ahead of the cab, where they powered a single crank shaft that was an integral part of the longitudinal drive shaft. The shaft was geared to one of the axles on each truck; power was then transmitted to the other axle by means of side rods on the face of the wheels.

The designer of the Heisler had intended to capture some of the features of the rod engine, minimizing the gearing. As such, the machine was the fastest of the geared locomotives, capable of speeds up to twenty to twenty-five miles an hour without shaking itself to pieces. But the Heisler's rocking motion made for rough riding and frequent overhauls.

Rod engines and geared engines are at their best when used properly for their advantages. On the main tram line with minimum grades, the rod engine would outperform; on the rough-tracked spurs, the geared locomotive proved its worth. For the Adirondacks, however, there were only three logging operations that employed this locomotive combination of both types. Possibly many felt their line was too short, or couldn't bear the expense of an extra locomotive, or in many cases had no difficult grades to contend with.

The one big exception was the Emporium Forestry Company, using an inflated roster of ten geared locomotives

28. Mac-A-Mac's new forty-two-ton Heisler upon arrival at Brandreth in 1912. The Heisler geared locomotive had a cylinder on each side mounted at a 45-degree angle toward the central drive shaft, which in turn was geared to all of the wheels. Courtesy of The Adirondack Museum.

and eleven rod locomotives. William Sykes apparently enjoyed purchasing used locomotives, although some of them never saw service on his woods operations. The company would have a Shay working the landings and accumulating a string of twelve to twenty-two loaded log cars. Then the loaded consist would be left at locations along the North Tram to be picked up by one of the Schenectady-built moguls. On many days a string of log cars would be left at the Dodge Brook wye for the Grasse River Railroad locomotive on the company's passenger operation from Childwold to Cranberry Lake. The little Grasse River engine would push the log cars in the front and pull the single coach and possibly some boxcars of lumber from the company's sawmill at Cranberry Lake. The main logging line, called the North Tram, extended for thirty miles to the north end of the timber ownership. It was a long six-hour trip at day's end for the Shay to return home from the far end with a few log cars in tow. The company laid over eighty miles of track during their forty-four years of railroad logging.

Equipment and Rolling Stock

Rolling stock on the logging railroads had many sizes, shapes, and origins. Secondhand equipment was passed down from one operator to the next; discarded cars from the mainline railroads were rescued from oblivion; shop-built equipment filled many specific needs. And then there were those operators who maintained shops capable of building their own log cars and other equipment.

Log Cars

Adirondack loggers used two types of log cars—flats and skeletons. The majority had flat cars, often with a set of rails fastened lengthwise along the top of the bed for the log loader to travel upon, moving from car to car. At least three operations, large ones, used the short skeleton cars, also with rails mounted on top of the bunks.

Secondhand flat cars were readily available for lease or

29. *Advertisement for logging locomotives and other equipment frequented the old* American Lumberman. *This one is dated August 1, 1903.*

30. *Skeleton log cars consisted of a pair of log bunks with rails on top for the loader to travel along. Each car held one tier of logs. Photograph taken next to sawmill log pond of Rich Lumber Company. Circa 1905.*

31. *Champlain Realty Company (International Paper Company) built these pulpwood rack cars for use on mainline railroads.*

purchase at a reasonable price from the large railroads. They had the disadvantage of being heavier than skeleton cars, especially when bark, dirt, and other material would accumulate on the deck to add weight for the return trip.

Skeleton cars were not that difficult to build in a well-equipped company shop. The Emporium Forestry Company had an outstanding shop at Conifer that constructed the 150 or more skeleton log cars used by the company. They featured a trip-chain device that was patented by company executive William S. Caflisch. Log cars were constantly being repaired or rebuilt by the shop crew, which even built its own wood-frame railroad trucks.

When the International Paper Company built their short logging railroad into Woods Lake, they purchased two hundred flat-bottomed gondola cars and brought them up to Woods Lake. A small sawmill was set up to supply the lumber for the crew, which then cut doors in the car sides and built racks around each car in order to contain loosely dumped four-foot pulpwood.

Frank A. Cutting, a hemlock bark dealer with an office in Boston, needed a special car to transport the thousands of cords of sheets of hemlock bark that he shipped to tanneries throughout New England. He designed and built one hundred railroad cars with open tops and widely spaced frames on the side and ends. The Brooklyn Cooperage Company railroad hauled many of Cutting's bark cars out of the Lake Ozonia area.

32. Frank Cutting's letterhead illustrated the style of car he had built to transport hemlock bark.

33. The Barnhart log loader was the most popular of the car-mounted loaders. The Brooklyn Cooperage Company crew was loading the company-owned flat cars with hardwood logs. Note the metal side stakes, which could swing down for unloading. Circa 1910.

Loaders

The most popular type of log loader used in the Adirondacks was the car-mounted log loader that could self-propel from car to car upon rails that were secured to the bed of the log cars. While sitting on one of the log cars, it would load the car behind it with five to seven thousand board feet of logs, in the case of hardwood. Then with a cable secured about three cars ahead, the loader would winch itself ahead onto the next car, using a patented slack pulley. The loader would then load the car it had just been sitting upon, continuing in that fashion until the entire string of log cars was loaded.

The most popular of these loaders was the Barnhart loader, named after its developer in Pennsylvania. It was a powerful and very practical machine that found ready ac-

34. A builder's photo of the American Hoist and Derrick Company log loader purchased by the Horse Shoe Forestry Company. Circa 1905.

ceptance by the loggers. Other makes of similar style loaders used in the North Country were the American and the Clyde. At least half of the Adirondack railroad loggers used these car-mounted loaders.

The Emporium Forestry Company had a number of Barnhart loaders, sometimes as many as four operating at the same time. It took skill to be a loader operator, and not many could equal Emporium's Ed Ressler's boast of loading eight hundred logs per day. Ed could be temperamental, but if he wasn't mad at the tong setter, he could swing the tongs out rapidly and yet lay them gently into the tong setter's hands. The log would then lift and come back in a flurry and yet set down gently in place.

Some of the logging operators still loaded logs the old way, rolling the logs onto the cars from tall decks or skidways. On the Mac-A-Mac railroad, the decks along the rail sidings were built high enough so that the landing crew could roll the sixteen-foot-long spruce logs over the top of the twelve-foot stakes mounted on regulation flat cars.

Railroad Shop

Any logging railroad had to have a shop of some sort; repairs were a constant factor. The outstanding railroad shop, however, was the one at Conifer, operated by the Emporium Forestry Company. The shop crew not only built their own log cars and railroad trucks, they could also completely re-

35. The Barnhart log loader. The housing and boom have been removed to show the arrangement and accessibility of the machinery.

36. The Robert Higbie Company's log loader was handling mostly pine logs with an occasional hardwood at the time of this photo. The wrapper chains were tightened by placing a log or two on top of the secured chain. Source: N.Y. Assembly Doc.

37. The Oval Wood Dish Company used a gasoline-powered, skid-mounted Raymond loader, placed on New York Central flats. Circa 1925. Courtesy of the Tupper Lake Public Library.

38. Loading log cars or sleds by hand required careful coordination by the two yard men working the skidway. Photograph at Brandreth by H. M. Beach. Courtesy of the Brandreth heirs.

build a locomotive. Jesse Blesh served the shop for many years as master mechanic, a position he had held at the company's Pennsylvania operations. At his retirement, the position was filled by Ray Zenger. Among the devices to be credited to the Emporium and its shop crew was a patented flexible stay bolt that separated the sheets in the locomotive fire box.

C H A P T E R   F I V E

Rails Through the Timber

Operations and Mishaps

THE LOGGING RAILROADS of the Adirondacks, like other northeastern operations, were not large in size, especially when compared with counterparts in other areas of the United States. Few of them exceeded twenty miles of total trackage; some were less than ten.

The exceptions to this smaller size became the most notable and most interesting of all of the Adirondack loggers. The Brooklyn Cooperage Company and the Oval Wood Dish Company each had over thirty miles of track laid. Then there was the premier Adirondack logging railroader, the Emporium Forestry Company. Including its common carrier, the Grasse River Railroad, the Emporium put down more than eighty miles of rails into the timber during its forty-four years of existence.

Sawlogs were often loaded by means of some type of steam-powered loader, frequently the car-mounted type that moved from car to car on rails. But there was another system used for loading log-length softwood pulpwood that was somewhat unique to the Adirondacks. It was called a jackworks. Logs would be floated down a stream or boomed across a lake to a location where they could be gathered in the water near the railroad track for loading onto cars. Steam-powered conveyors hauled the logs out of the water on an endless chain and conveyed them up onto a ramp from which they could be rolled by hand onto the log cars alongside the ramp or deck. A number of these jackworks were set up in the Adirondacks.

39. The winter cut of spruce logs was stacked railside in a Rich Lumber Company operation, ready for rail transport. Circa 1915.

The International Paper Company had what was possibly the only Adirondack logging railroad that hauled four-foot-length pulpwood from the woods. To do this, they had a unique loading operation at their Woods Lake operation. The spruce pulpwood logs were winter-hauled on sleds to Twitchell Creek in twelve- or sixteen-foot lengths and stacked streamside. In spring the logs were floated down to one of the jackworks, each of which was set up with a slasher mill. A slasher mill had a large circular saw that cut the logs up into four-foot lengths. The four-foot wood was then dumped back into Twitchell Creek again for floating down to another jackworks, which conveyed the short sticks out of the creek, dumping them into open-top rack cars.

Brake systems on log trains rarely met the standards for mainline railroad operations, unless they had to run their cars over the mainline railroads. Many of the Adirondack loggers had only hand brakes on the cars, which often slowed down operations a bit. When a loaded train reached the head end of a steep downgrade, the engineer would have to stop the train and have the brakes hand set on every car.

40. The Mac-A-Mac Corporation jackworks in operation at the North Pond flow grounds, Brandreth. The logs were pulled up or jacked from the water on an endless chain powered by the stationary steam engine in the background. The flat cars were loaded by hand. Circa 1914. Photograph by H. M. Beach. Courtesy of the Brandreth heirs.

The brakeman would use a "Jim Crow" bar or maybe even an old pick handle to tighten the brake chain as much as possible.

The brakeman had a particularly difficult list of chores on a logging railroad. Most operations did not carry a caboose; thus there was no place for shelter during adverse weather. Footing on a carload of logs was treacherous, and he had to be able to move along the logs with dexterity. Many fell from the perch, some to their deaths.

In June 1901 the Moose River Lumber Company was hauling some spruce logs to their McKeever sawmill over the New York Central line. On June 24 brakeman Nelson Avery was working some car shifts at Clearwater when he fell under the cars. He died from the injuries the next day.

Derailments were not uncommon on the hastily built logging trackage, often built with rails that were already well worn. The locomotives all carried rerail irons for frequent use.

When John Hurd built the Northern Adirondack Railroad, the tracks were definitely substandard, at least in the early years. Ferris J. Meigs, president of the Santa Clara Lumber Company, with which Hurd was also associated, relates a story in his unpublished history of the company to show how bad the conditions were on Hurd's railroad. On one particular occasion, a lone passenger car was coupled onto the end of a string of empty log cars when suddenly the log cars began to derail one after another on a curve, bouncing and dancing along on the ties. The engineer continued on, apparently unaware of the situation, until a passenger from the coach was able to run up ahead along the slowly moving train and warn him. It is said that the engineer then had the brakeman in the coach set the brakes while on a straight way, and the drag caused the cars to jump back on the rails, one after another. Exaggeration, possibly, but there were many derailments on Hurd's road, rarely serious because train speed seldom exceeded twenty miles per hour.

Rails didn't receive the close inspection given on the large railroads. After years of hard use, the rails on the Emporium had received considerable wear on the main tram, in some places spreading out beyond the standard-gauge width. This spread usually didn't present a problem for the wide "tires" or drive wheels of the Shays. However, when Shay No. 40 was sold to a logging line in West Virginia in 1950, it had to be delivered on a flat car; the wheels were too wide to go through the track frogs on the main line. This track condition once prompted company president George Sykes to remark to the president of the New York Central Railroad, a personal friend, that the Emporium's sixteen-mile Grasse River Railroad may not be as long as the New York Central's eleven-thousand mile system, but it was definitely wider.

Even the temporary sidings built by the mainline railroads for the pickup of logs or lumber cars could be treacherous. On March 8, 1902, an extra New York Central train

with seven empty flats and a caboose was backing into a siding at Underwood. Underwood was a small freight station located at a sawmill site about a mile south of Tupper Lake Junction. Running at a speed of about twenty miles per hour, the locomotive tender began to sway on the uneven, frost-heaved track. The tender soon left the rails, pulling the remainder of the train with it. The locomotive also tumbled, fatally injuring fireman J. R. Cummings and severely bruising engineer D. Kelly.

One of the cautions a locomotive engineer had to keep in mind was the danger of low water level in the boiler. The heat from the fire box could quickly burn through the crown sheet without a covering of water above; the results would be disastrous.

In 1908 Shay No. 2 on the Rich Lumber Company railroad was pushing a string of empties up a grade toward High Falls when a low water supply caused the boiler to explode at a location just below Dobson's Trail. The engineer was killed, as was a jewelry salesman riding on an empty log car in front of the locomotive. When the boiler exploded, the smoke-box door on the front of the engine flew off, and the brass number plate struck the salesman a fatal blow, decapitating him, according to one report. For hours afterward men searched for his valuable jewelry case, but no one ever admitted finding it. The Shay was rebuilt and renumbered as No. 6.

Sometimes it was the unexpected over which the engineer had no control. That was the situation one day at the village of McDonald on the Ottawa Division of the New York Central. McDonald village was where John McDonald's Bay Pond railroad junctioned with the New York Central. Both of the Bay Pond locomotives would customarily double up to highball a string of about ten carloads of pulpwood logs, an engine on each end of the string. On approaching the sidings at McDonald, engine number 2 in the front would cut off at another siding to allow the pusher engine to place the cars on the mainline siding for pickup by the New York Central freight engine. On this particular day, the air brakes wouldn't hold as the pusher engine attempted to stop the cars. The lead log car jumped a track frog and left the rails, dumping off two carloads of logs, which just missed the company office forty feet from the track. It was a day long remembered by the office personnel.

Imagine the surprise that confronted engineer George LaFountain one day while at the controls of the Mac-A-Mac railroad Alco-Schenectady 2-6-0. He had just brought a long log train over a slight rise when he witnessed a runaway team of horses start across the trestle in front of him. One of the horses had his legs drop down between the ties, which then pinned him in place. George yelled at his son Ernie, who was firing, to jump, dumped his air, and then "took to the birds" himself. But Ernie froze. Fortunately, the train ground to a halt a few feet short of the horses, after which the animals were hauled free and allowed to tear across the trestle.

Train crewmen traditionally considered themselves immune from any labor not directly involved with operating the train. There were often good reasons for the engineer to remain at his post, and even on the cruder logging operations it was rare to see any train crew member involved in

41. An unusual style of log-loading deck located at Brandreth. The ten- to twelve-foot high deck, two thousand feet long, was constructed of peeled spruce logs, and the log sleds were hauled up to the top for unloading. The approach ramp on the other end, surmounted by the horses with loaded sleds, was probably a gentler grade than the exit ramp. Circa 1914.

42. The boiler explosion on Rich Lumber Company's Shay No. 2 in 1908 killed at least one person, a passenger on a flat car in front of the engine. Nothing is known regarding the engine crew, but the evidence does not give a positive picture. Company employee Hank Belcher is on the right. Courtesy of Carl Abramson.

43. It was an embarrassing incident for engineer Bill Cudhea when the Oval Wood Dish Company Heisler ran out of control down the steep grade on Mount Matumbla (Blue Mountain) and ended up as shown. Courtesy of J. T. Giblin.

any other type of labor. But not all logging bosses subscribed to that work ethic. William S. Caflisch, one of the founding patriarchs of the Emporium Forestry Company, was a gruff individual who wouldn't hesitate to jump right in and work physically with the men. There was an occasion when he was shoveling gravel with the crew along the shoulder of the railroad bed. The locomotive of the work train was standing idle, and Caflisch, noticing the fireman had nothing to do, called up "Pardner [he called everyone Pardner], there's an extra shovel down here." The fireman's refusal, and his reply that he wasn't hired to do that kind of work, did not sit well with Caflisch. When the fireman returned to the office that night, he found himself without a job. Only considerable pleading got him his job back, and he eventually was promoted to an engineer's job.

The working hours of a log train crew were not easy ones. And working conditions were certainly far different from those of the crew on a fast passenger or even a wayside freight run. From most indications, however, the log train crewmen loved their work and prized their seniority.

CHAPTER SIX

Company Villages and Common Carriers

Adirondack villages were few and far between in the late 1800s and early 1900s, the era of the large-scale timber harvest. For the most part, the Adirondack villages then existing were the locations of permanent wood-product industries or service providers for the summer tourist industry. When the major timber operators acquired large timber tracts and wanted to set up mills, they had to create, in some cases, entirely new villages. Employees had to have residence in proximity, and services had to be provided. Thus, at one time, the Adirondacks hosted a variety of company towns built to support the lumber industry during its prime years. The lumbermen's villages were usually quite temporary.

Allied with these company villages that were established in remote locations were the railroads, usually built and owned by the company. The village had to have an access by rail to the mainline railroad, as well as having a means to transport the logs out of the woods. Some of the smaller railroad operations, however, were able to avoid creating a village by building a mill near an existing railroad settlement or hauling their logs on the mainline railroad to a more distant mill.

Few of the lumber industry villages exist today; most of them can hardly even be recognized as ghost towns. Nature has healed the sites and permitted few reminders to remain. In other cases a scattering of hunting camps on a poorly maintained road might serve as a site locator. But there are exceptions. Both Wanakena and Conifer have continued to exist as communities, though but shells of the once-thriving sawmill towns. Others such as Santa Clara and McKeever continue to hold on with a few residents in older structures barely surviving from the prosperous days.

When the Rich Lumber Company moved into the Adirondacks in 1902 and built the village of Wanakena, it had already had the experience of building three different company settlements in Pennsylvania. However, Wanakena became the company's pride. At its peak the village had a railroad depot, a restaurant, two hotels, a boardinghouse, a large general store, and a clubhouse with a library, reading room, and bowling alley, and homes for nearly five hundred inhabitants. Most of the buildings were company owned.

One of the structures not owned by the company was the large general store. It was owned and operated by Clayton Rich, brother of company owner Herbert C. Rich, and Wally F. Andrews, the same pair that had set up stores at the Pennsylvania operations. On a siding along the side of the store, flat cars were loaded with supplies for transport to the logging camps—orders such as fifty bushels of potatoes and one hundred pounds of coffee that had to be hand ground at the store.

The obvious location for the establishment of any new sawmill village would be at a site adjacent to one of the established mainline railroads that passed through the Adirondacks. Thus, there were a number of logging and sawmill settlements made along W. Seward Webb's Mohawk and Malone Railroad (New York Central) and along John Hurd's Northern Adirondack Railroad. Today it is difficult to find any evidence of their existence.

John McDonald, a well-known railroad logger, built two such villages, Brandreth and Bay Pond. Although each consisted of a railroad depot, general store, engine house, post office, company office, boardinghouse, and about seven family homes, they only lasted for eight or nine years. The sites now reveal very little indication of former habitation.

Lumberman Patrick A. Ducey built an even larger town, which likewise suffered an early demise. In about 1882, Ducey set up a large sawmill in Franklin County, said to be the largest in the state at that time, as it cut about 125,000 board feet per day. He then proceeded to create the village of Brandon. Situated on the Northern Adirondack Railroad, the settlement contained almost a hundred buildings, in-

The General Store

The general store at Wanakena had almost anything a person might need for life in this remote community. If the item wasn't in stock, an order from the catalog would have it there within a few days.

A frequent visitor to Wanakena, Herbert F. Keith, provided an apt description for the store in his book *Man of the Woods:*

"In the basement were fifty-gallon barrels of vinegar, salt pork, sour pickles, sweet pickles, olives (large and small), molasses, corned beef, and many other liquid and brine-prepared staples. The main floor contained many barrels of cookies, crackers, sugar, and oatmeal. Large tinfoil-lined chests of tea and plenty of cheeses, tubs of butter, lard, and other items were also found here. The clerks were so adept at scooping or cutting amounts you ordered that a set of scales seemed almost unnecessary. The patent medicines and drug department consisted of hundreds of bottles. The store cat always slept among them and yet never tipped any over. At the front of the store was a large rack displaying shoes from the latest style of button shoe to the lumberjack's caulked boots. Only one shoe of each kind was shown, never a pair, and occasionally a French Canadian would be seen walking around town with two different kinds of shoes on. Some of the Canucks helped themselves to the sample shoes just as to a cracker out of the barrel and finally had to be told the shoes were not samples to take home.

"A long nicely varnished counter extended part way down the left side as you entered the store. The top was marked off in yards and fractions of a yard with little round-headed brass nails so it was convenient to measure the yard goods. This counter was the only clear spot in the whole store, and sometimes the train crew and some of the other workers would sit on it in their dirty overalls. One of the clerks patiently wired all the nails together underneath the counter where they stuck through and then hooked up a good strong spark coil which could be operated by a switch on the opposite side of the room. One day when six or seven of the town's prize sitters were comfortably seated at the counter, the switch was thrown. Several of the women customers were bowled over in the rush. From that day on, the dress goods counter was as clean as a whistle.

44. The Rich Lumber Company passenger coach parked alongside the general store must have been idle. The attention of the little tots is entirely on the photographer. Courtesy of Bill Gleason.

"Sometime during the cold hard winter the dry goods salesman would arrive with his twenty large trunks full of sample goods, and of course the word of his coming was passed around. His arrival was a big time for all and especially for the ladies. A temporary counter was placed down the center of the store where the salesman could lay out his wares. The store customers would look at the display and tell the clerks what they wanted to buy. These goods were left out for three to five days so that the lumberjacks and other workers from the surrounding camps could all have a chance to order what they wanted. It was quite a sight to see the salesman pack all those samples away in his specially built trunks. The commission he got from the goods which were ordered justified the twenty-five-cent cigar he smoked as a sort of celebration.

"The top floor of the store was the clothing department. It was well stocked with an assortment from the latest style of men's suits to blankets and mattresses. The remnant counter was well patronized, and many school kids wore snow suits of assorted colors. Some of the heavy items sold at the store were wagons, sleighs, harnesses, oats, hay, coal, feed, ice, and stoves. The store carried about sixty thousand dollars worth of stock. There was a large porch built across the front end, and a covered platform ran its entire length. Beside the platform ran a railroad siding."

45. John McDonald's village on the New York Central Railroad was known as Brandreth Station. Shown are the company store and office along with the New York Central depot. Circa 1915.

46. Warm blankets for the loggers were a feature at the Mac-A-Mac company store at Brandreth Station.

cluding four hotels, two churches, four general stores, hardware and drug stores, a school, and a town hall. Population boomed at twelve hundred. Ducey didn't build his own logging railroad, but as with many other lumbermen, he depended on Hurd's Northern Adirondack for his existence. Within fifteen years the virgin timber was depleted; the village died. Only vacant sand plains remain. It's a mystery where the houses once stood.

A post office was usually part of the lumberman's settlement, though often of limited function. When the International Paper Company established a settlement at the Woods Lake flag stop in 1916 for their short pulpwood railroad, they had barely enough residents to warrant a post office. It opened on November 7, 1917. In 1918 Postmaster Fletcher reported an average of twenty dollars of cancellations

47. A multiple-saw sash gang mill that could saw a number of logs instantaneously. Courtesy of the Tupper Lake Public Library.

monthly. He had to hang out a mail bag each day for the passing New York Central mail train, even if the bag was empty.

The Mills

Large production sawmills came into their own in the Adirondacks shortly before the advent of the railroad logging era. Volume production began back in the mid-1800s when the mills became equipped with a multiple arrangement of sash gang saws that were able to consume prodigious quantities of pine and spruce timber. The sash gang was a frame or sash with a multiple set of saw blades held in a vertical position. The frame would oscillate vertically to produce the sawing action as the log or squared timber was moved through. A frame could have as many as thirty-six blades set in it.

Each machine was referred to as a "gate," and a large mill might have as many as five or six gates. There might be a slabbing gang, a stock gang with thirty or more blades set to produce one-inch-thick lumber, a gang with fewer saws that were set further apart to produce thicker plank, plus possibly a couple of up-and-down saws. A couple of circular saws might also be found. Such an arrangement of variously sized sash gang saws was very noisy and kept the mill building in constant vibration. But a six-gate mill could easily cut fifteen million board feet yearly with two shifts and a period of shut-down for low water, their source of power.

The Glens Falls industrial area had a number of these large sawmills equipped with gang saws to rapidly consume the Adirondack spruce that the loggers floated down the Hudson River. When Patrick Ducey built his large Adirondack sawmill at Brandon in 1886, he installed a number of eight-blade gang saws to process the large white pine found along the St. Regis River. The nearby Everton Lumber Company had two mills on another branch of the St. Regis River, employing a combination of circular and sash gang saws, and turned out seventy thousand board feet of softwood lumber daily in the late 1880s. Soon afterward, a new type of saw was developed for sawmills—the band saw. The large band saw was quickly adopted as a practical piece of machinery for the main head-rig in the sawmills. Huge sawmills were built that required a large river or a railroad to keep them supplied with logs. Specialty mills such as cooperage factories were constructed, consuming fifteen million board feet and more yearly. The railroads were kept busy.

48. All three types of saws were displayed in the filing room of a Tupper Lake sawmill: circular saw, band saw, and sash gang saws. The well-dressed filer was named Barden. Courtesy of the Tupper Lake Public Library.

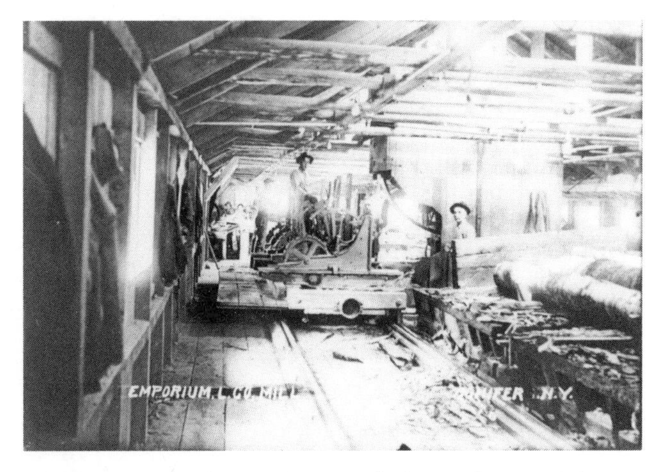

49. The Emporium Forestry Company had an early version of a band sawmill at Conifer. Behind the sawyer on the right, the large band saw wheel and housing can be seen. The log carriage on the left had two riders who repositioned the log after each cut. Circa 1912.

John Hurd, builder of the Northern Adirondack Railroad, put Tupper Lake on the map when his railroad reached that location in 1889, and he built in the following year a sawmill surpassing anything that had ever been in New York State. Known as the "Big Mill," it was capable of sawing from twenty to thirty-three million board feet yearly.

Band-saw mills gained rapid acceptance once they became developed for lumber production. The Moose River Lumber Company built a band mill in McKeever in 1894 that was sawing sixteen million board feet of spruce yearly. The Emporium Forestry Company built two efficient band mills to saw hardwood—their original mill at Conifer and a later mill at Cranberry Lake. The Conifer mill cranked out ten million board feet of hardwood yearly for thirty-seven years, but the expensive Cranberry Lake operation doubled that to almost twenty million.

Sawmills were not the only wood-product industry that supported these new Adirondack forest villages. Sometimes the company town would support a complex of wood-product manufacturers. The original or principal company ownership might entice and make attractive a situation for other manufacturers to set up shop on the location. These other manufacturers would have to be consumers, of course, of the species or grades of logs not used by the company that founded the village. Thus, a number of different types of wood users would be lined up along the railroad track in their remote company-town location.

With a complex of different mills, the obvious advantage to the company owning the tract would be that markets were created for species of trees otherwise not harvested on the company-owned timberland. And, of course, a larger industrial base was provided to support the village. The advantage for the smaller secondary industries was that a source of logs was provided, delivered right to the mill by rail. Of course, the health and future of the smaller mills was dependent on the health of the principal operation. A cause of failure often noted with such villages was an insufficient supply of timber to last the anticipated life of the village. Reasons for the failure were: the quantity of timber proved to be less than reported by timber estimates; the timber volume was intentionally misrepresented by the seller; the quality of the timber proved inferior and thus unusable; or severe fires swept through to wipe out part of the timber resource. However, there were some successful and profitable industrial villages created in the Adirondack north woods, even if they were of short duration.

When the Goodyear Lumber Company purchased a large tract of timberland in Jerseyfield, along the southern edge of the Adirondacks, they were able to create an industrial complex that it was hoped would last for at least ten years. At the terminus of the logging railroad in Salisbury, the Goodyears arranged for the construction of a wagon-hub mill (using yellow birch); a veneer mill (using high-grade birch and maple); and the largest mill in the complex, a cooperage mill (using the lower grades of hardwoods). However, the timber supply was apparently exhausted in less than ten years, and the mill owners gradually moved their operations elsewhere.

The ultimate in relatively short-term Adirondack industrial complexes was achieved by the Rich Lumber Company at Wanakena. Profiting from their experience in creating full utilization operations in Pennsylvania, the Rich family created a complex that has possibly never been equaled in the northeastern United States. Lined up along one edge of the log pond were five different industries that could together use about any log brought down by the company railroad. There was the principal industry, a large sawmill consuming spruce and pine sawlogs; a whip-butt mill (using birch logs); a barrel-heading mill (using low-grade hardwoods); a shoe-last mill (using hard maple); and a veneer mill (using high-grade yellow birch).

None of the Wanakena mills was owned by the Rich Lumber Company; the sawmill was manufacturing lumber under contract for Rich Lumber, and the other mills were purchasing logs delivered by rail. The entire operation was short-lived; within eight years the timber supply was gone, partially depleted by severe forest fires.

Common Carriers

When the new villages were created by a lumbering firm, they were faced with the need to provide employees for the mill, for the railroad, and for the logging camps. If the village or the mill was located remote from a mainline railroad, as they oftentimes were, there arose the necessity of providing transportation for passengers as well as for the mill products going out to market. What better system of transportation was there than to convert part of the company railroad system into a "common carrier." A common carrier designation would allow them to purchase rights-of-way over other ownerships if needed as well as to carry revenue passengers. And in a few cases even the entire logging line was converted into a regular common carrier after the initial logging chores were completed.

The best known of the lumber company common carriers in the Adirondacks was the Grasse River Railroad, built by the Emporium Forestry Company as a sixteen-mile rail link between the New York Central at Childwold station and the village of Cranberry Lake, passing through the company town of Conifer en route. The Grasse River existed for forty-

50. Grasse River Railroad No. 43, a Brooks-built 0-6-0, takes on water from a brook during a passenger run to Childwold in the 1920s. This small locomotive with forty-eight-inch drivers was built in 1883 and was later sold to a Pittsburgh steel mill.

51. Grasse River Railroad No. 2 passes the Conifer Depot during a December 23, 1949 snowstorm, running extra with freight cars for the NYC interchange at Childwold. Photograph by Philip R. Hastings.

two years under the ownership of the Emporium and its successor, the Heywood-Wakefield Company. In its prime years, during the 1920s, the short line was running two trains daily in each direction. So fascinating and unique was the lumber company's little Grasse River Railroad that it became the subject of rail fan interest for many years as a provider of many fond memories such as those recounted in the chapter on the Emporium Forestry Company.

When the Gould Paper Company interests built a nineteen-mile logging railroad into their extensive timberlands on Tug Hill, just west of the Adirondacks, they incorporated with the authorization to carry passengers as well as logs and other freight. Chartered as the Glenfield and Western Railroad, their common carrier existed for about twenty-eight years.

The common carrier built by the Rich Lumber Company had a shorter life. Chartered as the Cranberry Lake Railroad, this line formed a vital six-mile link from the com-

52. The Grasse River Railroad junctioned with the New York Central at Childwold. The company's Mack rail bus was on hand to meet New York Central train No. 2 at Childwold in 1949. Photograph by Philip R. Hastings.

53. The afternoon train had just arrived at Wanakena, discharging a full load of well-dressed passengers. It was a warm summer day as the vacationers began to stroll up the boardwalk toward the Hotel Wanakena. Circa 1910.

pany mill village of Wanakena to the outside world, where it junctioned with the Carthage and Adirondack Railroad at Benson Mines. The life span of the Cranberry Lake Railroad was shortened to twelve years when the timber supply became exhausted in 1914.

Of even shorter duration was the seven-mile Newton Falls and Northern Railroad, built by lumberman Robert W. Higbie after he had obtained a large tract of timber from the Newton Falls Paper Company. A sawmill and village were created on the South Branch of the Grasse River and named New Bridge. Its eleven-year existence came to an end in 1919. There were other early logging pikes that eventually changed their character, by design, and became bona fide passenger lines. Included as such would be Hurd's Northern Adirondack Railroad and Webb's Mohawk and Malone Railroad (both later part of the New York Central system) and the earlier St. Regis Falls and Everton Railroad.

CHAPTER SEVEN

Logging the Old-Growth Timber

Logging is a strenuous occupation that underwent very few changes during the 1800s and early 1900s. By its very nature, the harvest of trees had resisted most mechanization efforts until only recent years. Back when the loggers first entered the Adirondacks from the settlements on the fringes and worked their way up the rivers, it was the ax, the horse, and the swift river waters that provided the mills with the needed raw material. That movement into the Adirondacks began in the mid–eighteenth century. By the mid–nineteenth century the loggers had reached the heart of the wilderness, and it was still the ax and the horse and the river currents to transport the logs—a hundred years of logging with no change in the tools used.

Even at the advent of the era of logging railroads, the late 1880s, the life and labor of a logger had changed very little. Occasional changes were to occur, but ever so slowly. Until about 1890 trees were felled and bucked into logs by ax alone, a tedious method. Then a new tool, the crosscut saw, was introduced. In the northern Adirondacks, lumbermen Patrick Ducey and Peter MacFarlane had moved in from Michigan in the 1880s and reportedly brought with them a new technique of sawing trees instead of chopping them. Ducey went on to build the sawmill village of Brandon and MacFarlane the sawmill village of Everton and the Everton Railroad.

54. The crosscut saw was introduced into the Adirondacks, it is said, in the late 1850s. Prior to that, the trees were cut into log lengths by ax alone. Loggers for the Robert W. Higbie Company are felling a large maple. Circa 1904. Source: N.Y. *Assembly Doc.*

Changes came slowly, however. Even after the crosscut saw was introduced into the Adirondack woods, it remained the custom for many years in some areas to fell the trees by ax only and then to cut the fallen tree into log lengths by means of the crosscut saw. Only the most expert of the axmen were assigned to fell the trees.

In the eyes of the public and the popular writers of the time, the life of the lumberjack was a romantic one. The lumberjack knew better, but he seldom traded it for an easier life during his younger years. As with loggers throughout the United States, the North Country lumberjack had to be not only strong physically but strong in self-reliance, hardy, and dependable as well. He had to be content living under conditions that most people would consider deplorable, for months at a time. The lumberman was a true character of American history whose deeds are part of rich folklore.

Gradual changes did come to the Adirondack loggers, but they came slowly. It was during the railroad era that the first mechanized competition appeared for the horse that had been such a hard-working stalwart for the logger from the very beginning. Military technology in World War I had demonstrated the practicality of crawler tracks or tanks, and after the war this concept made gradual inroads into the North Country woods in the form of early tractor models.

For the first one hundred years or more, the Adirondack logger sought only the mature softwood trees. The market at the time would accept only large logs, but there were plenty

55. The white pine grew large around Cranberry Lake, and these six selected logs made an outstanding sled load measuring 5,781 board feet. The load was pulled by a single team of well-matched horses. Photograph by H. M. Beach.

Table 1
Timber Values in 1911

	Tree	*Price ($)*	*Standard*
Stumpage			
	Spruce	.60–1.00	market log
	Pine	1.00–1.25	market log
	Hardwoods	.60–2.00	market log
	Hemlock	.60–1.50	market log
	Hemlock bark	2.00–3.00	cord
	Poplar, basswood	2.00–2.50	cord
	White birch	.50–1.00	cord
Delivered to Mill or River Bank			
	Spruce	2.00–3.00	market log
	Pine	2.00–3.00	market log
	Hardwoods	1.50–2.50	market log
	Hemlock	1.50–2.77	market log
	Hemlock bark	6.50	cord
	Poplar, basswood	6.50	cord
	White birch	4.50	cord

Notes: "Market log" is a log of specified size that was used as a standard in determining volume of logs; it was used instead of a log rule in the early years in the Adirondacks. The Glens Falls standard was a log 13 feet long, 19 inches in diameter; Saranac standard was 13 feet long, 22 inches in diameter. Five market logs was considered to be one thousand board feet. For hemlock bark, 2,200 pounds equaled one cord.

of old-growth trees to choose from. The forests were not cut heavily.

However, the late 1800s were to spawn some market shifts that resulted in some drastic changes in the handling of the Adirondack forests. And the technology of logging railroads was to play a part.

The Logging Camp

Another of the aspects of a logger's life that changed but slowly was the logging camp. The camp was a necessity, a seasonal home in the wilderness for workers who probably had traveled far from home to pursue their chosen woods employment. Large groups of experienced woodsmen made the trek south from their native Canada, and newly arrived immigrants from abroad came north looking for work. A few loggers even drifted over from the New England area and the Maritime Provinces. Few of the Adirondack lumberjacks were original residents of the region.

Before 1914 and the First World War, there were more than 150 camps operating in the Adirondacks during the logging season, employing over seven thousand men. And what a mixture of nationalities there were to be found working together! In 1890 one-third of the Adirondack loggers were foreign born, immigrants from Ireland, Norway, Sweden, Russia, Lithuania, Romania, Germany, and Italy. That was not true of the northern part of the Adirondacks, however. Both the Brooklyn Cooperage Company and the Santa Clara Lumber Company found that their woods labor in that area was 75 percent French Canadian. After the Great Depression hit in 1929, the demand for timber took a dive,

56. Logging camp of Santa Clara Lumber Company in the Cold River country. Loggers would frequently pose in a boxing stance for a photographer, even with a pipe in the mouth, or climb on each others shoulders. Courtesy of the Tupper Lake Public Library.

and by the winter of 1932–33 the number of logging camps in the Adirondacks dropped to about twenty. Logging employment became scarce for the itinerant lumberjacks.

A mixing of nationalities in the camps was often a practice to be avoided, however. Feuds could be dangerous, and the difficult task of diplomacy lay in the tactful hands or tough fists of the camp foreman. Separate work areas were often set up; possibly even camps in separate locations were needed to keep nationalities separated.

The Rich Lumber Company had separate camps for the Italian woods workers just west of Wanakena village. But the company found that they even had to have separate camps for the immigrants from northern and southern Italy. They were bitter enemies who sometimes didn't hesitate to settle their arguments with a sharp knife.

The Racquette River Paper Company set up two large camps near the Oval Wood Dish Company camps at Kildare, which were manned primarily by Russian immigrants who had left Russia at the time of the Bolshevik Revolution. Although communication was quite difficult, they proved to be excellent loggers who learned quickly.

57. The older logging camps were constructed of logs with sheets of spruce bark covering the roof. Notice the long building divided into two sections, the men's sleeping quarters separated from the kitchen and eating room.

In the earlier years of the nineteenth century, the logging camp was, as one might expect, quite crude. A single low log structure, possibly sixty feet long, it had a dirt floor, and the spaces between the logs were chinked with moss. The bunks were crowded into one end of the building, and the eating facility plus the kitchen were in a separate room on the other end. A small barn housed the horses. Perishables and vegetables might be stored in a small building built into a bank for cool temperatures. Or a storage house might be built over a stream or spring, using double log walls with a foot of dirt between to create a cool temperature by condensation.

Crude as these camps were, they might hold as many as forty to fifty men. It wasn't unusual to find a few pigs in residence also, feeding on scraps, and even a cow. Frank Cutting had a large year-round center camp on the Brooklyn Cooperage Company railroad near Lake Ozonia that was used to supply his other logging camps. It was customary there to see fifty to one hundred pigs, twenty cattle, and twenty sheep, as well as chickens.

By the advent of the era of railroads on the log jobs, the camps had improved in layout and comfort. Two-story buildings were common, often of lumber construction, with wood floors and tar paper on the roof. The bunkhouse was separate from the mess hall, although some camps put the bunk room on a second floor above the eating area. Bunks were double-deckers, even triple at times, and with heat supplied by a cast-iron stove in the center of the room, the man in the top bunk had to suffer the heat while the man below almost froze during the bitterly cold nights. There was little ventilation of the rank nighttime air in the building, but the men got plenty of fresh air all day long to make up for that.

Hygiene was little regarded in most logging camps, especially during the winter months, when the men seldom bathed. Beds could be noticeably filthy. Food was the cleanest part of the outfit.

An office building contained the living quarters for the camp foreman and camp clerk plus the company scaler when he stayed overnight. Usually the "van" or company store was in the same building and supplied all of the essential elements such as wool clothing, patent medicines, pain killers, and tobacco. A large barn would always be nearby to

58. The odor from the wet socks mixed with the wood smoke would create a unique air in the bunk room of the logging camps. Source: N.Y. Assembly Doc.

house the horses and to store the hay and oats. Such a camp would support from sixty to possibly one hundred men.

There probably was never a logging camp immune to the most unwelcome of visitors: body lice and bed bugs. The little critters made their presence most uncomfortable for the logger needing a good night's rest, and once they arrived they were determined to stay.

Scrubbing with yellow soap helped a little, but one of the Sunday camp chores was boiling the "ticks," the canvas straw mattress coverings, in the sixty-gallon copper "cootie kettle" to make an attempt at killing the little critters. Afterward, the tick would be filled again with fresh straw from the barn. Clothes were also boiled in the kettle to destroy the troublesome bugs. If the problem persisted, the bunk boards could be wet down with kerosene, which, despite the odor, would provide some relief, for a couple of weeks anyhow. Mosquitoes and punkies were discouraged with a liberal rubdown of lard mixed with a small amount of roofing tar.

Accidents in the woods were frequent, both to men and to horses. Not uncommon was the incident at one of the camps of a contractor working for Oval Wood Dish Company in 1915. A loaded log sled left the road on a turn and threw the teamster into the woods. He suffered a fractured skull and died soon after.

Bad cuts and other injuries were handled as best the camp boss was able to do. A logger once described his own first-aid technique as follows: first "chew up a wad of tobacco, then spit it onto a large witch hobble leaf and finally tie the leaf over the cut with a dirty handkerchief." Pine pitch was said to be a remedy for about everything. Few knew what a tourniquet was. About the only medicine that Oval Wood Dish Company kept in their camps was a large bottle of iodine. Pour it on the cut and wrap it.

Many of the camps had telephones, but little would be gained by calling for a doctor in the event of severe injuries. The camps were too far from town. If the injury was bad enough, the man was placed on a log sled for a bumpy ride down off the mountain to a point where he could be placed aboard the next log train headed out. Then he would be shipped out to Tupper Lake for the main office to worry about. There was even less that could be done in the fall of 1918, however, when the worldwide plague of Spanish flu hit the logging camps. Many loggers lost their lives.

Food in the logging camps was legendary and usually

A Woodsman's Tale

Life in the lumber camps spawned some strange tales that were passed on as factual events. Note this tale about a logger working near Spruce Lake in the Adirondacks, as related in the *Malone Palladium* of September 3, 1903.

"Some time ago, William McCoy was working with John Duffy getting out some long poles to repair a chute, which is used to slide logs down the mountainside. Duffy went to cut a limb with an upward swing, when the ax slipped from his hands and went flying through the air. It struck McCoy, and its keen edge shaved off a part of his cheek.

"William Henry, a nurse, of Utica, took McCoy in hand. After having partially stopped the flow of blood he went out to the stable, took a little fawn that some of the boys had captured two days before, shaved the hair for about nine inches off the animal's side and then carried it to the camp. While some of the woodsmen held the fawn, Henry cut the skin, peeled it off and applied it immediately to the face of McCoy. Having filled it in place firmly, he rubbed over it a thick coat of balsam gum, and over that he placed tight bandages. The cheek stopped bleeding and the graft was found to be a success. Soon afterward, however, McCoy noticed when he drew his hand across his cheek that hair was growing on the grafted skin. In a few days more the hair had grown so thickly that its color and nature were plainly visible. It was the hair of the fawn growing, and moreover, it was spotted like that of a fawn. He did not dare shave for fear of breaking the skin and allowed the hair to grow until the fall of that year. Then the spots disappeared and the "blue" coat of a full grown deer took its place. When spring came round, he saw that the hair of the cheek was falling out, and fine red hair was growing. At last the blue, or winter coat, was entirely gone, and the red summer coat took its place. In fact, he and the other woodsmen, to their merriment, saw that the hair varied and changed precisely as does the coat of a deer."

59. The woods crew is poised for the evening meal in the mess hall of a Santa Clara Lumber Company logging camp. Note the variety of ages.

lived up to their fine reputation of supplying the needs of the hungry woodsmen in copious quantities. The camp boss had to be sure the men were fed well if the loggers were going to be kept at the camp for the season. At the sound of the "gut hammer" or gong, the waiting crew would file to places probably set with inverted agate plates topped with agate cups upon tables covered with a brown oilcloth. The sugar might be put out in tomato cans, the inverted spoons standing up in tins.

Staple food items included such things as meat, beans, potatoes, tea, and coffee. Carrots, cabbage, turnips, and onions would likely be found in the cool storage, and there would be a barrel of vinegar and a barrel or two of salt pork. Breads, pies, and cookies were baked at the camp and were usually a delicacy. Pancakes were served about every morning. Loggers had to be fed well, and the word soon passed around regarding any particular camp or company that didn't feed well. The loggers might refer to such a camp as a "whistle stop" or a "greasy sack outfit."

Occasionally the cook was a woman, especially in the earlier years. Although female camp cooks were not common, a tough woman could hold her own in the verbal battles with the lumberjacks. She had to gain their respect or there would be a disruption of the work at hand.

There seemed to be an uncontested rule in logging camps, no matter where one went in the nation, that silence was to be maintained in the mess room during a meal. The only sound was the rattle of iron forks and spoons on the agate or tin dishes and cups intermingled with the burps and slurps and an occasional request to pass some food down.

Strong drink was not permitted in the logging camps, not even beer. But Rich Lumber Company made an exception with their Italian camps. Beer had always been an essential element of the Italians' homeland diet, and they wouldn't work without it. In view of the hard work they rendered, this exception was a wise compromise.

Most of the camps established by the railroad loggers were built along the rail lines. It was a matter of convenience; supplies were delivered on flat cars or on old boxcars. Camps sited in more remote areas would have to be supplied by wagon over a tote road from the nearest convenient rail location.

The Logging Crew

The woodsmen were customarily divided into crews that would work together. Brooklyn Cooperage Company divided their men into a crew of nine: four sawyers or fellers to drop the trees, two swampers to cut out a path for the skidding team and to limb the fallen trees, one teamster with the skidding horses, and two rollers or canthook men to pile the logs. The logs were generally piled up on a skidway awaiting snow so that they could then be sledded to the railroad loading points. Or if close enough to the railroad and if the con-

60. The loggers tools are displayed by a crew cutting for the Rich Lumber Company. Having notched the tree with double-bitted axes, they are about to make the back cut with the crosscut saw that the one logger is holding on his shoulder; a maul and a wedge might have been used to keep the saw from binding in the cut.

61. The skid horse was a hard-working and intelligent animal. These horses were skidding single spruce logs to the skidway to be piled for loading on sleds. Source: N.Y. Assembly Doc.

62. The scaling crew measured, tallied, and stamped each log. Stamping each log with the owner's brand indicates they were probably to be put in a river for a log drive. Note that all logs are of the thirteen-foot Adirondack standard length.

tour of the land afforded the opportunity, the logs were skidded singly downhill to the rail-loading location.

Santa Clara Lumber Company used a crew combination of five men and one skid horse. Tree felling was done the conventional way, notching by ax about waist high on the side toward the fall and felled with a crosscut saw. The company would set quotas for the crews as high as fifty logs per day. In the central Adirondacks in the mid-1800s it was said that an average production for a good axman chopping the tree into log lengths was seventy hardwood logs, possibly eighty. With softwood, a good chopper could put forth a hundred logs per day.

All of the softwood logs cut in the Adirondack forests in the 1800s were cut thirteen feet in length. It was a custom peculiar to the Adirondacks because elsewhere in the state softwood logs were cut sixteen feet if possible, as well as in shorter lengths when necessary. The reason for adopting a thirteen-foot Adirondack standard is unknown. Even William F. Fox, the New York State forestry commissioner in the 1800s who wrote the 1901 history of the lumber industry in New York, confessed that he did not know the origin and that it was a one-hundred-year-old custom at his time. The thirteen-foot standard continued until at least 1920, in some areas into the 1930s.

After the logs had been piled on a skidway, the scaler and his assistant would "scale" or measure the logs for board-foot content. The scaler was an employee of the jobber or company and traveled around to the various logging camps.

If the jobber was harvesting hemlock bark, commonly done in the years before World War I, the crew might consist of only three men. Operating during the peeling season of mid-May to mid-August, the bark was taken off in four-foot-long sheets. After taking off the first four-foot-high ring of bark from the base of the standing tree, the hemlock tree was felled and the trunk girdled with an ax at four-foot intervals. Another man behind him would pry the bark off with a bark spud, and the sheets of bark would be stacked for loading onto a sled. Although the hemlock tree was sometimes cut into logs after the bark was removed, it was sometimes just left to rot where it lay. The market for hemlock logs was not always the best, hemlock not being a preferred construction lumber.

63. Hemlock bark harvesting by immigrant Italian wood cutters. Note piles of bark sheets next to peeled logs.

64. Large quantities of hemlock bark were moved by the logging railroads. In this scene on the Mac-A-Mac Railroad, a New York Central locomotive waits for loading at a jackworks. Note the long stacks of hemlock bark. Circa 1915.

65. Bark was being loaded on a boxcar at a siding deep in the woods. The bark sled had no metal nails or fasteners. From the collections at College Archives, SUNY ESF.

After the fallen tree was bucked into log lengths, the logs were first ground-skidded a short distance to a point where they could be stacked onto a pile for sled loading. In the older days the logs were jacked up onto the pile with a six-foot pole made of a tough wood such as ash. One end of the pole was flattened into a wedge shape. Later a spiked iron socket was fitted on the end of the pole, and eventually the peavey became a standard tool for rolling and prying the logs.

Log Sledding

As the logs were being cut and stacked on a skidway, other crews would be cutting out the haul roads for the log sleds to take the logs down to a location where they could be banked for a later river drive or piled for loading on the railroad. The main haul road would be laid out in the shortest route to the railside landing that the terrain and grade would permit. Spotting the road location required a skillful eye that could avoid problems of grade for the heavily loaded sleds.

Branch roads would reach out from the main haul road to various log skidways. Frequent turnouts along the main road were necessary for the horse-drawn sleds to pass each other when going in opposite directions. Or the jobber might build a separate go-back road for the returning empty sleds to use, laid out along a shorter route with minimal concern for grades because the sleds would be empty.

Reaching the landing alongside the railroad tracks, the logs would often have to be piled again where they could be picked up by the machine loading the log cars.

Although the tree felling and skidding began in early fall, possibly even in summer, the sled hauling could not begin until about the first of January. The heavily loaded sleds required frozen ground and a good snow cover. And they required good roads, as level as possible. Small trees four to six inches in diameter were cut and piled crossways to fill in the troublesome dips or to build a little elevation where unmovable rocks stuck up in the road. Side-hill or dugway roads were cut into a hillside where a road slabbed across a hill to drop gradually in elevation. Swamp areas were corduroyed with small trees laid crossways. After the snow came, more leveling would be done by filling the low spots with snow, packing it down, and allowing it to freeze. Considerable work was needed to keep a busy sled road in good shape. And, of course, thaws obviously increased the work load for the "road monkeys" who had that responsibility.

When it came time to start the sled hauling, the road would be plowed or scraped fairly level. Ruts from the sled runners would soon wear into the road surface. Except on downgrades, these ruts were kept icy to decrease friction for the sled runners.

Keeping the roads iced became a full-time job for a couple of men and a team of horses, working through the night. A large water-box structure, possibly fifteen feet long and six feet high and wide, was mounted on sled runners and used to sprinkle the roads during the dark, freezing hours. The tank was filled at a water hole dug out in a nearby creek or swamp. A large barrel was used as a dipper; it was filled with

66. The skidways are filled with fine spruce logs ready for loading into sleds on a Santa Clara Lumber Company job, about 1910. The hillside has been cleared of merchantable spruce but the hardwood remains. Courtesy of The Adirondack Museum.

water and pulled by the team up a set of skids to the top of the tank, where it emptied. Traveling along the sled roads, the teamster's helper would pull the plugs on the back end of the water box, sending a stream of water into each sled-runner track. A build-up of ice in the center area between the ruts would be avoided because that's where the horses walked. A tankful of water would generally last for at least a half mile. If the sprinkler head froze, as it was wont to do, a fire would have to be built under the tank to thaw the opening. Kerosene torches provided some illumination along the road, if needed.

A sled road on a downgrade presented a different situation; then the chore was to slow down the sled. Either friction on the runners had to be increased or some type of braking system had to be used to prevent the sled from overtaking the hooves of the fast-stepping horses. Hay or sand would be spread in the ruts by the "road monkey" assigned to care for the sled road. Sand would be kept in a sand shed or dug from a nearby frozen bank, which he would have to thaw by building a fire with some dry rotten wood, roots, or pine knots that were filled with pitch. Some teamsters would wrap chains around the sled runners or clamp a U-shaped piece of iron on the front of the runner that could be swung down into the rut. Or they might drag some logs behind the sled.

An excessively steep sled road made it necessary to "snub" the log load down by means of a cable or a rope wound two or three times around a tree at the top of the grade. Some used a mechanical brake known as a Barienger brake. It was an interesting lesson in horse-sense to watch a team with a sled being held back with a braking system. The first time down the steep slope, the team would be terrified,

67. A fresh snow had fallen and the sled roads had been leveled for hauling out the spruce logs cut the previous fall on a Santa Clara Lumber Company job in the Cold River country. Circa 1910. Courtesy of the Tupper Lake Public Library.

68. The sprinkler sled used to water the sled ruts where ice was wanted. The barrel on top was used to fill the box with water. Photograph by G. A. Whipple. From the collections at College Archives, SUNY ESF.

69. Hay has been spread in the sled runner ruts on this steep descent to slow down the sled. Source: N.Y. Assembly Doc.

knowing that the heavy load behind them could quickly overtake and collide with them. But they soon learned that the load was being safely held back, and they would lean forward in their harnesses and relax as they were lowered down.

There were occasions when laying out the sled roads that having an uphill grade was unavoidable. Usually this meant a reduced log load, but the loggers had an answer for this dilemma. A high topping-off skidway would be built at the height of the grade to top off or increase the load for the remainder of the trip.

The skidways for loading the sleds were built crib-work

70. The Barienger brake was used to slow the descent of loaded sleds on steep grades. The operator with lever in hand would apply friction to the drums around which the cable was wound. Photograph taken near Cranberry Lake, where the Emporium Forestry Company was lowering Linn tractor sleds down a hillside. Circa 1920.

style on a slope where the logs could be rolled across a level top to drop down onto the sled below. It was common to see skidways built in steps as a doubleheader for building tall loads on the sleds. The taller step behind would be used for rolling the logs over to top off the large load.

The type of sled used would vary with the terrain and length of haul. For a relatively short trip down a slope, the log would either be ground-skidded or taken on a bobsled, called in some places a "muley sled." The bobsled was composed of a single set of iron-shod runners connected in the front by an iron "roll rod." A heavy wooden beam was bolted across the center of the runners. Then above that a spike-studded bunk was fastened to the beam at the center of the bunk, allowing the bunk to pivot with the log load. The logs were loaded with one end resting on the bunk and the other end dragging on the ground to create a braking effect.

For longer hauls, the loggers used a "two-sled" or a variation of it. This sled consisted of two separate, independent sets of runners with ten-foot bunks on top, connected only by a pair of diagonally crossed chains. When loaded, and this type of sled could carry as much as five or six thousand board feet loaded to a height of eight or ten feet, the weight of the logs would keep the two sets of runners functioning as a single sled. The crossed chains between the two sets of runners

Table 2
Logging Costs in 1911

Product	*Work*	*Cost ($)*	*Standard*
Hardwood	cutting, skidding, hauling 3-mile haul to skid	.60–.70	market log
Poplar	cutting ($.75) peeling & stacking ($1.25) hauling ($2.00–3.00)	4.00–5.00	cord
Hemlock bark	cutting & peeling ($2.00) piling & hauling ($2.50)	4.50	cord

Lumberjack wages

Year	*Wages per Month ($)*
1870	25–30
1890	35–45[a]
1900	28[b]
1915	26
1919	60–100[c]
1920	90–104
1921	26–40

Note: Room and board was usually included. Pay was for a six-day work week, ten hours a day.
[a]Rated on ability to cut 60, 70, or 80 logs per day.
[b]A woman cook was paid $30.
[c]World War I caused a manpower shortage. Board was deducted.

71. The tall sled loads of logs with high vertical sides were built up by the use of frames such as this to hold the logs until the load was firmly chained. The location is George Bushey's job for International Paper Company at Woods Lake in March 1921.

72. The spring-pole technique of tightening the wrapper chain shows well on this sled load of spruce on a Newton Falls Paper Company operation. A twist is made around the butt end of the pole, which is then bent down and secured under strain by another chain. Pity the poor teamster if the chain holding the pole let go. W. N. Kellogg was woods superintendent for the company. Circa 1917. Courtesy of Bill Kellogg.

would jack the rear end of the sled into a wide turn on the corners. Only the last bunk was spiked.

The wrapper chain around the sled load of logs was tightened by a spring binder, which could be a lethal weapon in careless hands. One end of a stout stick was placed under the chain and then swung forcefully around to twist and tighten the chain. Securing the pole so that it did not suddenly whip back was essential. While the sled was being loaded, the runners would sometimes sink into the icy ruts and freeze in place. Many heavy blows on the runners with a maul would be needed to assist the horses to get the sled moving.

Horses were treated with exceptional care, and in some respects with more kindness than that extended to men. But one can appreciate the need for such concern. The teamster had the responsibility of care, of course; he tucked them in at night and went to check on them first thing upon rising. The teamster was up possibly by 2 AM with breakfast at 3 AM and on the road with his team in the bitter cold of the early morning, well before the first rays of light filtered through the trees. By 5 AM the sled was loaded and ready to move.

Most of the horses were Belgians, often obtained from farms in Ontario for $80 to $110 each; in the years after World War I, the price was over $300. A good logging horse knew the special commands of the teamster and was able to sense various hazards quickly. A well-trained skid horse could even work alone without an escort, twitching a log from the cutting crew down to the man at the skidway and returning without anyone walking along with him. If the log hung up en route on an obstruction such as a rock, the horse knew enough to "gee and haw" in different directions on his own until the log came free.

If a horse fell in deep snow, it became quite difficult for him to get back on his feet with the harness in place. He would lie still, as trained, until the teamster unhooked the straps and chains. Most horses would readily walk across a railroad trestle, carefully stepping on the ties. Working eleven hours a day during the season, a horse could be expected to last about six years.

Old-Time Lumberjack Competition

The loggers rarely indulged in any type of competition of their cutting skills. It wasn't a popular pastime, and there was little time allowed for it in the busy logging camps. But there was one exception that became popular and widespread—the putting together of a champion-sized sled load of logs. The feat was considered to be quite a challenge, and the word about record loads soon spread from camp to camp. Of course, once constructed, the load had to be hauled by the horses to the normal destination.

When Fred Johnson was running Carlson's Camp on the Mac-A-Mac logging railroad in 1914, he arranged for the landing crew of F. Morrisey and Thomas McGarren to select the logs and then put together a record sled load. They rolled 298 logs onto one sled, a mammoth load sixteen feet high and sixteen feet wide that scaled $21^{3}/_{8}$ cords and must have weighed about thirty tons. Teamster Joe LaPort hauled it out with a single team over carefully iced roads. Fortunately, arrangements had been made for photographer Henry M. Beach to be on hand for the occasion to record what must have been a record feat.

In that same year, Nathaniel Ingram's team was hauling logs on a job near Sacandaga Lake, and his horses sledded an oversized load of 157 logs. The fact was publicized to the extent that it upset the crews in the nearby camp of George Perkins. Teamster Selah Page vowed he could do better, and he was urged to try. The loading crew loaded 220 logs on his

73. This massive load of peeled spruce logs on Carlson's Mac-A-Mac Corporation job at Brandreth in 1914 was considered to be a record load. The scaler at the rear of the load measured a load width of sixteen feet and a height of sixteen feet, scaling $21\frac{3}{5}$ cords for the 298 logs. A single team drew the estimated weight of thirty tons on well-sprinkled sled roads. The sled was expertly loaded by F. Morrisey and Thomas McGarren, and the team was handled by Joe LaPort. Photograph by H. M. Beach.

74. A claimed record load of $12\frac{5}{8}$ cords of four-foot-long spruce pulpwood by the Beede Brothers in Keene Valley. The single bunk rack was thirty-two feet long.

sled, used thirteen toggle chains, and claimed this to be a record never exceeded.

Innovations and Mechanizations Appear

Changes in the woods came slowly and frequently resulted only when a practice was introduced from areas outside of the Adirondacks.

When the Emporium Forestry Company moved their operation into the Adirondacks from Pennsylvania, they brought with them a scheme apparently untried in New York. Pennsylvania loggers would build a U-shaped slide constructed of logs or lumber, hook a few logs together in the slide, and have the horses pull the train of logs along the greased slide surface.

75. The log slide patented by Frank Sykes in use by Emporium in the Dead Creek area. The hardwood logs were attached end to end by means of grapple hooks and pulled along the oiled slide by a team of horses. Courtesy of the Sykes family.

In 1914 Emporium woods superintendent Frank Sykes built and patented his own portable log slide or chute made from lumber, which could be transported and assembled in eight-foot sections. Each section used three separate pieces that would form a trough with tapered sides and with slots on the ends to piece the section together with ties along the bottom. A train of logs would be pulled long distances along the slide to a railroad landing, one horse pulling on each side of the slide. In later years a Cletrac tractor replaced the horses. The uphill and level portions were kept well greased, the company buying slide oil by the barrel.

The concept did not go unnoticed in the Adirondacks. Oval Wood Dish Company later employed a log slide near Derrick (Camp 5) on the New York Central Railroad. The slide went slightly uphill and was thus kept well greased.

Gravity slides down the mountain terminating with a dump into a lake or river were occasionally used. A trough was built with the proper gradient to allow free movement of the wood, and a stream of water would be fed into the slide at the top. A jobber operating near Benson Mines built a three-mile slide in the 1890s to convey pulpwood down to the valley, where it could be loaded on the railroad. His creation

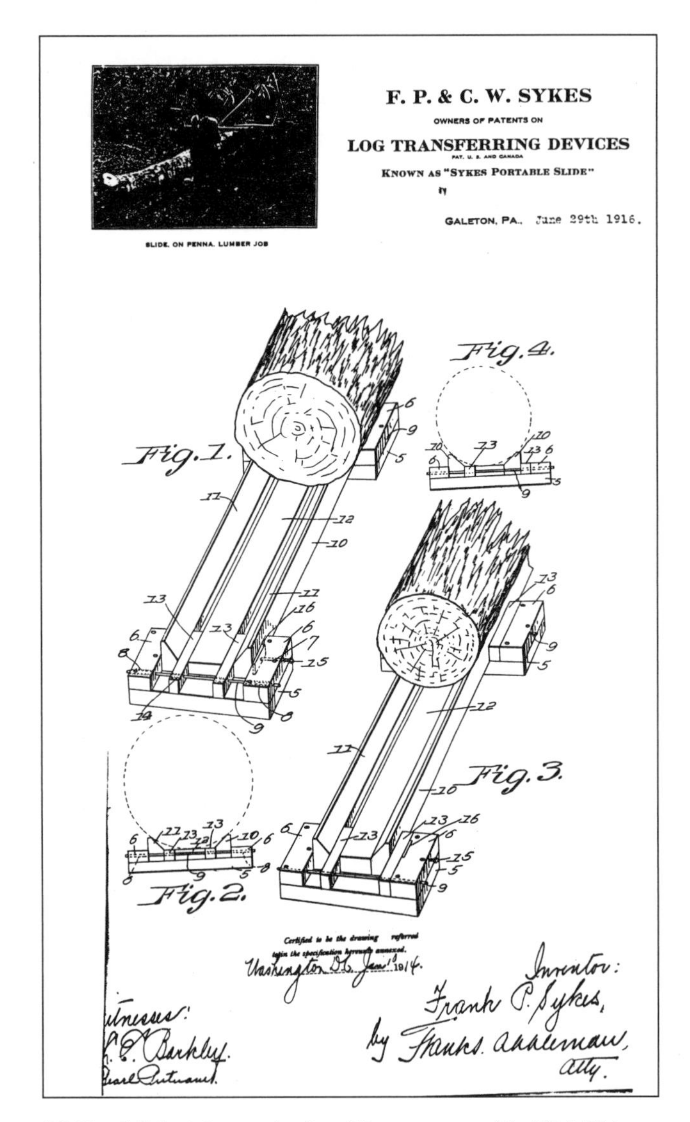

76. *Frank Sykes's innovative log slide was patented in 1914. This drawing was submitted with the patent application.*

77. *The teamster may have been having trouble keeping the log in the trough of Sykes's patented slide. Photo taken 1920 by R. J. Hoyle at an Emporium Forestry Company job near Silver Lake. From the collections at College Archives, SUNY ESF.*

78. *The J. and J. Rogers Company slide was a water trough that carried four-foot pulpwood down from Spruce Hill into the Ausable River.*

could move sixty cords per hour. At one point in time, there was said to be a pile of stacked pulpwood a thousand feet long, twenty-six feet high, and forty feet wide that had been moved off the mountain by the trough.

J. & J. Rogers Pulp Company of Ausable Forks had an even longer water slide at about the same time. Stretching for seven and one-half miles, it dumped the four-foot wood into the Ausable River, which then carried it down to the mill.

When jobber George Bushey was cutting logs for the International Paper Company in the Indian Mountain Pond area south of Cranberry Lake, he used a long slide built of peeled hardwood logs to dump the long logs into Chair Rock Flow on Cranberry Lake. The slide extended for two miles around steep Indian Mountain. The spruce and pine logs were peeled, fastened together with short chains, and pulled along by horse power. As the string of logs approached the point of the slide where it dropped off steeply, the horses were shifted to the rear of the line to push the string from that end. Then as each log approached the steep drop-off, a man knocked off the chain and the log sped down the steep chute and flew off the end into the air before landing in the flow area. The logs were then boomed and towed down to the outlet of the lake, where they were sent on down the Oswegatchie River. It took two days to haul the half acre of boomed logs the distance of six miles down the lake.

79. It took many logs to build the supporting crib work for the log chute. Photo by W. A. McDonald in 1915. From the collections at College Archives, SUNY ESF.

80. The snubbing drum set up by the Santa Clara Lumber Company to raise and lower the log sleds down a steep grade was formerly part of a steam log skidder. Photograph by H. H. Tryon, 1916. From the collections at College Archives, SUNY ESF.

Steam mechanization made it possible to pull floating logs out of the water also. As reported in a number of the individual railroad histories in the next part of this book, mechanized jackworks were set up on the banks of a river or a lake to jack the logs out of the water and load them onto railroad cars. The logs were either taken out of the water on a conveyor or by a steam-powered boom device.

Steam-powered cable logging systems never gained a large foothold in the northeastern United States, including the Adirondacks. The terrain is much less challenging than that of the Northwest, and the timber is not of such prodigious size. Occasional attempts at cable logging were tried in the Adirondacks, however.

Oval Wood Dish Company had a Lidgerwood steel tower rail-mounted log skidder that used a high-lead slack-line system. Its unknown how long it was in use, but in 1918 it was set up on the slopes of Mount Matumbla when logging out of Kildare. Logs were loaded on the flat cars with a gasoline jammer.

The Santa Clara Lumber Company employed a Lidgerwood system for a while on Mount Seward and other steep slopes. The logs thrashing around in the air as they were cabled into the landing by a trolley traveling along a high-lead cable were so destructive to other young trees that the company suspended the use of the system. The company then modified the system to use ground leads for cabling the logs to the landing. But ground-cable skidding on the rough terrain was found to be impractical as well as too destructive to young growth to suit the company. Again, horse power won out in the Adirondacks.

However, the practical-minded woods boss for Santa Clara found a good use for the steam engine on the skidder. The engine and a cable drum were mounted at the top of a mountain slope and used to haul up the horses and empty log sleds. Even the horses rode up in comfort on a special sled.

TRACTORS ENTER THE WOODS

The Gould Paper Company is credited with being the first to use tractors in the Adirondack woods. The company owned large tracts of pulpwood on the watersheds of the Black River and the South Branch of the Moose River, where they were conducting some large timber harvests in the early 1900s. Although Gould Paper did operate a logging railroad over on the Tug Hill plateau, west of the Adirondacks, the Adirondack pulpwood was being driven long length down the Moose River to the pulpmill at Lyons Falls.

Hauling sleds loaded with five cords of pulpwood from the Black River valley over to the Moose River involved a long grueling trip for the horses, and it was only one trip a day for many teams. The sled road had to surmount the high ground of Ice Cave Mountain, which separated the watersheds; a steady, stiff grade of four miles had to be pulled from the valley to that summit.

Five cords of pulpwood will weigh about fifteen thousand pounds, and when combined with a rugged sled that weighed in the vicinity of two thousand pounds, it made a heavy load to pull uphill by two horses that may have weighed a total of thirty-five hundred pounds. As many as 150 teams were working the haul in the season of 1917–18, with several of the teams stationed at difficult slopes to double up on the sled. It was a stupendous haul that season under the direction of woods superintendent John B. Todd, but it was apparent to the company that some changes needed to be made if all of the wood was to be taken out of the cut along the Black River.

In a story well known in the annals of Adirondack logging history, John Todd met with a Mr. Linn, who had requested a meeting with the company to discuss a new type of caterpillar or endless-track machine that he had devised. At the Utica meeting, Linn expounded the advantages of his new machine, which was propelled by a crawler track in the rear and steered by sled runners in the front. It was similar to the older steam Lombards but powered by a gasoline engine instead of steam.

The machine price of $6,000 deterred John Todd from a purchase agreement, but at Linn's suggestion a demonstration at Linn's expense was allowed. A tractor was brought onto the Moose River operation and aptly demonstrated, as he had predicted, that it could do the work of twenty teams. Todd bought three Linn tractors for that hauling season of 1918–19, and the company soon had a dozen or more of these machines, which were made in Morris, New York. The later, larger models were able to haul as many as twenty loaded sleds in train fashion along a well-made sled road. Gould continued to use Linn tractors to log the more remote pockets of timber.

81. The Linn gasoline tractor was a pioneer among Adirondack tractors, using crawler tracks in the back and sled runners on the front for steering. They were made in Norris, N.Y., and were received favorably by Adirondack loggers.

The standard type of crawler tractor also made its introduction into the Adirondack timber at the same time. World War I had shown the practicality of endless crawler tracks on rough terrain, and a few stripped-down surplus tank models made their way onto logging jobs. Emporium Forestry Company was quoted a price of $3,000 by the War Department for a Renault tank with armor plate removed.

The Holt Manufacturing Company (later absorbed by Caterpillar Company) offered a five-ton and a ten-ton crawler tractor for the loggers. George Bushey, a successful logging contractor for many years with the International Paper Company, preferred the Holt over the cheaper-to-operate Linn tractors, but found the five-ton Holt to be too light. On the Woods Lake operation he ran three ten-ton Holts, finding that if the job called for two machines he needed three on hand to allow for breakdowns.

Bushey still had a good stable of horses, because it was his maxim that if the haul was no more than six or eight miles with no uphill, horses did the job more cheaply—provided, that is, that the job took no more than two years to complete.

Other operators used a different standard. A tractor was not practical, they said, on a haul short enough for a team of horses to make two trips per day, hauling eight to twelve cords per load. The tractor would require a haul of at least seven miles and the capability to haul forty to fifty cords per trip. Otherwise, horses did it cheaper.

A ten-ton Holt cost $6,500 in 1923, and it required about thirty sleds to keep the tractor properly supplied. Sleds cost about $200; thus each tractor outfit cost $12,500. Operating costs ran $1,000 to $1,500 yearly for each machine and sleds.

Another well-known North Country logging contractor, Clarence Strife, used Holts at Brandreth after World War I. But Strife was unhappy with them on long hauls; they had a tendency to jackknife on steep slopes because they were not able to hold back the load sufficiently. The Holt also gave operation problems in deep, wet snow. Strife, therefore, used the Holt for yarding and the Linn for long hauls.

In 1919 the Cleveland Tractor Company tried unsuc-

82. The ten-ton Holt was one of the earliest tractors used by Adirondack loggers. There was fresh snow on the ground on this cold day in March 1921, when George Bushey's tractor hauled out an unusually large load of 75.77 cords at the Woods Lake operation, International Paper Company.

cessfully to gain a solid place among the Adirondack loggers. They tried to sell their one-and-one-half-ton tractor to the International Paper Company for use in Woods Lake, but resident manager T. J. Wilbur quickly turned it down because of its light weight.

These few basic changes in logging tools and techniques were the only ones evident all during the primitive logging years and on into the era of railroad logging. The scene continued to change very little until the years immediately following World War II.

CHAPTER EIGHT

Forest Fires and the Railroads

The early years of railroading in the Adirondacks, especially at the turn of the twentieth century, paralleled the age of the calamitous forest fires that struck a heavy blow to the north woods. A couple of outstanding years of adverse weather conditions, an increase in flammable ground fuel because of heavy logging, plus a much larger force of men working in the woods created some explosive situations. And the railroads unquestionably became one of the leading causes of a large increase in forest fires. Not the only major cause, however; some fires were ignited far from any existing railroad.

Serious losses occurred in the forests and in the villages dependent upon woods activity. The state legislature was slow to act at the time to alleviate the sufferings, but firm action eventually did occur. Special regulations were promulgated and directed exclusively at the railroads operating within the Adirondacks. Also resulting were new operating policies set forth by the active logging railroads themselves.

Public attention was especially drawn to the horrors of forest fires started by the railroads during the early years of the Northern Adirondack Railroad and the Chateaugay Railroad in the northern section of the Adirondacks. By 1890 the burned areas along John Hurd's Northern Adirondack stretched for miles on both sides of the track. Passengers traveling the seventy miles from Plattsburgh to Saranac Lake on the Chateaugay saw only rocky, blackened land with few visible live trees. Fires started by the railroads were devastating.

The visible horrors of fire were particularly noticeable in the northern Adirondacks because of the terrain and soil types found there. Outwash plains are quite common in this area, characterized by sandy soils that retain very little water and that support the types of plant growth that become very dry in times of drought. Burns were thus able to become severe and would leave the area in a desolate condition for many decades afterward.

After the destructive fires of the early 1890s, there came the disastrous year of 1899. That was one of the driest years on record, spawning one of the largest numbers of forest fires known in the records. Railroad locomotives were reported by the New York superintendent of forests as being responsible for 24 of the reported 322 fires started in the Adirondacks. The village of Tupper Lake was said to have been almost destroyed by fire on July 29–30.

But 1899 was only a prelude to the greater conflagration yet to come. The month of May 1903 turned out to be the driest May since 1826, with unusually high temperatures baking the ground before the trees had leafed out. The explosive conditions were either not recognized by the railroad operators or were simply ignored. The law at that time had required certain precautions for the railroads, such as spark arresters, but the laws were frequently not adhered to or enforced. Railroad management continued to send heavy train loads through the Adirondack woods that taxed the locomotive capacities, spewing out sparks from the heavy exhaust. Locomotive firemen continued to be permitted to dump white-hot coals on the roadbed when they stoked the boilers. Add to this the lack of sufficient fire patrols and the indifference of the section men, and one had a portrait of impending disaster. It happened. Four hundred and sixty-four thousand acres burned over in New York State that summer, most of that in the Adirondacks. It was claimed that 121 fires were started by locomotives.

The inexcusable negligence by the railroads contributed to a large share of the blame for these fires. At least half of the railroad fires were sparked by locomotives of the New York Central and the New York and Ottawa. Plotting the burned areas on a map also reveals that a number of them were likewise started along the logging railroads.

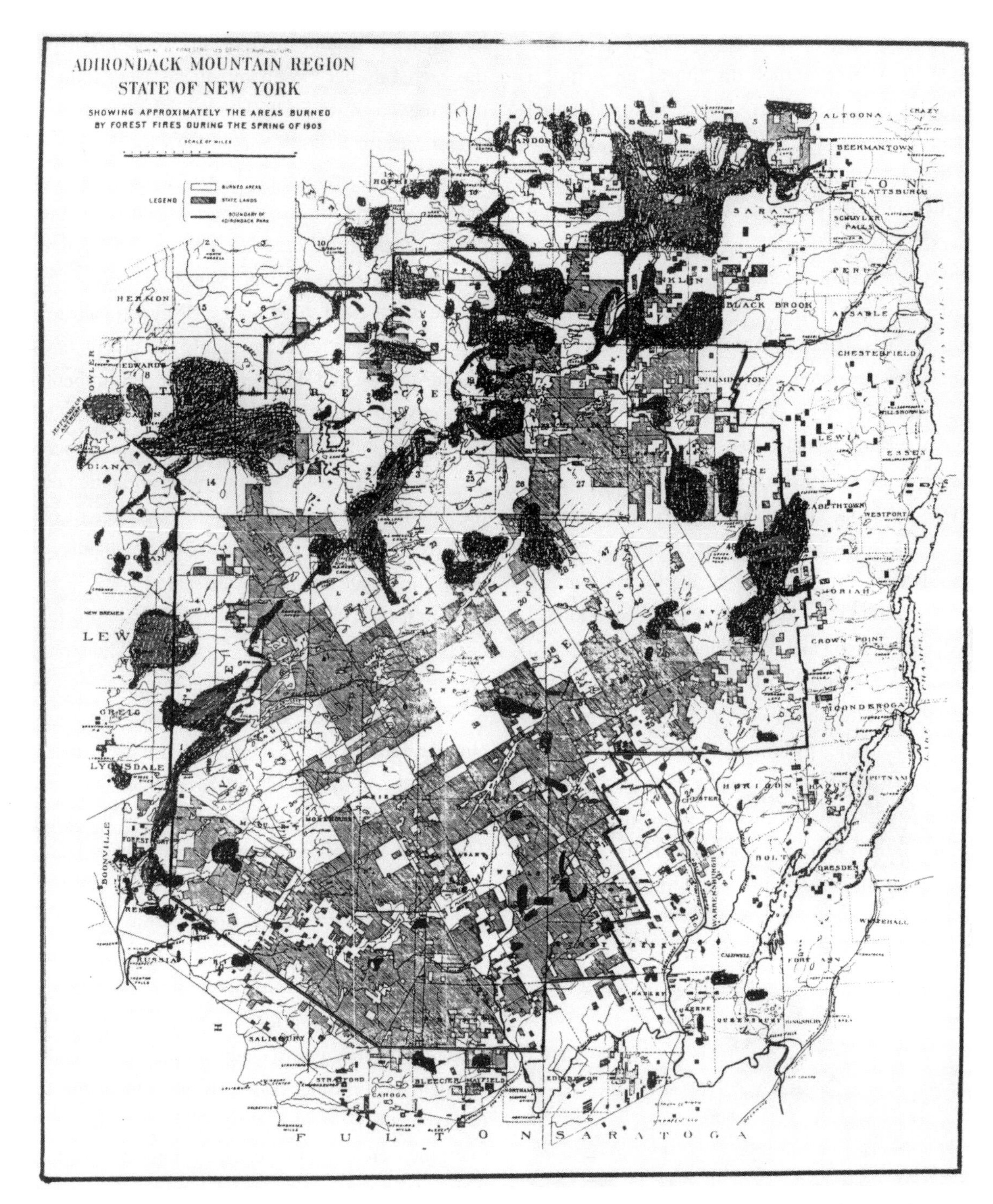

83. *The areas burned by the forest fires of 1903 are shown by the heavy black areas, the lighter shading indicating the state-owned land at that time. The Blue Line, as it existed in 1903, is shown by the heavier line that delineates the Adirondack Park. Note how the burned areas often followed the railroad lines.*

Nehasane Park, W. Seward Webb's spacious summer estate, suffered twelve thousand acres burned over in that year of 1903. The magnificent buildings were saved only because fire equipment was rushed up by rail from Herkimer and Ilion, powerful steam pumps mounted on freight cars.

A. A. Low's Horse Shoe Park sustained a large fire loss, and Low filed for damages. In December 1904 the New York Central and Hudson River Railroad awarded him $21,500 for losses in the fires of 1899 and 1903.

Not many years later, there was a repeat of that holocaust that was even more catastrophic. In 1908 the weather conditions again turned extremely dry, this time reaching a critical stage in the late summer. The New York Central had reportedly just acquired a number of new mainline engines, and they had ended up assigning the older replaced engines to the Adirondack Division without the proper fire safety devices.

Over three hundred thousand acres were scorched that year, at least 20 percent of which were reportedly started by the railroads. Webb's Nehasane Park got hit again, and A. A. Low was said to have counted over twenty fires in the vicinity of his park on one late August afternoon.

The most catastrophic blaze started the second week in September, set by a New York Central locomotive a half mile southwest of Long Lake. One hundred and fifty fire fighters had dug miles of trench hoping to contain the burn, when suddenly the wind shifted to the south, and the fire leaped out of control. The blaze raced uncontested toward the small village of Long Lake West (later known as Sabattis) on the New York Central, north of Nahasane. With the flames in sight, all of the horses were turned loose and more than seventy residents fled by train to Tupper Lake. It is said that another train with a firefighting crew aboard had tried to make it through, but the heat from the fire had bent the rails too much for the train to travel over. All buildings were destroyed plus about thirty thousand acres of timberland, the most serious fire ever in the Adirondacks. Some sources set the loss of Long Lake West village at $335,000.

This same fire swept westward through cut-over timberland as far as Cranberry Lake. The residents of Wanakena, the Rich Lumber Company village, were understandably quite nervous with the constant smell of smoke and the circulating rumors. The company ran the log trains only at night when the dew was on the ground. Barrels of water were placed at every wooden trestle. A steam-operated water pump was mounted on a large flat car, and water tank cars were quickly made available with hundreds of feet of hose on the car. A train stood ready at Newton Falls to evacuate the residents of any village should the emergency arise. As the fire approached Wanakena, the Rich Lumber Company mills were shut down, and the men were sent out to join the woods crew in battling the threatening fire. Other residents dug holes in the ground to bury their valuables before heading for the railroad station. By September the village of Wanakena was practically surrounded by flames, but fortunately the village and mills were spared.

84. The destroyed remains of Sabattis after the disastrous fire of September, 1908. Note the condition of the side track and the remains of several freight cars. The New York Central main line shows in the background. Source: New York State Annual Report.

Forest fires left behind a terrible scene on the timberlands of the Rich Lumber Company. Their property included the sandy plains along the Oswegatchie River, which once supported some magnificent white pine stands. Logging slash in this area had created a hot fire bed. It has been said that the industrial life of Wanakena village was cut noticeably short by destructive fires such as the one in 1908. The forest fires of 1908 burned over 308,000 acres in the state, almost all of it in the Adirondacks.

By October of that year the state was goaded to action, and hearings were held that month on the problem of the fires caused by sparks from locomotives. A survey had shown that in the forty days prior to September 29, over four hundred fires had been started adjacent to the railroads running through the Adirondacks. Forty percent of the fires were caused by the railroads that year. The spark arresters used at that time were quite ineffective, using a quarter-inch mesh, which did not stop the pea-sized sparks. Railroad engineers had contended that they could not get a full head of steam when using a finer mesh, but something obviously had to be done.

Thus in 1908 new regulations were adopted that required the railroads to burn oil instead of coal within the Adirondack Preserve during the hours of 8 AM to 8 PM and between the dates of April 15 and October 31. By April 15, 1910, all locomotives operating in the Adirondacks were to have oil-fired equipment installed, and the use was mandatory.

85. A fire drill was in order for the NYC fire train. Circa 1910. Note the water pump on top of the locomotive boiler.

87. A New York Central fire crew equipped with fire torches and back pumps. The torches were used to burn over the hundred-foot cleared right-of-way each spring. Note the velocipede cars, which were used by the patrolmen in the twenty-nine sections of the Adirondack Division. The fire-tank cars are in the background. Source: American Forests, *July 1928.*

86. New York Central Railroad officials were viewing the testing of steam operated pumps mounted on steel flat cars. Each of the tanks carried seven thousand gallons of water and each flat was equipped with one thousand feet of $1\frac{1}{2}$" hose, shovels, and other fire-fighting equipment. Steam was supplied by the locomotive. New York Central had five of these fire-tank cars on the Adirondack Division. Source: American Forests, *July 1928.*

88. Webb's Nehasane Park financed their own fire train because of large fire losses on the Park property. The fire pump was driven by a steam boiler set up in a boxcar, supplied by a 6,187-gallon water tank on the flat car. Circa 1908.

The Emporium Forestry Company soon switched over all of the motive power on the company's common carrier Grasse River Railroad to fuel oil, but all of the Shays on the logging line remained coal burners and were thus grounded for daytime operation for a few months.

A number of the railroads installed firefighting equipment in cars and located the cars at strategic points. The logging railroads in particular took this precaution. The New York Central provided special fire trains consisting of about five flat cars, each loaded with a seven-thousand-gallon water tank and pump with hose. The railroad also provided crews to clean up flammable material along the lines. The Delaware and Hudson did the same along the Chateaugay, but as per custom, John Hurd's negligent Northern Adirondack failed to act on the early regulations directed by the Public Service Commission.

The commission later reported that no fires of any consequence caused by a locomotive had occurred in the Adirondacks after the new regulations went into effect in 1909.

The Emporium Forestry Company was still burning coal in the 1920s. In April of that year, the company's Grasse River Railroad issued a stern directive to the employees that all possible precautions had to be taken to prevent fires. Locomotive ash pans and front-end nettings had to be maintained in safe condition. Ash pan slides could not be operated in an open position unless proper netting was in place. When cinders were dumped, the employee was responsible for seeing that all fires were extinguished before leaving, and hot clinkers could only be thrown off at the regular cinder pile.

The above directive had been issued because early in 1920 the state had provided temporary relief to the Grasse River Railroad as well as the New York Central Railroad and the Delaware and Hudson Railroad, permitting use of coal fuel during the daytime hours. There was an oil fuel shortage at the time. The railroads in turn were required to provide effective spark arresters and ash pans, intensive patrols, fire trains, and available oil-burning locomotives for service if necessary. Emporium employee Ray Zenger fondly remembered following along the track on a speeder fifteen minutes behind the locomotive to look for hot spots. He took along his fishing pole and made short stops at a few good fishing holes along the way.

By that point in history, the railroads were operating quite well in reducing the number of fires. Hazardous years continued to occur, dry years such as 1921, 1930, and 1934, but fire losses were kept at acceptably low figures.

PART TWO

INDIVIDUAL LOGGING RAILROADS

In the following part of the book, a history of each Adirondack logging railroad is given, including what is known about the locomotives and other equipment used by the railroad. A map is included for each railroad, prepared from information supplied by old maps, recollections of old-timers, and other railroad enthusiasts. The base maps were adapted from U.S. Geological Survey quadrangles.

The logging railroads are arranged geographically in the text and grouped together according to the mainline railroad to which they had made connections. There were three mainline railroads with logging lines attached: the New York Central's Adirondack Division (Webb's railroad), Hurd's Northern Adirondack (later the New York Central's Ottawa Division), and the Carthage and Adirondack (later part of New York Central). The other mainline railroads, such as the Delaware and Hudson Railroad, had no logging railroads branching off of them.

Histories are also included for the New York Central's Adirondack Division (Webb's Mohawk and Malone) and the New York Central's Ottawa Division (Hurd's Northern Adirondack) because of the vital part that they both played in the development of lumbering in the Adirondacks. Neither was a logging railroad by definition, but both hauled large quantities of logs and lumber in the early years. This was especially true of the Northern Adirondack Railroad, which had its beginnings as a log hauler.

The reader will notice that a couple of the logging railroads covered in this section were not true logging railroads by strict definition. However, since logs were a major freight item while serving the lumber industry, they are included.

Not included in the book are wood industry yard railroads and the occasional short private industrial railroad built from a sawmill to a mainline siding to transport the sawn product.

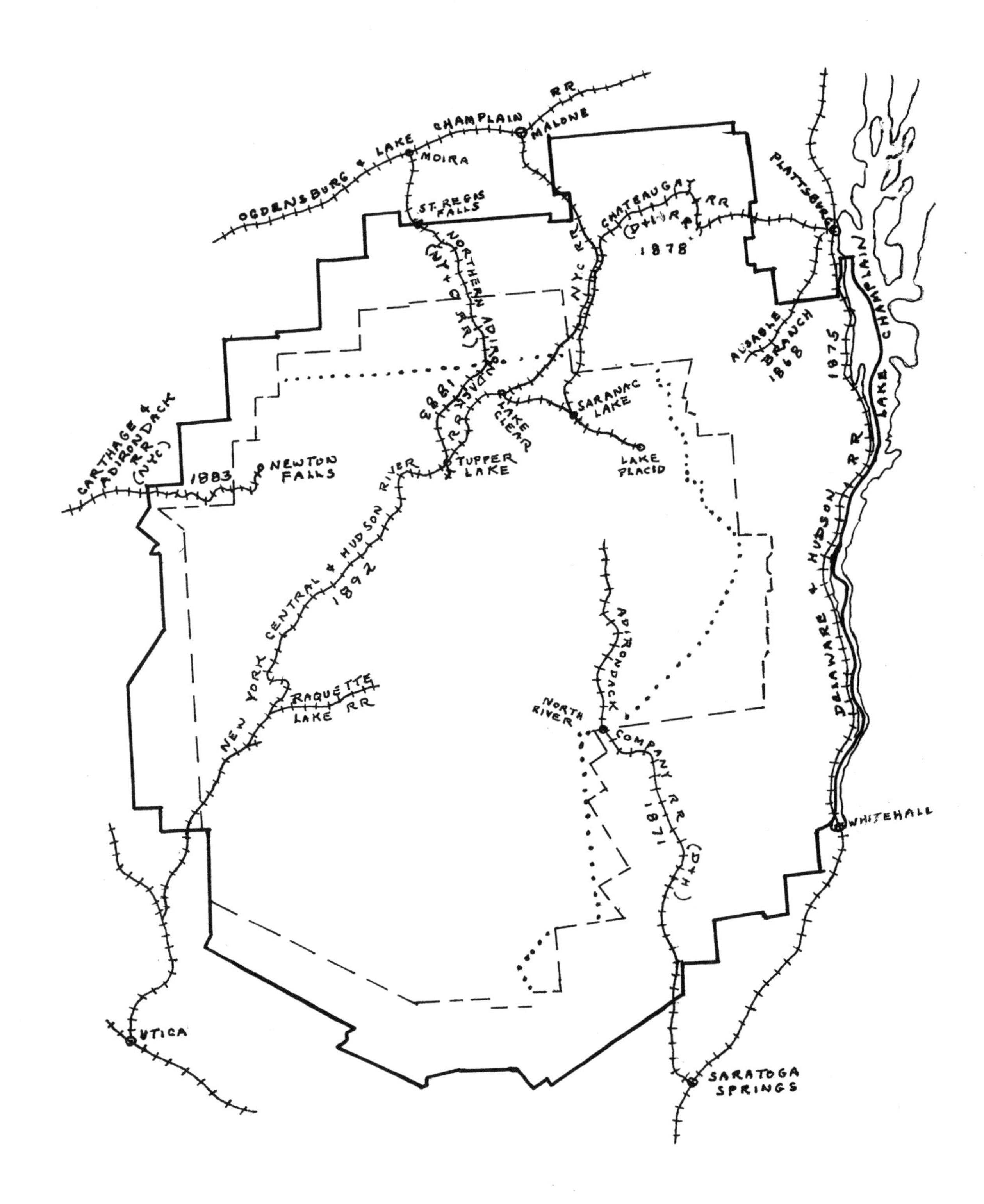

MAIN LINE RAILROADS IN ADIRONDACKS

ADIRONDACK "BLUE LINE"

- • • • • • • IN 1890'S, ORIGINAL
- — — — EARLY 1900'S
- ——— PRESENT

DATE OF CONSTRUCTION SHOWN

SCALE MILES 0 10 20

Map 1.

SECTION ONE

Logging Railroads That Connected with the New York Central and Hudson River Railroad, Adirondack Division

THE NEW YORK CENTRAL'S Adirondack Division, originally built by W. Seward Webb in 1891–92, was the first and only railroad ever to completely cross the Adirondack region. This railroad became a major factor in opening up the Adirondacks to development in the early part of the twentieth century, especially the logging and lumber industry.

The Adirondack and St. Lawrence Railroad, as it was originally chartered and labeled, was not a logging railroad as such; it was far more than that. However, the development of this railroad is briefly covered here because of the historic effect it fostered upon Adirondack forest management and the number of logging pikes it spawned. Webb's railroad opened a vast mountainous region to public transportation with its through route from the Mohawk Valley to Montreal. A brief history of the railroad is given in this section, followed by histories of each of the logging railroads that connected with the New York Central. The logging railroads are listed in the geographic order in which they appeared on the New York Central map, from south to north, not necessarily in the chronological order of their construction. Maps 2 and 3 show the location of the logging railroads in connection with the main line railroads.

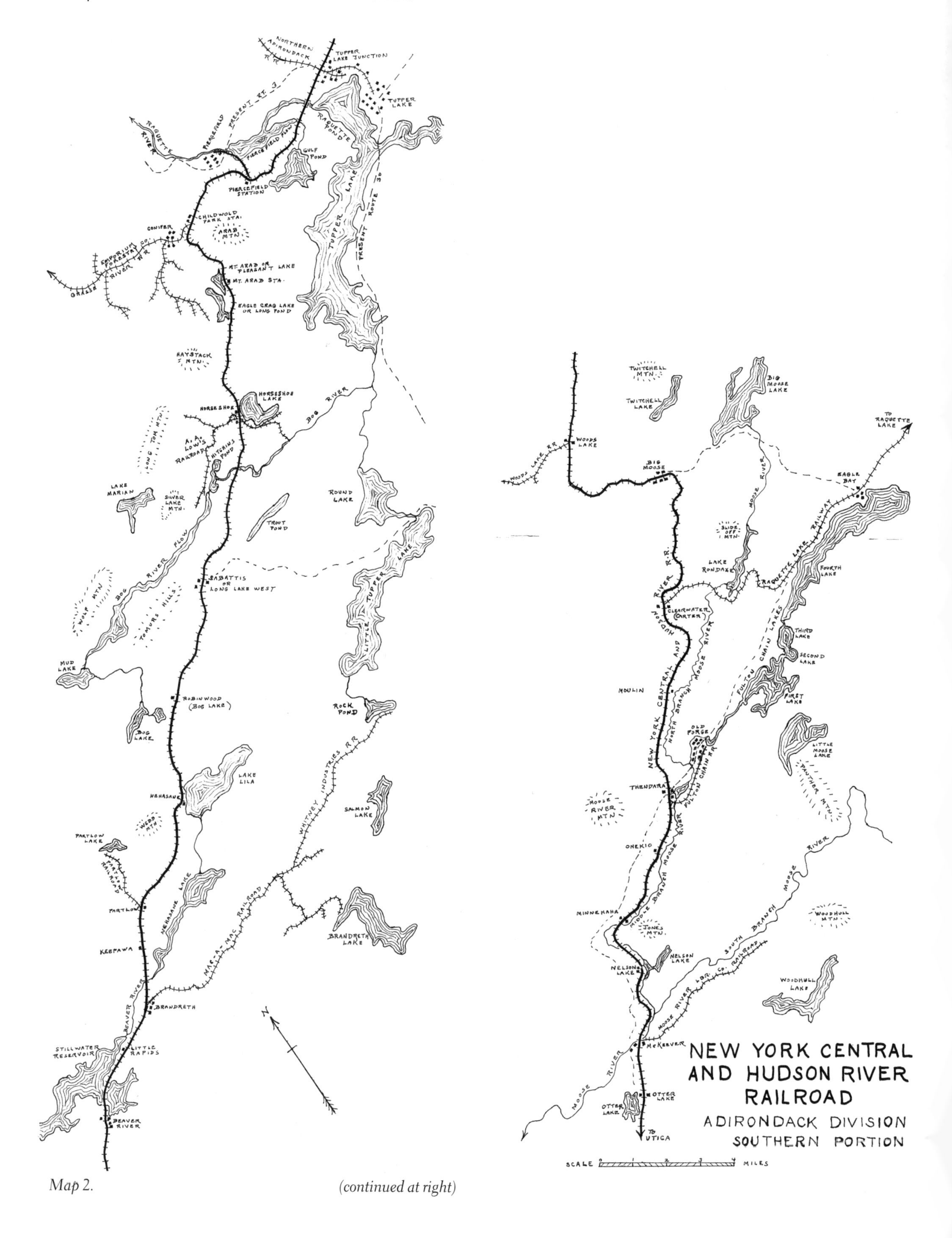

Map 2.

(continued at right)

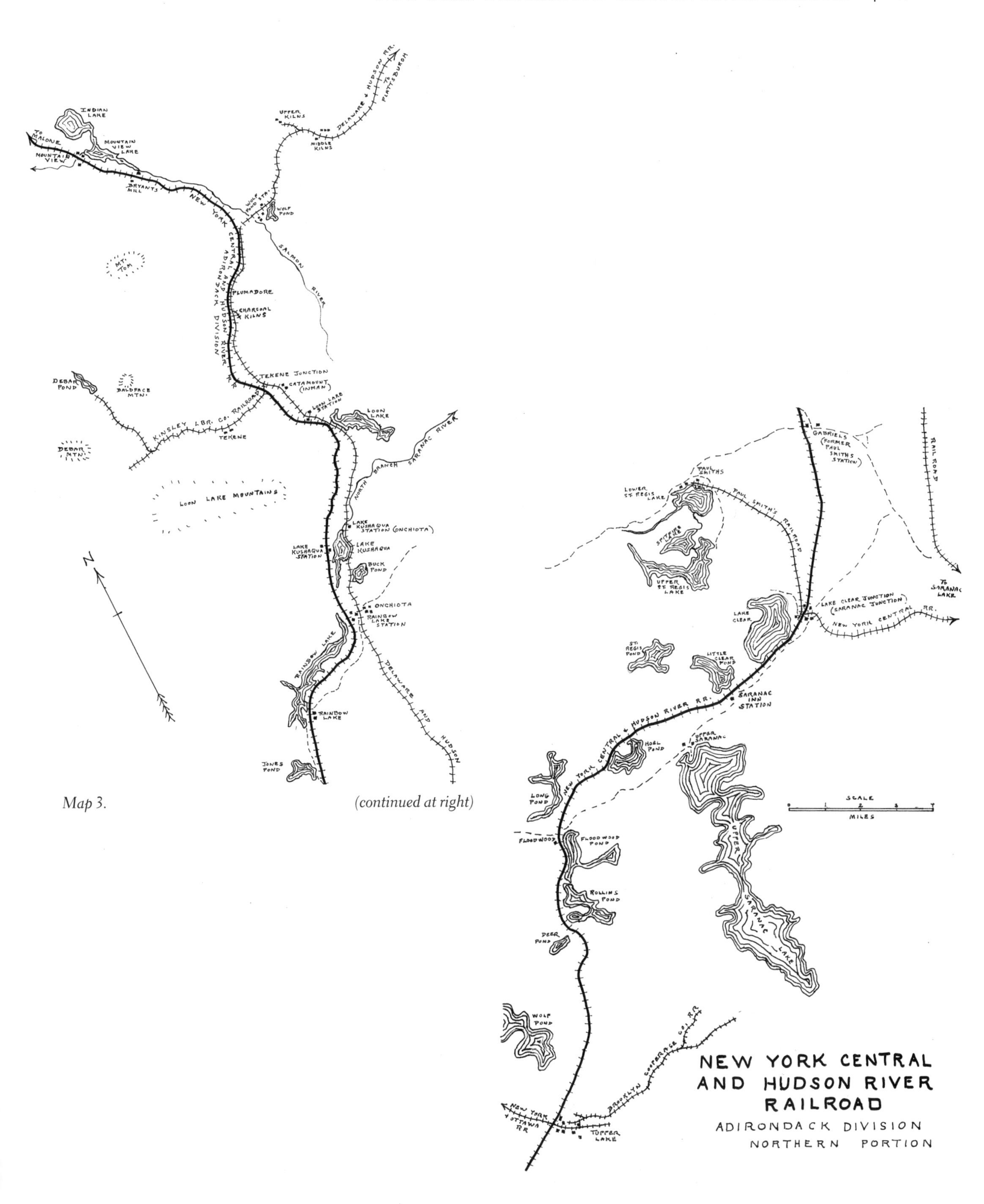

Map 3.

(continued at right)

CHAPTER NINE

Adirondack and St. Lawrence Railroad, 1890–1892; Mohawk and Malone Railroad, 1892–1893; New York Central and Hudson River Railroad, Adirondack Division, 1893–1972

The most controversial railroad ever built into the Adirondack region was Dr. William Seward Webb's Adirondack and St. Lawrence Railroad, more commonly known in the early years by the record name Webb gave it, the Mohawk and Malone. Webb's railroad was a monumental 106-mile task built at Webb's own expense and at an unusually fast rate of construction.

Public sentiment was strong in opposition to this only railroad ever to pass completely through the Adirondacks. But Webb was decidedly an entrepreneur to be reckoned with—and not to be disagreed with.

As far back as 1876, the Delaware and Hudson Railroad had completed a line along Lake Champlain from New York City to Montreal and had captured most of the traffic from New York through to Canada's largest city. In 1890, however, this position of supremacy was challenged by Dr. Webb, a man of personal wealth as well as a son-in-law of William H. Vanderbilt.

Even by 1890, much of the Adirondacks was still somewhat inaccessible for the many wealthy summer visitors. For example, travel to the picturesque Fulton Chain of Lakes meant a twenty-four-hour trip from Utica. First the vacationer had to endure the ride to Boonville on the Rome, Watertown and Ogdensburg Railroad with an overnight stay upon arrival. The next leg was a stage ride to Moose River village, followed by a ten-mile jaunt on the wooden-railed Minnehaha Railroad, then a steamboat passage through the locks to Fulton Chain Landing and finally two to three miles on the trail to Old Forge. Webb's railroad was to make a large change in that travel itinerary.

The New York Central had been showing interest in establishing a railroad through the Adirondacks even back in the 1880s. Their primary interest at the time was not so much the prospect of local traffic as it was the desire to acquire direct access to Montreal in competition with the Rome, Watertown and Ogdensburg Railroad. Of course, the rich, untapped timber resources were also an attraction. Thus, the New York Central encouraged the wealthy Webb to ally with them in undertaking the task of building rail through the Adirondacks.

Before any action materialized, however, the New York Central was able to lease the competitive Rome, Watertown and Ogdensburg on the western edge of the Adirondacks, and it then withdrew any backing of Webb for a railroad

New York Central & Hudson River R.R. Co.
Grand Central Station.
E. Van Etten,
General Superintendent.
New York, Dec. 27, 1899.

89. Letterhead.

90. William Seward Webb, who had the determination and vast financial resources to build the first railroad through the Adirondack region, photographed at his Nehasane Lodge. Courtesy of The Adirondack Museum.

through the Adirondacks. But Webb was fully aware of the potential before him, so he proceeded to build the entire line at his own expense and supervision. In September 1890 Webb acquired the narrow-gauge Herkimer, Newport and Poland Railroad, standard-gauged it, and with this sixteen-mile pike as a starting point, put hundreds of laborers and surveyors to work. Webb formed the Adirondack and St. Lawrence Railway Company in December of that year. Construction crews were put in the field to work from both directions and to do so rapidly. For reasons of his own, Webb wanted this railroad built as rapidly as it was possible to do so.

As Webb's surveyors cut their way through the Adirondack forests, the checkerboard pattern of state ownership continually created a barrier to thwart his chosen path. He had sought permission from the state to cross through public land but frequently met with drawn-out delays. When news of Webb's intention reached the many residents of urban areas who considered the Adirondacks to be their choice resort, a mighty protest was raised. Indignation over supposed desecration of the forest was vehemently voiced in the press. The Delaware and Hudson Railroad, opposed to the project because Webb's railroad had the potential of mounting strong competition, created a strong lobby effort from their Albany office. From a position less than ten years previous of being publicly castigated for laying waste to huge areas of forestland and blackening it with charcoal kilns as they built new railroad lines, the D&H Railroad suddenly became the voice of forest conservation and didn't want the Adirondacks spoiled.

Seeking to gain allies for his new project, Webb made strong overtures toward the Adirondack Park Association, an organization formed in April 1890 to preserve Adirondack forests and promote the establishment of an Adirondack state park. To gain their support for his seeking permission to cross state land, he pledged that his railroad would not convey out any lumber sawn from logs less than twelve inches in diameter and would not transport any charcoal made from green timber. One wonders if any person ever questioned how such a promise could ever be fulfilled or ever verified for compliance.

Unsure of ever gaining permission from the New York Forest Commission and the New York State Land Board to cross over state-owned forestland, Webb tried a new approach. He began to purchase more Adirondack properties.

Public Fear of Logging

Charles S. Sargent's comments on the June 10, 1891, issue of *Garden and Forest* echoes the fearful misconception of many urban dwellers.

"If the woods were crossed and recrossed with railroads, settlements would spring up at their intersection, the deer would be frightened from their ranges, the fish in the streams and lakes would be caught for city markets, and the wildness would be chased away with the game. The place would still have attractions, but it would no longer be a wilderness."

This occurred, of course, prior to the 1894 amendment to the New York State Constitution, which prohibited the cutting of any trees on the Adirondack Preserve, and thus it would have been legally possible in 1890 to cut through state land. But Webb was impatient and began to add immense acreage to his holdings. In 1891, for example, seventy-one thousand acres were purchased from the Adirondack Timber and Mining Company for $373,343. Webb's private ownership, it is said, exceeded two hundred thousand acres at this point in history.

Thus, with so much land owned by Webb, his surveyors were in many situations able to work around parcels owned by the state. Eventually Webb did receive a favorable ruling from the state regarding land swapping. Although the constitution prohibited the sale of any public land, it did not prohibit exchanges of land parcels of equal value. Webb's wealth and influence had prevailed again. It wasn't until November 1894 that a constitutional convention instituted the "forever wild" provision on all state-owned lands in the Adirondacks.

Construction of the railroad proceeded rapidly at the strong insistence of Webb. Beginning in the spring of 1891, four thousand men were quickly at work clearing and grubbing out a hundred-foot-wide swath through the forest, leveling a roadbed that was to have no curves greater than four degrees and laying the seventy-five-pound rail as fast as it arrived. Costs were not spared; six-spike fish plates were used and ties laid at three thousand to the mile. Webb himself was frequently on the scene, arriving unannounced to ask the contractors why they had not progressed further. It mattered little that his urgent demands kept increasing the construction costs and that he was driving some of the contractors crazy.

When Webb realized that the railroad would be passing through the vicinity of Tupper Lake, he made an offer to buy John Hurd's struggling Northern Adirondack Railroad. It needed considerable upgrading, but cost and time would be saved if he could obtain it. Six hundred thousand dollars were offered, but Hurd refused, calculating he could hold out for more. Webb called his bluff and began making plans for a separate and competing line through the same region.

Webb then proceeded to build his own railroad line north from Tupper Lake, parallel to Hurd's line for many miles before swinging to the northeast to reach Malone. Hurd filed for bankruptcy soon afterward and lost everything.

Soon after the above drama, Webb offered to build his railroad through the city of Malone if the city would provide $30,000 for construction costs. It was felt that he would probably have built to Malone with or without their contribution, but the city fathers of Malone took no chances. They agreed.

The managers of Webb's railroad knew that the early

91. When William Seward Webb's Adirondack railroad was first formed in 1890, the locomotives were lettered Adirondack & St. Lawrence Company, as noted on No. 16. Courtesy of the Tupper Lake Public Library.

92. In 1892, when Webb consolidated his railroad companies into the Mohawk and Malone Railroad, the engines received new lettering and a paint job. Engine No. 50, Compound, was photographed at Tupper Lake Junction. Courtesy of the Tupper Lake Public Library.

MOHAWK & MALONE RAILWAY CO.

Adirondack & St. Lawrence Line

LOCAL TIME TABLE—NORTHBOUND.

TO THE ADIRONDACKS, MONTREAL, ETC.

Dist. f'm Herkimer	STATIONS. April 23d, 1893.	Adirondack and Montreal Express. *47	Day Express. †21	Mail. †1	Mail. †8	Freight. †61
	Lv. NEW YORK, (N. Y. C.)	*7 00 pm	†11 59 pm		†10.00 am	
	" Poughkeepsie, "	9 15 "	‡ 3 10 am		10.00 "	
	" Albany, "	11.15 "	† 8.25 "		1.35 pm	
	Ar. Herkimer, "	*1.20 am	†11.07 "		†4.17 "	
	Lv. BUFFALO, (N. Y. C.)	*3.50 pm	†4 40 am		†8 50 am	
	" Rochester, "	6 15 "	6.50 "		10.30 "	
	" Syracuse, "	8.45 "	9.40 "		1 30 pm	
	" Utica, "	10.25 "	11.20 "	†6.00 am	3.25 "	
	Ar. Herkimer, "	*10.50 "	†11.45 "	†6.30 "	†3.55 "	
0	Lv. HERKIMER	1.30 am	11.50 am	7.00 am	5.00 pm	2.35 pm
3.3	" Kast Bridge		f11.59 "	f7 07 "	f5.09 "	2.46 "
5.2	" Countryman's		f12.03 pm	f7.11 "	f5.14 "	2.54 "
7.0	" County House		f12.07 "	f7.15 "	f5.18 "	3.00 "
8.7	" Middleville	d1 47 am	12.11 "	7.19 "	5.22 "	3 16 "
11.2	" Fenner's Grove		f12 17 "	f7.24 "	f5 27 "	3.30 "
13.0	" Newport	d1.56 am	12 21 "	7.28 "	5.31 "	3.45 "
16.4	" Poland	d2 01 "	12.28 "	7 35 "	5.39 "	4 04 "
[illegible]	" Gravesville			f7 43 "	f5.49 "	
23.6	" TRENTON FALLS		f12.45 pm	f7.50 "	f5.56 "	4 50 pm
25.5	Ar. PROSPECT		12.49 "	7 54 "	6.00 "	5.15 "
	Lv. (Junction Hinckley Br.)				6.05 "	6.10 "
27.6	Ar. REMSEN	2 22 am	12.55 pm	8 00 am	6 10 "	6 30 "
	Lv. REMSEN					6.45 "
31.6	" Honnedaga		f1 03 pm			7.08 "
35.6	" Forestport		1.12 "			7.28 "
42.4	" White Lake		1 28 "			8.02 "
47.5	" Otter Lake		f1.42 "			8.30 "
49.4	" McKeever		f1 45 "			8.40 "
56.0	" FULTON CHAIN	3 20 am	2 06 "			9 30 "
69.0	" Big Moose		f2.30 "			10 25 "
[illegible]	" Beaver River		f2.48 "			11.15 "
[illegible]	" Little Rapids (Private Stations.)					
[illegible]	" Ne-ha-sa-ne (Private Stations.)					
[illegible]	" Bog Lake		f3 17 pm			12.25 am
[illegible]	" Horseshoe Pond		f3 37 "			[illegible]
[illegible]	" Childwold		f3.53 "			[illegible]
[illegible]	Ar. TUPPER LAKE JUNC					[illegible]
						† No. 68
	Lv. TUPPER LAKE JUNC	5.10 am	4 10 pm			5 25 am
[illegible]	" Saranac Inn	x5.36 "	f4 38 "			6 30 "
[illegible]	Ar. LAKE CLEAR	5.48 "	4 47 "			[illegible]
	Lv. (Junction Saranac Br.)					
[illegible]	" Paul Smith's	x5.52 am	f4 59 pm			7 25 am
[illegible]	" Rainbow Lake		f5 02 "			[illegible]
[illegible]	" Lake Kushaqua		f5.12 "			[illegible]
[illegible]	" Loon Lake	x6.14 am	5 20 "			8.35 "
[illegible]	" Mountain View		f5 42 "			9.20 "
[illegible]	" Owl's Head		5 46 "			9.40 "
[illegible]	" Whippleville		f6.01 "			10.10 "
[illegible]	Ar. MALONE (Union Station)	7.05 am	6.10 "			10.30 "
	Lv. Malone (St. L. & A)	7.10 am	6 15 pm			
	" Huntingdon "	7.45 "	6 51 "			
	" Coteau Junc. "	8 24 "	7 30 "			
	Ar. Montreal (G. T. Ry.)	*9.20 "	8 30 "			
	Ar. Ottawa (C. A. Ry.)	†11 [illegible] am				

HINCKLEY BRANCH.

From the Main Line.				Dist. bet. Sta.	STATIONS			To the Main Line.			
		Way F'gt †78	Way F'gt †71					Way F'gt †72	Way F'gt †74		
		pm 5.25	pm 2.30	.0	Ar.	HINCKLEY	Lv.	pm 2.45	pm 5 45		
		5 15	2 20	2.8	Lv.	Prospect	Ar.	3 00	6 05		

SARANAC BRANCH.

From the Main Line.				Dist. bet. Sta.	STATIONS.			To the Main Line.			
*57 Daily	†55	†58	*51 Daily					*50 Daily	†52	†54	*56 Daily
pm 9.20	pm 6.00	am 11.15	am 6.10	.0	Ar.	SARANAC LAKE	Lv.	am 5.15	am 10.20	pm 4 20	pm 8.35
9 00	5.35	10 50	5.50	5.6	Lv.	Lake Clear	Ar.	5.40	10.40	4.45	8.55

* Daily. † Daily except Sunday. ‡ Daily except Saturday.
f Train stops when signaled, for passengers.
d Train stops to leave passengers holding tickets from Utica or points beyond, and from Albany or points beyond. x Train stops to leave passengers from Herkimer or points beyond and on signal to take on passengers for Malone or points beyond.
Freight trains carry passengers in caboose cars.

MOHAWK & MALONE RAILWAY CO.

Adirondack & St. Lawrence Line

LOCAL TIME TABLE—SOUTHBOUND.

FROM THE ADIRONDACKS, MONTREAL, ETC.

Dist. f'm Malone.	STATIONS. April 23d, 1893.	Day Express. †86	N. Y. Express. *44	Mail. †2	Mail. †4	Freight. †64
	Lv. OTTAWA (C. A. Ry.)		†3.25 pm			
	" MONTREAL (G. T. Ry.)	†7.00 am	*5.30 pm			
	" Coteau Junc. (St. L. & A.)	8.00 "	6.35 "			
	" Huntingdon, "	8 40 "	7.13 "			
	" Malone, "	9.15 "	7.45 "			
0	Lv. MALONE (Union Station)	9.20 am	7 50 pm			2 05 pm
4.0	" Whippleville	f9.27 "				2.20 "
10 6	" Owl's Head	9.40 "				2.45 "
12.9	" Mountain View	f9.45 "				2.55 "
24.9	" Loon Lake	10 09 "	x8.32 pm			3.50 "
29.1	" Lake Kushaqua	f10.16 "				4.10 "
34.2	" Rainbow Lake	f10.26 "				4.35 "
36.6	" Paul Smith's	f10 32 "	x8.50 pm			4.50 "
41.7	" LAKE CLEAR	10.42 "	9.00 "			
	" (Junc. Saranac Branch)					5 35 pm
44.7	" Saranac Inn	f10.47 am	x9 05 pm			6.05 "
59.7	Ar. TUPPER LAKE JUNC.					7.20 "
						†62
	Lv. TUPPER LAKE JUNC.	11.20 am	9 30 pm			7 00 am
66.2	" Childwold	f11 34 "				7 25 "
73.4	" Horseshoe Pond	f11.49 "				7.57 "
82.4	" Bog Lake	f12 08 pm				8.35 "
86.0	" Ne-ha-sa-ne (Private Stations.)					
92.9	" Little Rapids (Private Stations.)					
95.5	" Beaver River	f12 38 pm				9.37 am
104 1	" Big Moose	f12 54 "				10.16 "
115 1	" FULTON CHAIN	1.18 "	11.30 pm			11.05 "
123.7	" McKeever	f1 37 "				11.46 "
125.6	" Otter Lake	f1.42 "				11 58 "
130.7	" White Lake	1.53 "				12.30 pm
137.6	" Forestport	2.07 "				1.12 "
141.6	" Honnedaga	f2.18 "				1.30 "
145.5	" REMSEN	2.26 "	12.20 am	8.25 am	6.45 pm	2.05 "
147.6	Ar. PROSPECT	2.31 "		8.30 "	6.49 "	2.20 "
	Lv. (Junc. Hinckley Branch)			8 35 "		3.05 "
149.5	" TRENTON FALLS	f2.38 pm		f8 39 "	f6.52 pm	3.15 "
152.7	" Gravesville			f8.45 "	f6.57 "	
156.7	" Poland	2.55 pm	d12.42 am	8.54 "	7.06 "	4.04 pm
160.1	" Newport	3.05 "	d12.48 "	9 02 "	7.12 "	4.20 "
161.9	" Fenner's Grove	f3.09 "		f9.05 "	f7.16 "	4.29 "
164.4	" Middleville	3 16 "	d12 56 am	9.10 "	7.21 "	4.42 "
166.1	" County House	f3.20 "		f9 14 "	f7 24 "	4.50 "
167.9	" Countryman's	f3 24 "		f9.18 "	f7 27 "	4.57 "
169.8	" Kast Bridge	f3 28 "		f9.22 "	f7.31 "	5.09 "
173 1	Ar. HERKIMER	3.35 "	1.15 am	9.30 "	7.38 "	5.25 "
	Lv. Herkimer (N. Y. C.)	†3.55 pm	*1 15 am	†10.00 am	†7 39 pm	
	Ar. Albany, "	6 50 "	3 20 "	12.55 pm	10.25 "	
	" Poughkeepsie, "		5 30 "	3 50 "		
	" NEW YORK, "	10 30 pm	7.45 "	6 41 "		
	Lv. Herkimer, (N. Y. C.)	†4.17 pm	*1.34 am	†10 03 am	†8 31 pm	
	Ar. Utica, "	4 46 "	2 00 "	10.35 "	9 05 "	
	" Syracuse, "	6.45 "	3.25 "	12 45 pm	11.10 "	
	" Rochester, "	9.35 "	5.45 "	3.47 "	1.25 am	
	" BUFFALO, "	11 40 "	[illegible] "	5 10 "	[illegible] "	

HINCKLEY BRANCH.

From the Main Line.				Dist. bet. Sta.	STATIONS.			To the Main Line.			
		Way F'gt †78	Way F'gt †71					Way Fg't †72	Way Fg't †74		
		p m 5.25	p m 2.30	.0	Ar.	HINCKLEY	Lv.	p m 2.45	p m 5 45		
		5 15	2.20	2.8	Lv.	Prospect	Ar	3 00	6.05		

SARANAC BRANCH.

From the Main Line.				Dist. bet. Sta.	STATIONS.			To the Main Line.			
*57 D'y	†55	†58	*51 D'y					*50 D'y	†52	†54	*56 D'y
p m 9.20	p m 6.00	a m 11.15	a m 6 10	.0	Ar.	SARANAC LAKE	Lv.	a m 5.15	a m 10.20	p m 4.20	p m 8.35
9.00	5.35	10.50	5.50	5.6	Lv.	Lake Clear	Ar.	5.40	10.40	4.45	8.55

* Daily. † Daily except Sunday. ¶ Daily except Monday. f Train stops when signaled, for passengers.

d Train stops on signal to take on passengers for Utica or points beyond and for Albany or points beyond.

x Train stops on signal to take passengers for Herkimer or points beyond and to leave passengers from Malone or points beyond.

Freight trains carry passengers in caboose cars.

93. Webb's Adirondack and St. Lawrence Railroad was completed in 1892. The following spring the line was taken over by the new corporation Mohawk and Malone Railway Company, and this timetable was issued.

success of the line would depend upon the lucrative summer tourist travel to the hotels on the Adirondack lakes. But they realized in the spring of 1891 that the new line from Malone down to the Saranac lakes would not be completed in time for the 1891 summer traffic. And that was revenue they wanted to tap into immediately.

It was decided, therefore, to build a fifteen-mile spur line from the terminus of the Northern Adirondack Railroad at Tupper Lake village over to Saranac Lake. Work began on this section in May and was quickly built. The portion from Tupper Lake to Clear Lake became part of the main line when the section coming down from Malone was completed.

By July 1, 1892, the construction crew working from the south had completed track laying to Thendara, and those working from the north had reached Tupper Lake Junction. But Webb continued his impatience, demanding a faster pace. The northern crew had relatively easy terrain to work with, sandy plains and gentle stream bottoms with very few steep grades. The average grade from Malone up to the level

94. Logs piled along new construction of Webb's railroad, possibly cut from the right-of-way clearing, are loaded by means of a Barnhart loader. Adirondack and St. Lawrence No. 16, a ten-wheeler, waits patiently on the temporary elevated trackage. Circa 1892. Courtesy of the Tupper Lake Public Library.

95. Small station stops along the Adirondack Division of the New York Central provided service for sportsmen and vacationers to reach their camps or motels. The location is Mountain View, a stop at Indian Lake in the north edge of the Adirondacks.

of the Adirondack plateau was only eighty feet per mile, or 1.5 percent.

In June 1892 Webb consolidated the different railroad corporations involved in the Adirondack project into one corporation, the Mohawk and Malone Railroad. The new name apparently never appeared on any rolling stock other than locomotives, however. For the remainder of Webb's short ownership, the equipment retained the name Adirondack and St. Lawrence Railroad. But the Mohawk and Malone name caught on with the public; the railroad was popularly referred to as the M&M for years after the New York Central took it over.

On October 12 of that same year, the two crews that were working from opposite directions came together, and the last spike was unceremoniously driven at a location a half mile north of the Twitchell Creek bridge, near Big Moose. The assistant project engineer supposedly attempted to drive the spike signaling completion but missed twice and snapped off the handle. The job was finished by a St. Regis Indian laborer.

Although regular trains began running through from New York City to Montreal on October 24, there was much work yet to be done. Twenty-seven work trains with about thirty cars each, and nine steam shovels, were kept busy for many months filling in temporary trestles, widening and stabilizing banks and cuts, and replacing temporary culverts.

During construction there were heard the voices of the ever-present skeptics who derided the project as foolish, soon to be "two streaks of rust through the woods." But these voices were soon silenced. The action of the state constitutional convention of 1894 made it impossible for any future railroads to be built through the Adirondacks, and Webb's railroad practically had a monopoly on the region. Reports came out in 1902 that the New York Central was considering double-tracking the entire length of the division, but, of course, it never happened.

To construct a railroad almost two hundred miles in length (Herkimer to the Canadian border), in part through the Adirondack wilderness, in that short a time was phenomenal for that era, to say the least. From the time the first survey was started in the spring of 1891, it took only about eighteen months before regular traffic began. The cost was estimated at close to $6 million. W. Seward Webb had no intention of operating a large railroad, however, and no doubt from the beginning, he knew who would take it over. The next year, on May 1, 1893, the railroad was leased to the New York Central and Hudson River Railroad, and it became their Adirondack Division. By 1913 all of the stock had been purchased, and it was merged into the New York Central ownership. To Adirondack residents, however, it was still Webb's M&M. In the histories that follow, all references will simply be to the New York Central.

96. The New York Central passenger service made a regular stop at the little Childwold depot, the location where the Grasse River Railroad interchanged with the New York Central.

Log Train Activity on the Adirondack Division

Logs were the initial freight item on Webb's Adirondack and St. Lawrence Railroad, and forest products remained important throughout the life of the division. Temporary sidings were built in a number of locations along the line, where cars loaded with logs or lumber could be picked up.

Typical is the operation conducted by jobber George Harvey between 1908 and 1913 near the small Woods Lake station and settlement. The softwood pulpwood was cut along the Razorback Pond Outlet on the west side of the New York Central tracks and landed at the Woods Lake station, where it was loaded on cars.

In those early years of the Adirondack Division, the forestland would literally become alive with activity about the first of May for several miles from the railroad. That's when the pulpwood-peeling season began; spruce and fir logs were at that time peeled on the ground right at the stump. In 1917 there were 60 logging camps operating along the Adirondack Division, providing employment for three thousand men. The entire Adirondack region had over seven thousand men working in a total of 150 logging camps. Webb's railroad had become a gateway to tremendous areas of merchantable timber and busy logging operations.

A log train through Herkimer village was a frequent late-afternoon occurrence as the New York Central brought in a train load of logs for the Standard Desk Company.

Derailments of log trains on the Adirondack Division were common, probably in part because of the older woods equipment rather than bad track conditions on the main line. It was a log train that left the track near Tupper Lake Junction on May 17, 1912, killing one of the train crewmen.

Human failures always played a role, however. A southbound New York Central log train on the main line was double-headed for the upgrade toward Big Moose in February 1930 when the engineer realized that a water stop was essential. He stopped the train on a slight downgrade just before reaching Beaver River settlement, and the two locomotives were cut off to continue on to Beaver River for water. They ran up light because Beaver River was located on an upgrade and the crew chose not to start the train again on an uphill pull. Upon returning to pick up the train they encountered their string of cars rolling freely toward them on the downgrade toward Beaver River. The hard collision killed brakeman John S. Kivlen, who, it was later determined, was responsible for the tragedy. He had failed to fix the angle cock on the stopped train properly, allowing it to roll downhill on its own.

The Adirondack Division of the New York Central became the trunk from which ten branch logging railroads were built between 1897 and 1936. The first one was the short line built into Lake Rondaxe by John A. Dix; the last was the railroad built into Whitney Park. The last logging railroad on the Division to expire was the Emporium's Grasse River Railroad, in 1957.

Passenger service on the division began to decline after World War II, and in the early 1950s the New York State Public Service Commission began to hold hearings regarding cessation of service. At a hearing in Utica in 1953, the Brandreth Lake Corporation presented testimony that the New York Central Railroad was the only access to Brandreth Park for the ninety to one hundred summer residents and for the thirty or more lumberjacks coming from Tupper Lake and elsewhere. Brandreth Park had been lumbered sporadically since 1912, when John McDonald had built the Mac-A-Mac logging railroad into the large ownership. At least 500,000 cords (250 million board feet) of wood had been harvested on the ownership during that period, and it all went out by rail from Brandreth Station. In the early 1950s the log trucks were brought in by railroad to work on the property and the harvested timber was still being shipped out by rail. Thus the opposition to New York Central's proposed discontinuance was understandable.

Discontinuance did not come in the 1950s, but times were changing. Passenger service ceased on the Adirondack Division on April 24, 1965, and freight service halted not many years later, in April 1972. Webb's phenomenal railroad was abandoned after eighty years of service.

Tourist service was restored on a portion of the southern end in the 1990s by the Adirondack Scenic Railroad. The company has leased the trackage from the present owner, the State of New York, and has plans to restore the tracks all the way to Lake Placid at a future date.

THE LOGGING CREW ARRIVED BY RAIL

The intensity of logging activity in the forests along the Adirondack Division, which lasted up until the end of the Second World War, sometimes provided an unusual mix of passengers. Tupper Lake was the hub of lumbering for the region and provided not only an outlet for the loggers' recreational needs but also a base for those seeking woods employees. There was an occasion in 1939 when George Ainsworth of the Forestport Lumber Company needed a crew to drive his logs down the Black River in the southwest corner of the Adirondacks. What he was given turned out to be more than he could handle. Here is the report in the June 4, 1939, *Utica Observer-Dispatch* magazine.

"One year Ainsworth went up 'Webb's Railroad' to Tupper Lake to gather a gang of log drivers. He put 40 men on the train to Forestport and they got drunk. They had a rousing party. They kicked the oil lamps off the ceiling. They broke the windows and let in the air.

They fetched the emergency ax out of the glass box and laughed at the 'hatchet.' They hurled it through one of the remaining windows.

" 'I'm going to take this bunch down to Herkimer,' the conductor told Ainsworth. 'That's all right with me,' replied Ainsworth, 'but my ticket says Forestport, and that's where I get off.' 'You can get off,' said the conductor, 'but your men go down to the law in Herkimer.'

"So the train stopped at Forestport and the conductor, letting Ainsworth pass, stepped in front of the 40 lumberjacks. The jolly crew climbed right over the conductor and tumbled out after Ainsworth. Not a man went to Herkimer. They went up river and drove logs."

C H A P T E R   T E N

Moose River Lumber Company Railroad

The construction of W. Seward Webb's Adirondack and St. Lawrence Railroad created a number of new settlements in what had been an almost unbroken wilderness. Among these was the settlement of McKeever, originally set up in 1892 as a camp for the railroad construction workers. It is believed the location was named after R. Townsend McKeever, once a private secretary for Webb and later an assistant superintendent on the Adirondack and St. Lawrence Railroad.

Located just below the confluence of the South and Middle Branches of the Moose River, McKeever was to become a logical location for a forest products industry. Population grew enough in this remote location for the railroad company to build a passenger depot there in 1893.

As Webb purchased enormous amounts of Adirondack land for the location of his new Adirondack and St. Lawrence Railroad, he accumulated huge areas of virgin timber in Herkimer and Hamilton Counties. A May 1891 purchase of seventy-one thousand acres from the Adirondack Timber and Mining Company included Township 8 in Brown's Tract, a valuable timber tract. The events that followed were to make it possible for the creation of a thriving lumber industry at a small settlement known as McKeever.

In 1886 the state had built a dam on the Black River at the foot of what is now Big Stillwater Reservoir to impound water for use by manufacturers below. The rising water eventually flooded much of Webb's Township 8 property and cut off access to other land owned. The dam was eighteen feet high and impounded water for twenty miles. Webb took the state to court.

The result of the 1896 settlement added considerably to Webb's coffers. The state paid Webb $600,000 for an agreed purchase of seventy-five thousand acres of timberland, plus damages incurred, and Webb retained timber cutting rights on the land for an eighteen-year period. New lumber settlements were to be born.

Webb provided logging contracts to three different firms for timber harvest during that eight-year period. A local operator by the name of Firman Ouderkirk was given a two-year contract for timber cutting, and by terms of the contract, he had to build a sawmill at Beaver River on Webb's new railroad to saw all of the logs.

A smaller one-year contract to cut timber was given to Patrick and Dennis Moynehan, and by terms of the contract

Table 3
Moose River Lumber Company Railroad

Dates	c. 1903–c. 1915
Product hauled	hardwood sawlogs and pulpwood (4-foot lengths)
Length	5.5 miles
Gauge	standard
Motive power	1 rod locomotive, make unknown
Equipment	approx. 30 flat cars, American log loader

97. *Sawmill at Beaver Falls. The New York Central track is in the foreground. Circa 1892. Source: N.Y. Assembly Doc.*

they were compelled to haul their logs on Webb's railroad to Ouderkirk's sawmill at Beaver River.

By far the largest contract was dished out to a political ally, John A. Dix. Dix was a former governor of New York State (1872–74 and 1910–12) and had been influential in aiding Webb during his acquisition of Adirondack properties. On May 1, 1894, Dix and his partners, Lemon and Edward Thomson, were given an eight-year timber contract to cut all of the softwood timber on a portion of Township 8. Here, again, Webb closely controlled not only how the land would be lumbered but how and where the logs and sawn lumber would be transported. Terms of the contract specified that no logs could be floated down the Moose River beyond McKeever village and that all sawn timber had to be loaded on Webb's railroad at the McKeever depot.

These logging contracts given to Ouderkirk, the Moynehan brothers, and John Dix were somewhat instrumental in the development of many Adirondack villages. They formed key links in founding the settlements of Brandreth, Beaver River, Woods Lake, Big Moose, Carter, Moulin, Thendara, and McKeever, all on Webb's M&M Railroad.

John Dix and partners built a band mill in 1894 at McKeever to manufacture the spruce timber and named the firm the Moose River Lumber Company. The location was also referred to as Thomson Mills. As so often occurred, the sawmill burned down about 1896, but it was soon rebuilt.

Production in that first year (1894) was 10.2 million board feet, 66 percent of it spruce and the remainder hemlock and white pine. Production in 1896 fell off to 5.5 million board feet because of the fire, but was boosted to 16.1 million board feet the following year. The contract for Webb's softwood timber expired in 1902, causing the mill activity to take a large drop.

98. The Moose River Lumber Company mill at McKeever. Circa 1905.

By the year 1903 the Moose River Lumber Company was able to gain control of a large acreage east of McKeever and south of the South Branch of the Moose River. The area was rich in hardwoods. A hardwood sawmill was geared up, possibly using the older mill, located between the New York Central tracks at McKeever and the Moose River.

Much of the labor force for the new hardwood mill moved over to McKeever from Big Moose, including mill superintendent John Potter. Potter suffered a wood splinter in his eye and lost sight in that eye while at McKeever. A later superintendent, Thomas Marleau, also from Big Moose, recalls the time when the large band saw came off the wheels and went careening down the mill floor. One of the workers, frozen in fear as it narrowly missed him, promptly marched to the office and quit on the spot. Some of the workers had also come to McKeever from Ducey's big mill at Brandon, which had closed down shortly before.

To move the hardwood logs to the sawmill, a 5.5-mile standard-gauge railroad was built about 1903, easterly through the property toward Woodhull Mountain, more or less parallel with the South Branch of the Moose River. The railroad serviced six logging camps, each with about forty men cutting and stacking hardwood logs railside. Albert Merritt was supervisor of logging operations.

The Moose River Lumber Company railroad was built to standard gauge for two principal reasons. First was the location of the logging railroad—on the opposite side of the New York Central right-of-way from the mill—requiring the use of the NYC main line for a short distance in order to cross over with the log train. Second, the original plan was to extend the railroad easterly as far as the Hudson River watershed, where the Thomson's reportedly had an interest in large acreages of timber land and where access was possible to pulpwood markets. This second goal proved to be quite unrealistic.

Nothing is known of the motive power used on the Moose River Lumber Company railroad other than it was a small "rod" locomotive of unknown make. Log loading was accomplished with an American loader that moved along the empty log flatcars with rails on the car bed. The loader had been purchased from the Horse Shoe Forestry Company in the early 1900s.

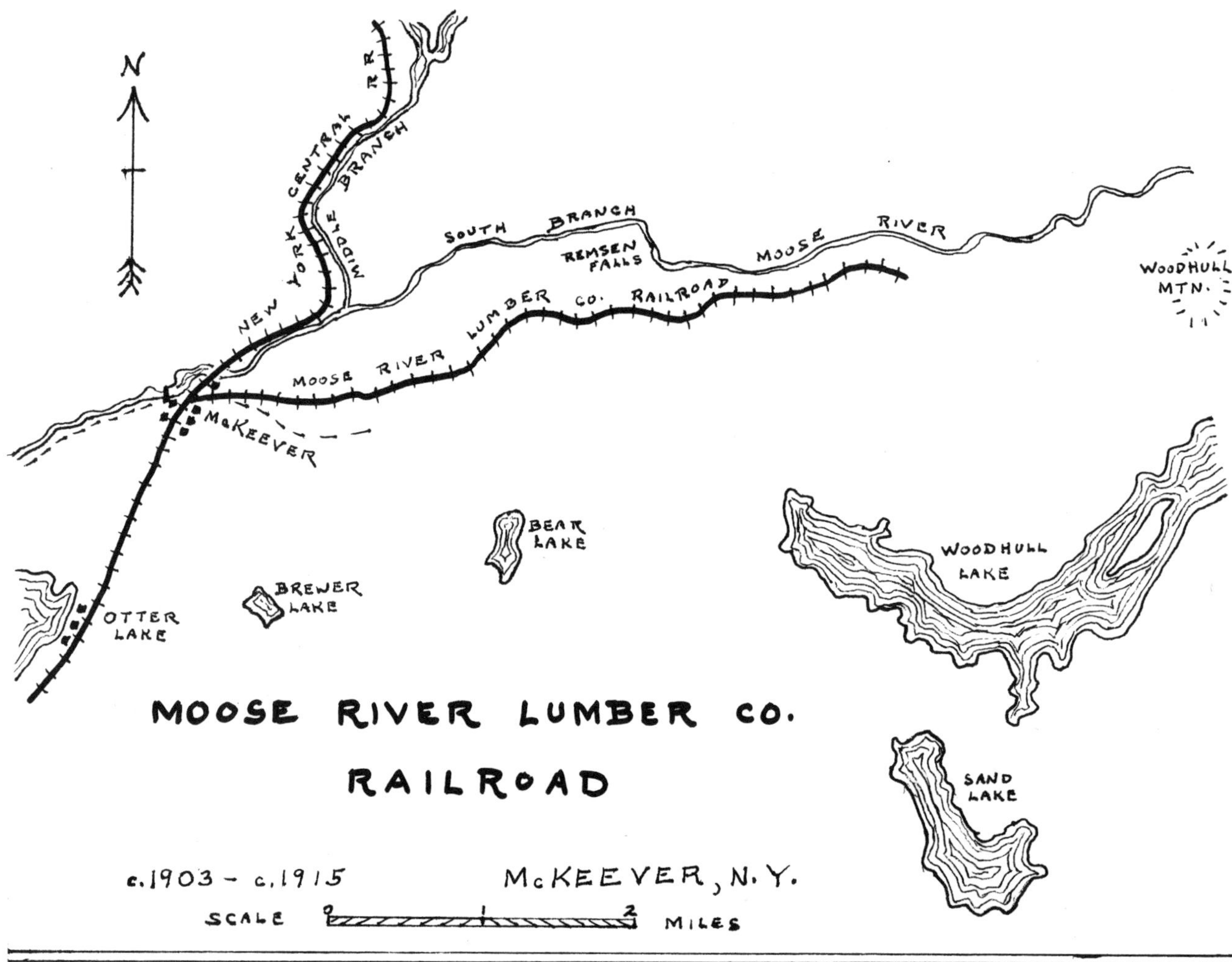

N
NEW YORK CENTRAL RR
MIDDLE BRANCH
SOUTH BRANCH
MOOSE RIVER
REMSEN FALLS
WOODHULL MTN.
MOOSE RIVER LUMBER CO. RAILROAD
McKEEVER
BEAR LAKE
BREWER LAKE
OTTER LAKE
WOODHULL LAKE
SAND LAKE
MOOSE RIVER LUMBER CO.
RAILROAD
c.1903 – c.1915
McKEEVER, N.Y.
SCALE 0 1 2 MILES

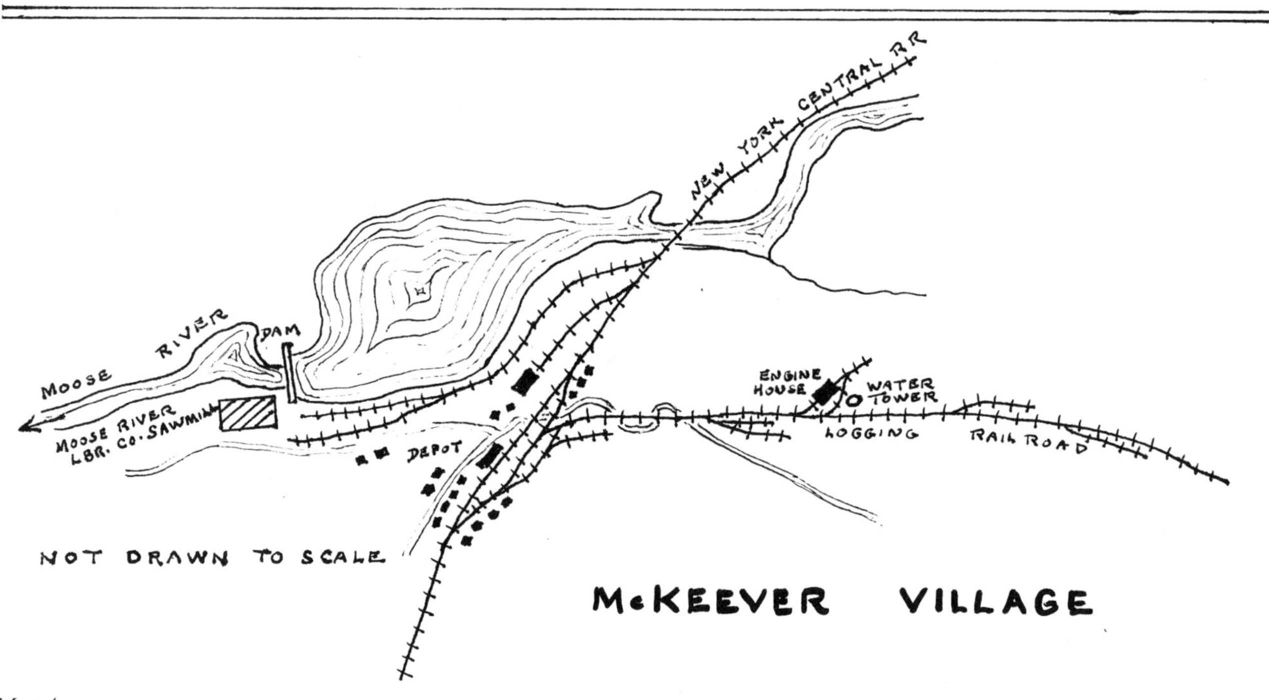

NEW YORK CENTRAL RR
MOOSE RIVER
DAM
MOOSE RIVER LBR. CO. SAWMILL
DEPOT
ENGINE HOUSE
WATER TOWER
LOGGING RAILROAD
NOT DRAWN TO SCALE
McKEEVER VILLAGE

Map 4.

99. Spring has arrived, and much of the winter cut of hardwood logs is waiting transport to the mill by the small Moose River Lumber Company locomotive. Circa 1905.

100. The log train was being loaded at a large hardwood log concentration yard. The American log loader still bears the symbol of the former owner, Horse Shoe Forestry Company. Circa 1905.

The sawmill was putting out about forty thousand board feet of birch and maple lumber daily. And the yellow birch timber was outstanding. One of the largest birch logs sawn was squared off at thirty inches on the carriage and sawn into twenty-six-inch-wide lumber "clean as a hound's tooth," according to the lumber grader Edwin Kling.

John Dix was one of the investors in the Iroquois Paper Company pulp mill, also located at McKeever. The logging railroad, therefore, would haul large quantities of four-foot pulpwood to the McKeever pulp mill. The pulp mill had four grinders operating, producing pulp for newsprint manufacture. At some point during this era there was also a small veneer mill built in the village.

The financial situation apparently became tough for the Moose River Lumber Company. In 1907 all of the real estate assets were conveyed to the Iroquois Pulp and Paper Company except for the sawmill building, the hardwood stumpage, and the logging equipment. Operations continued, however.

In 1913 a contract was made with the Sagamore Lumber Company of Buffalo to supply Sagamore with hardwood lumber (70 percent birch) at the following prices per thousand board foot: "#1C & Btr. = $3.50, #2C = $3.00, #3C = $2.50." But the financial situation worsened, and the Moose River Lumber Company went into bankruptcy in October 1915.

McKeever remained a small but active wood industry village for many years afterward. The Gould Paper Company had its woods headquarters there for their logging and river-driving operations on the Moose River. In fact, logging activities along the entire Adirondack Division of the New York Central was so intense that by 1917 there were sixty logging camps and three thousand men busy in the area.

With the closing of the sawmill, the properties were eventually sold to the Gould Paper Company, which continued its woods operation out of McKeever until about 1940. In 1947 Rice Veneer Company opened a veneer operation on property purchased from Gould Paper, selling it to the Georgia Pacific Company in 1956. With the closing of that veneer mill in 1960, the forest products activity of McKeever came to an end.

CHAPTER ELEVEN

John A. Dix Railroad Logging Operation and the Raquette Lake Railway

The Raquette Lake Railway, short as it was, had what was probably the most distinguished board of directors of any railroad in the entire United States—Collis P. Huntington, J. Pierpont Morgan, Alfred G. Vanderbilt, William C. Whitney, Henry Payne Whitney, W. Seward Webb, and Senator Chauncey M. Depew. But grand as its list of owners and passengers might have been, the railroad had its beginning as a crude logging railroad.

After W. Seward Webb had completed his Adirondack and St. Lawrence Railroad and leased it to the New York Central, he sold off some portions of his vast Adirondack ownership. In 1896 Township 8 in Herkimer County was sold to the State of New York, with Webb reserving the right to cut timber for eight years. The tract was well forested with old-growth timber.

As mentioned in chapter 10, Webb kept tight control over who cut the timber and where it was marketed, and he awarded a large timber contract to his political ally John A. Dix.

By 1896 John Dix had five million board feet of logs harvested in the Big Safford Lake area and piled on skidways at Rondaxe Lake. The plan was to float the logs down the Moose River to John Dix's sawmill at McKeever, the Moose River Lumber Company. But a problem materialized before the river drive ever began. Both banks of the North Branch of the Moose River near Carter were owned by William DeCamp, and he demanded a toll for passage of the logs. Dix wouldn't pay it; he then took the case to court and lost the decision. However, John Dix was too clever to accept defeat that easily.

Dix quickly hired Webb's lawyer, Charles Snyder, to work out a deceptive maneuver. Snyder was able to get Thomas Clark Durant to purchase a railroad right-of-way across DeCamp's land, ostensibly for the sole purpose of a rail line to Raquette Lake. DeCamp agreed without realizing that the source of the purchase money was John Dix.

Dix then used the newly purchased right-of-way to build a three-mile railroad, running easterly from the New York Central to Lake Rondaxe. The junction with the New York Central was at a remote location known as Clearwater,

Table 4
John A. Dix Railroad

Dates	c. 1896–1897
Product hauled	spruce sawlogs
Length	approx. 3 miles
Gauge	standard
Motive power	railroad operated by the New York Central and Hudson River Railroad

101. The general store at Carter (Clearwater) as viewed from the south end of the depot. Circa 1914. Photograph by Gerald Smith.

Table 5
Raquette Lake Railroad Common Carrier

Dates	1899–1933
Product hauled	passengers, spruce sawlogs
Length	18 miles
Gauge	standard

Motive Power

No.	*Builder*	*Type*	*Builder No.*	*Year Built*	*Weight (tons)*	*Source*	*Disposition*
1, 1st	Rome	4-6-0	314	1887	53	NYC 1902	scrapped 1914
1, 2d	Schenectady	4-8-0	4700	1898	72	NYC 1914	scrapped 1916
1, 3d	Schenectady	2-6-0	5506	1900	80	NYC 1916	scrapped 1947
2, 1st	Schenectady	4-6-0	2442	1887	53	NYC 1902	scrapped 1912
2, 2d	Schenectady	4-8-0	4532	1897	72	NYC 1912	scrapped 1915
2, 3d	Baldwin	2-6-0	17638	1900	80	NYC 1914	scrapped 1941

changed later to Carter Station in 1912. Logs began to roll over the rails to McKeever.

The loaded log cars were hauled over the New York Central from the Dix timber holdings to the McKeever mill by New York Central engines. Dix's timber contract with Webb was to run out in 1903, but even before that date he recouped his cost of building the railroad by selling the pike to Durant and associates. They had larger things in mind for John Dix's little railroad.

Raquette Lake Railway

At the turn of the century, picturesque Raquette Lake had a small number of summer mansions belonging to some of the wealthiest tycoons in the nation. One of them, Collis P. Huntington, had purchased the expansive Camp Pine Knot from the Durant family. He entertained dreams of riding all the way up to his camp location in his own private railroad coach. The strain and tiring inconvenience of changing trains and using public facilities could be avoided, and the tedious journey on the small steamers along the lakes from Old Forge to his vacation mansion would be no more, if a railroad could be extended to Raquette Lake.

Thus, with the operating Adirondack Division of the NYC not very many miles away from Raquette Lake, Huntington urged William West Durant to consider organizing a railroad branch to Raquette Lake. Durant was the son of Dr. Thomas Clark Durant, the mastermind behind the financing and construction of the first transcontinental railroad in the United States (the Union Pacific) and was a large landowner in the Adirondacks.

The proposal appealed to William Durant because of his large timber holdings near Raquette Lake, which could possibly be harvested with railroad access. And it was also genuinely believed at that point in history that the restrictions preventing timber cutting and development on state-owned land would soon be removed under public pressure. The hoped-for result would be the opening up of large state ownerships around the lake for timber harvest and freight for the railroad.

102. No. 2, a Schenectady-built 4-8-0, was purchased second-hand from the New York Central to be the second locomotive owned by the Raquette Lake Railway.

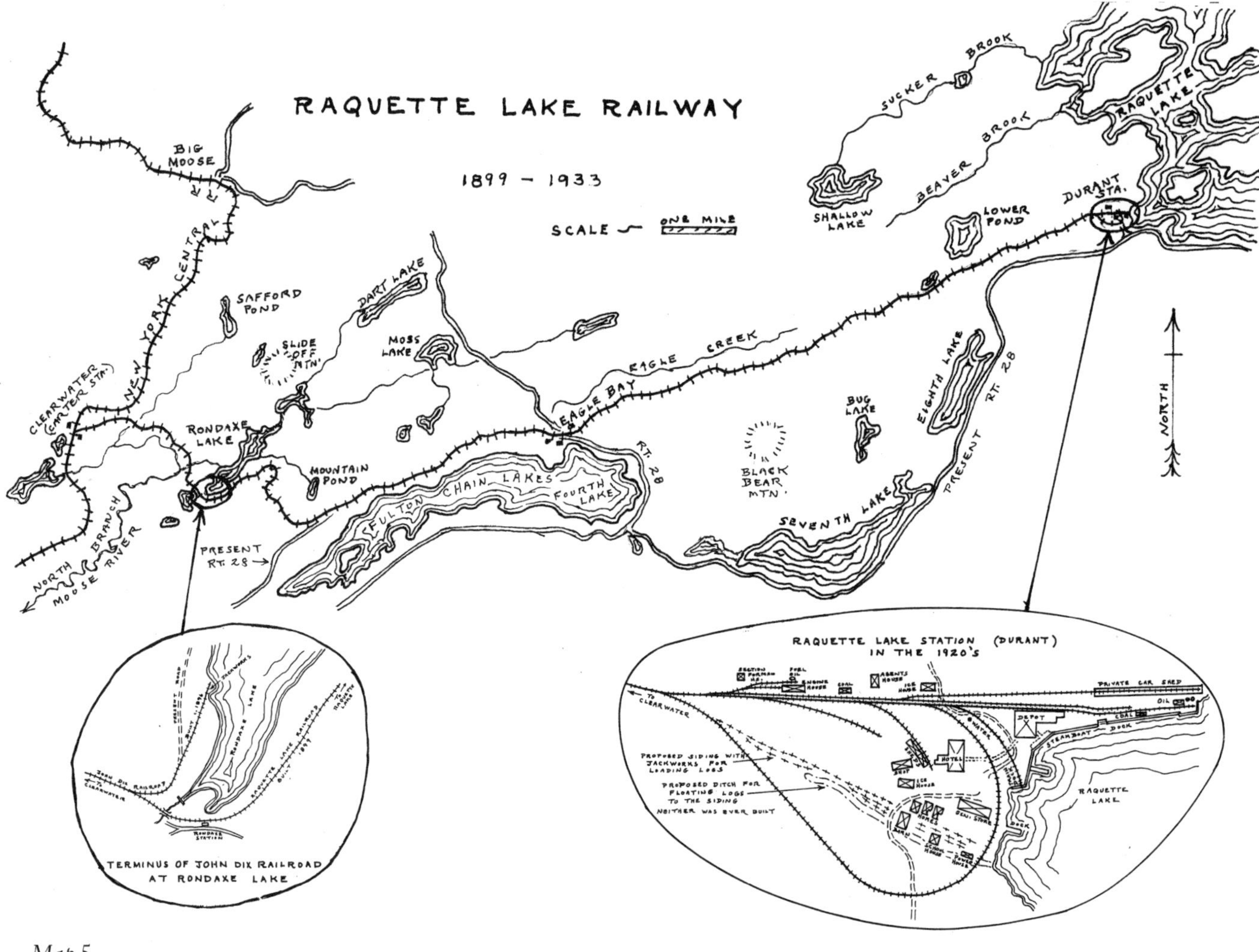

Map 5.

For the best access to his land, Durant had wanted to lay rail over to Sucker Brook Bay in the northwest section of the lake, but that meant crossing state-owned land, and the state constitutional act preventing the cutting of any trees on public Adirondack land was by then in effect. The surveys, therefore, put the railroad in an easterly direction to a terminus at the southwest corner of the lake. Opponents appeared, decrying the proposed railroad as a fire hazard to the camps and of benefit only to the logging interests. In spite of the opposition, it went through.

That very distinguished group of investors and board of directors was formed, and the railroad was built in a five-month period in 1899, using the spur built by John Dix as a beginning. Total length from Clearwater Junction to Raquette Lake was eighteen miles. Collis P. Huntington was able to run his private railroad car to the Raquette Lake dock for the first time in September 1899. The line was opened to the public in July 1900.

The New York Central began hauling log trains on the Raquette Lake line from Eagle Bay as early as the winter of 1899–1900. But the Raquette Lake Railway was not part of the New York Central system in the early years. It was privately operated, usually running two trains each way daily to provide convenient service to their patrons at the village of Raquette Lake and the camps in the area. Six locomotives were owned over the thirty-three years of operation; up to three of them were operating at any one time.

Expectations continued that the state-owned timberlands around Raquette Lake would be opened up before long. The general manager of the Raquette Lake Railway, Edward M. Burns, made the public statement that he estimated seven hundred million board feet of spruce would

103. The Raquette Lake Railway terminated on the shore of Raquette Lake, at a location first known as Durant.

soon be available from public land. In anticipation, therefore, the layout for the terminal at Raquette Lake included a "ditch for floating logs" alongside a railside loading location. In 1901 the U.S. Department of Agriculture forestry section developed a proposal with working plans for cutting the timber on state-owned Townships 5, 6, 40, and 41, complete with plans for an extended logging railroad. Nothing came of it, however; it was merely an economic development proposal. The federal government, incidentally, owns no timberland in the Adirondacks.

Forest products remained the principal freight item on the eighteen-mile run to Clearwater Junction. In 1911 there was 13,692 tons of outgoing freight, 92 percent of which was listed as forest products of some type, primarily logs. The train crewman with the most dangerous job on the log train was, of course, the brakeman; he had to be adept at finding his footing over slippery, moving, uneven log surfaces. On June 24, 1901, brakeman Nelson Avery had the misfortune of falling from a log car on a train bound for the sawmill at McKeever. The train was being shifted at a switch near Clearwater when Avery slipped and fell under the moving wheels. He died later that day.

In 1917 the New York Central acquired ownership of the Raquette Lake Railway and continued operation for another sixteen years. The railroad made its last run on September 30, 1933, thirty-three years after being built by that highly distinguished board of directors.

C H A P T E R   T W E L V E

Woods Lake Railroad

One of the opportunities opened up in 1892 by W. Seward Webb's new railroad spanning the Adirondack region was accessible meccas for sportsmen. Remote flag stops became established along the railroad to service the wilderness vacationers and sportsmen coming from urban areas.

One of those small, established train stops was the little depot of Woods Lake, located between Big Moose Station and the Beaver River settlement on Stillwater Reservoir. There were families living at the site as early as the 1890s. The Woods Lake flag stop became a popular destination for many deer hunters. Fourteen deer carcasses were shipped out from this depot during the hunting season of 1900. In 1915, however, the Woods Lake flag stop came to life and grew into a small settlement, a short-lived one, but busy while it lasted.

The International Paper Company had purchased Township 6 in the Town of Webb from the Adirondack Forest and Lumber Company in 1898. Then at the time that World War I was raging, International Paper decided to harvest the softwood pulpwood on that portion of their property along the Twitchell Creek drainage, south of Stillwater

International Paper Company
30 Broad Street
New York
Department of Woodlands
J. P. Riley, Manager
March 20th, 1919.

104. Letterhead.

Flow. The area was close enough to a number of different company-owned paper mills, which could receive the wood.

Paper companies had become major landowners in the Adirondacks by that point in history, but none of them had ever found a place for logging railroads in their softwood harvesting methods in the Adirondacks. International Paper Company's Woods Lake operation was to become the one and only exception. Their railroad was short—the Adirondacks' shortest logging railroad, with only about three miles of track—but the railroad had its own locomotive and rolling stock.

The company was actually receiving a large portion of their pulpwood shipments over the mainline railroads during this era, including many trainloads over the New York Central. For this operation, however, it was not considered economical to attempt to sled all of the wood up to a siding along the New York Central for loading into cars. Most of the Twitchell Creek cut lay downstream from the railroad line, and wood driven down the small stream would have to be removed at a point quite inconvenient to any siding on the NYC. Thus came the decision to build a small company railroad into the wood removal point on the stream. A cut-up mill and conveyor system known locally as a "jackworks" would be used to pull the wood out of the stream and load it on railroad cars.

Brisk activities began in 1916 to transform what had

Table 6
Woods Lake Railroad

Owner	International Paper Company
Dates	1916–c. 1926
Product hauled	softwood pulpwood (4-foot)
Length	3 miles
Gauge	standard
Motive power	Baldwin 2-6-2 saddletank, class 10, 26 1/4 B1, builder no. 14494, built 1895 for International Paper Co.
Equipment	200 pulpwood rack cars

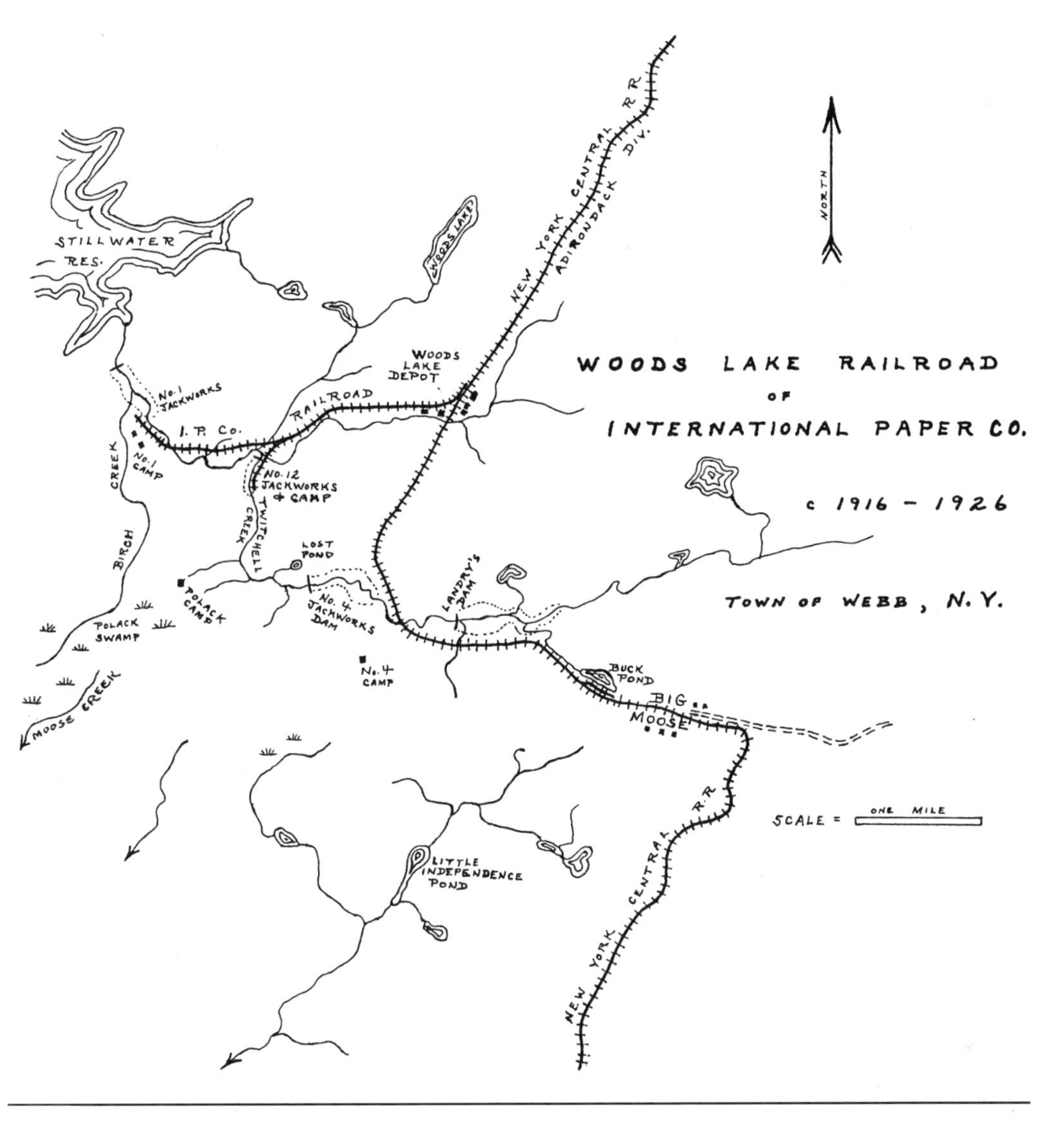
WOODS LAKE RAILROAD
OF
INTERNATIONAL PAPER CO.
c 1916 – 1926
TOWN OF WEBB, N.Y.
NORTH
SCALE = ONE MILE
NEW YORK CENTRAL R.R.
ADIRONDACK DIV.
STILLWATER RES.
WOODS LAKE
WOODS LAKE DEPOT
RAILROAD
I. P. Co.
No. 1 JACKWORKS
No. 1 CAMP
No. 12 JACKWORKS + CAMP
BIRCH CREEK
TWITCHELL CREEK
LOST POND
POLACK CAMP
POLACK SWAMP
No. 4 JACKWORKS DAM
No. 4 CAMP
LANDRY'S DAM
BUCK POND
BIG MOOSE
MOOSE CREEK
LITTLE INDEPENDENCE POND
NEW YORK CENTRAL R.R.

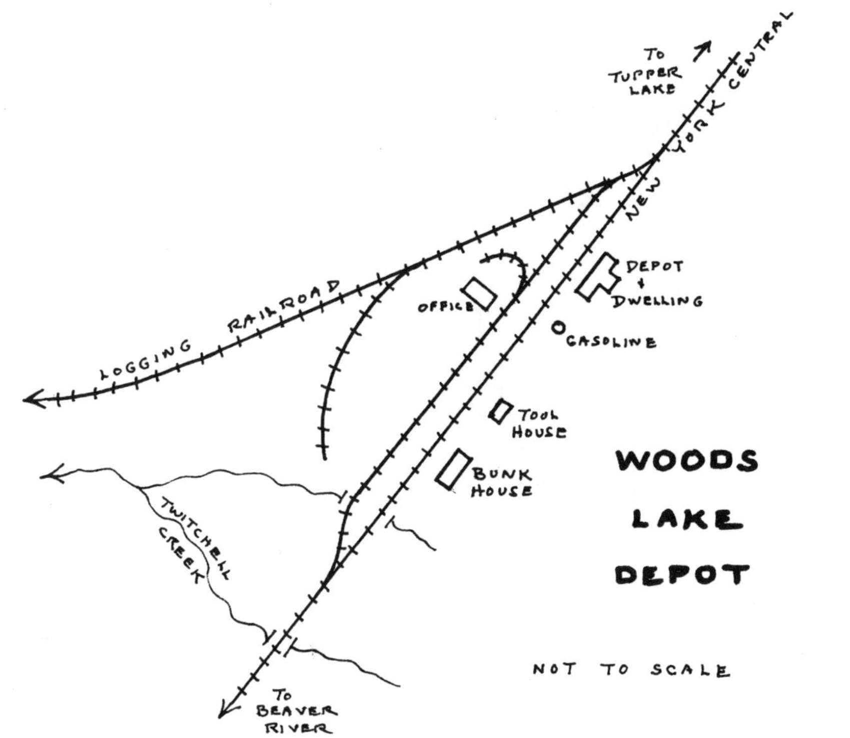
TO TUPPER LAKE
NEW YORK CENTRAL
LOGGING RAILROAD
OFFICE
DEPOT + DWELLING
GASOLINE
TOOL HOUSE
BUNK HOUSE
WOODS LAKE DEPOT
TWITCHELL CREEK
NOT TO SCALE
TO BEAVER RIVER

Map 6.

CHAMPLAIN REALTY COMPANY
DIVISION OFFICE 111 WHITE BUILDING, UTICA, N. Y.
WOODS LAKE, N. Y.
G. F. UNDERWOOD, PRESIDENT
JOHN P. RILEY, VICE PRESIDENT

105. Letterhead.

been the remote deer hunters' flag stop into a bustling logging village and depot of noticeable proportions. Before long there were enough residents to warrant a post office, which opened on November 7, 1917. The postmarking stamp was received on January 28, 1918, and about $20 of cancellations were recorded monthly in that year. Postmaster Charles Fletcher would hang out the mail bag each day for the mail train, even it was empty. Fletcher was also the clerk for the International Paper Company.

School classes were held in a railroad car for the few permanent families in residence. Besides the supervisory personnel that International Paper moved in, there was a section foreman and maintenance crew for the New York Central in residence.

In an effort to reduce operating costs, a seemingly endless project of all paper companies, International Paper released its cooks at the Woods Lake settlement in 1920 and hired a Mr. Peasley and his wife to cook together for a combined salary of $125 per month. The Peasleys added two more children to the school enrollment. Another resident left the hard way the following year—a company cow escaped the fenced pasture and was killed instantly by a northbound train, the meat unsalvageable.

International Paper built their "short line" railroad in 1915–16, forming a three-mile connection from the Twitchell Creek jackworks to the Woods Lake depot on the NYC. The sole function of the Woods Lake railroad was to move the loaded pulpwood cars from the jackworks over the short distance to the siding at Woods Lake depot, where they could be picked up by the New York Central.

Arriving in 1916 was a small company-owned locomotive that was sent down from an International Paper Company operation in Maine. It appears that this engine, a Baldwin-built 2-6-2 saddletank, was used by International Paper on a company logging railroad, the American Realty

106. An exceptional sled load of spruce put together by George Bushey's crew at Camp 4, above No. 4 jackworks. The scaler stands to the right of the sled, and a number on a log end indicates a load of 15,150 board feet. Circa 1921.

Company's South Bog railroad in Rangeley Plantation (1901–c.1906).

The locomotive had been built new for International Paper by Baldwin twenty years earlier and had doubtless seen some hard service on company railroads in Maine. En route to Woods Lake in 1916, therefore, the little saddletank was sent to the Rome Locomotive Works for repairs. It was outshopped in June 1916, from Rome, but not in perfect condition, as future operations were to reveal. The little railroad began operation that spring in 1916.

The Woods Lake operation during its early years was managed by William O'Brien, resident superintendent, who in turn reported to T. J. Wilbur, resident manager for Champlain Realty Company with an office at Utica. Champlain Realty Company was an operating subsidiary of International Paper Company, and as such, Wilbur reported to John P. Riley, its manager of the Department of Woodlands in New York City. T. J. Wilbur died in August 1921, and O'Brien moved up to succeeded him as resident manager at Utica the following year. In 1923 O'Brien wrote to Riley requesting a salary increase; he got a raise to $375 per month.

Wilbur kept in daily touch by telephone with resident supervisor O'Brien at Woods Lake and was constantly nagging him to cut operating costs. Thus, O'Brien received a reprimand one winter for paying the exorbitantly high price of $37 per ton for hay (FOB car) when he had previously been given a direct order to cut expenses.

The entire logging and railroading operation was eventually placed in the hands of a principal contractor, George Bushey. Other contractors that had worked the area were John Davignon, George Vincent, and George Harvey. Vincent also owned a hotel at Thendara that catered to the lumberjacks seeking recreation and gladly cashed the loggers' checks. Bushey had a renowned and respected name in the North Country and had contracted for International Paper as early as 1905, when he cut and river-drove forty-five thousand cords of pulpwood for the company. In those days, Bushey remembered, the pulpwood was all cut by ax, no saws.

George Bushey was truly a contractor of the "old school"; he worked hard, worked his men hard, and knew how to move wood under the adverse conditions imposed by Adirondack winters. He was a very competent and efficient woodsman, and it is said he argued constantly with the company superintendent, O'Brien, although he considered International Paper to be honest in their dealings with him. Bushey's wife, Kate, cooked in one of the log camps.

Bushey's contract with International Paper would call for him to deliver the wood streamside and slash it into four-foot lengths, provide a boardinghouse, load the railroad cars, and operate the company railroad, paying the cost of the locomotive fuel, repairs, and crew. His contract for the 1923–24 season called for him to receive $12.50 per cord for peeled spruce, fir, and hemlock and $10.50 per cord for rough (unpeeled) wood and for pine. Bushey logged twenty thousand cords along Birch and Moose Creeks that winter.

The softwood logs were cut into twelve- and sixteen-foot lengths in the woods for sledding. Then a cut-up mill at the jackworks slashed the logs into four-foot lengths before loading into the cars.

In addition to the indispensable horses, Bushey also used tractors for the long sled hauls and was one of the earliest jobbers in the Adirondacks to use them. His preference was the ten-ton Holt, which he knew how to use properly. He had three of them at Woods Lake, despite the tragic experience of one rolling over onto this brother Eddie, killing him in December 1920.

A new ten-ton Holt tractor cost $6,200 in 1921, and International Paper held the chattel note for Bushey's machines. Tractor sleds with chains cost $200 for the best ones, and a contractor would need at least fifty or sixty for a tractor

107. Contractor George Bushey and crew with a Holt tractor in March 1921. The train of five sleds contains 797 pieces of sixteen-foot spruce, scaling 75.77 cords.

108. The tractor sleds were loaded with the aid of a frame that allowed the landing crew to build a high, straight-sided load. March 1921.

job logging ten thousand cords per season. The large Holt would haul four sleds, loaded with about forty cords.

There were four jackworks located along Twitchell Creek or on tributaries, each with a cut-up mill, except at Landry's Dam. The saws and the jackworks conveyor were powered by steam engines and in some locations by gasoline motors. The principal railroad loading site was the Number 1 jackworks, located where the Tree Farm public campsite is now built, but loading was also done at Number 12 jackworks and at the Buck Lake jackworks. Loading the railroad cars was hard work, involving tossing the four-foot wood as it rapidly dropped off the conveyor, and the contractors had trouble keeping the men from walking off this tedious job.

The crew at No. 1 jackworks was loading fifteen to twenty cars per day in January 1921, the best day being twenty-four cars. The cut-up mill was capable of cutting over two hundred cords of four-foot pulpwood daily from the long logs floated down.

The Number 4 dam and jackworks was not used to load the cars. The long logs were floated down Twitchell Creek, jacked out at Number 4, cut into four-foot lengths and dumped back into the creek for floating down to Number 12 for loading on cars. Landry's Dam had a small shack at the end where attendant Joe Landry lived and had a telephone to inform him when to allow more water and logs through the dam.

The jackworks and logging camps were connected by single-wire magneto telephones with headquarters at Woods Lake depot. The largest logging camp, at Number 1 jackworks, had a capacity of 112 men plus horses under foreman Ed Smith.

In June 1921 a traveling salesman by the name of R. R. Morrow rode the company rail motor car into Camp 1 to see George Bushey. On his return trip the car jumped the track, throwing him to the ground and causing slight injuries and torn clothes. His request for $100 to pay for doctor and clothes bill was granted, but the company made sure that any future riders thereafter signed waivers.

Most of the railroad cars used to move the Woods Lake pulpwood were owned by the International Paper Company. In 1920 the company purchased two hundred flat-bottomed gondola cars and brought them to Woods Lake for refurbishing as pulpwood cars. Doors were cut in the car sides and racks built around the car to contain the loosely dumped four-foot wood.

A Lane No. 0 portable sawmill was set up at Woods Lake in May 1920 to saw out $1^{1}/_{4}$" x 8" beech for use in building the car racks. Larger spruce with red rot unacceptable for a ground-wood pulp mill was sawn into timbers, but the mill had constant trouble that winter when the fifty-horsepower portable sawmill boiler was unable to keep up the steam pressure. The unprofitable sawmill was shut down at the end of 1923.

The Baldwin locomotive turned out to be a constant headache for the crew at Woods Lake. Either it was sent to Woods Lake by International Paper in a well-used condition or it took some abuse on the little three-mile line; it was in constant need of repair. Engineer Elias Dowling, who was remembered as being a terrible grouch, constantly complained about the fitness of the Baldwin. Despite his disposition, he apparently had a point, because there were breakdowns soon after the commencing of the Woods Lake operation.

In 1918 the International Paper Company temporarily leased locomotive No. 35, a geared Climax, from the Emporium Forestry Company for $15 per day, with Emporium supplying the engineer, of course. Emporium had a large sawmill and logging railroad operation at Conifer, New York, operating a fine locomotive repair shop to service the many locomotives they kept busy logging the Adirondack woods. Another of the locomotive engineers at Woods Lake was Elmer Martin.

New tires (metal wheel rims) were purchased in 1919 for all ten locomotive wheels on the Baldwin. The total cost seems unbelievable—$53.77. It was a different story the next year; the repair bill from the Rome shops was $6,408. Major repairs were called for again in the fall of 1923, done this time at the Emporium shops in Conifer.

The repaired locomotive was returned to Woods Lake on December 28, and within less than two weeks disaster struck. On January 3, it ran off the track above Camp 12 and broke the frame completely through. Off again went the cripple to the Emporium shops, and this time Emporium locomotive No. 33, another Baldwin, was leased to fill the gap. There obviously was a good relationship with Emporium Forestry Company because in January and February 1923 Emporium had in turn leased International Paper's Baldwin after a disastrous engine house fire at Conifer, which disabled part of the Emporium locomotive roster. In 1924, Emporium proposed an arrangement to cut hardwood logs and haul them over the International Paper Company railroad while it was still in operating condition, but nothing came of it.

Forest fires were of serious concern during this era of steam, and a state law required the presence of a fire patrolman on the railroad from April 1 to November 15. Frank Onroful patrolled the three-mile stretch in 1920 as well as sections of the New York Central.

Pulpwood from the Woods Lake operation was sent to various company pulp mills in New York—Piercefield, Watertown, Hudson Falls, Brownsville (two-foot lengths), Corinth, and Fort Edward. The busy season was, of course, late winter and spring, moving out the winter's cut. In June 1923 there were 31,500 cords of peeled spruce, fir, and hemlock floating in the backed-up waters of Twitchell Creek, and the conveyors were busy loading from 70 to 135 cords daily into the cars. There were peak periods when as many as twenty to twenty-four cars (295 cords) would be loaded daily.

And then there was the usual and continual squabble over scale or measurement; it seems the paper companies can never escape it. The standard rack car with wood thrown in loosely (unpiled) and filled to the height of 7$^{1}/_{2}$ feet would hold from 12$^{1}/_{3}$ cords to possibly 14 cords of peeled spruce. A cord occupied about 192 cubic feet of space when loaded in this fashion. But arguments never ceased, especially over deductions for bad wood. Being ground-wood pulp mills, the plants were not able to utilize sticks with red rot or partial sapwood decay. And plant management had a continual battle with woodland management over acceptable standards. A familiar story.

M. O. Wood, manager of International Paper's Piercefield mill, put in a loud complaint concerning wood being shipped in from the Woods Lake area in 1919. It seems the wood had been cut in 1917 and left lying around in the bogs for two years, deteriorating to the point where one cord in fifteen was not fit for pulp. The company scaler Reynolds went up to Woods Lake and soon became embroiled in a dispute with contractor George Vincent over the definition of cull wood. Vincent had his men "pull the fire" under the locomotive boiler and left the job.

Another familiar problem was the failure of the New York Central Railroad to return the empty cars when needed. International Paper Company had eighty-two private rack cars (IPCX cars) in use in the district, plus the open-top rack cars available from NYC. Time and again, the New York Central would promise to supply all the cars needed but would fail to deliver.

The Woods Lake operation lasted from 1916 until 1926, approximately ten years. Pulpwood was also loaded in 1920–21 into cars at George Vincent's operation at Buck Pond, near Big Moose, using a steam-powered conveyor to pull the wood from the pond to cars on a siding next to the NYC main line. Brief consideration was given by Interna-

109. The Woods Lake station as it appeared in the 1970s. To the right of the New York Central track can be noted the location of the former siding where the International Paper Company locomotive accumulated the loaded pulpwood cars.

tional Paper to building a company railroad line into Little Independence Lake, southwest of Big Moose, but the plan was abandoned because of the poor market at the time. International Paper then hoped to sell the railroad equipment to a hardwood operator, but that didn't work out, either.

The Woods Lake settlement soon became nothing more than a caretaker's residence. Superintendent William O'Brien had already left, having been promoted to resident manager at Utica in January 1922. The rails were pulled up about 1932, but many of the buildings at Woods Lake depot and at No. 1 jackworks were left standing for years afterward. George Bushey retired not many years after the closing of Woods Lake, at an early age of fifty-five, but lived to see his ninety-fourth year.

The last full-time resident at Woods Lake station was caretaker Noah Fournier, who became ill in 1945 at the age of eighty-five and had to leave. The attractive old stationhouse stood alone for many years, used on occasion by the company, and the abandoned roadbed now serves as a truck road winding through the young spruce growth. Out at the other end, at the former No. 1 jackworks, the International Paper Company has built a public campsite for the Stillwater Tree Farm, located where once terminated the shortest logging railroad in the Adirondacks.

CHAPTER THIRTEEN

Mac-A-Mac Railroad

Scattered among the almost six million acres of forest land in the Adirondacks, among the immense acreage of state-owned land and smaller acreage of paper company-owned land, are a number of relatively large private holdings or "parks." These are large, privately owned forest properties purchased in the latter half of the nineteenth century by affluent individuals for, in most cases, private resort use. A remnant of these remain, often in skeleton form and often in corporation form involving an army of heirs.

Forest management on these private parks formerly took a back seat to summer luxury and wilderness enjoyment. But then there arose mounting expenses, growing numbers of heirs to placate, and inheritance taxes to satisfy, paid sometimes by a sudden, intense effort of timber harvesting that shows yet today in occasional large areas of unproductive land. Not entirely, however; some of the parks have wisely applied practical forest management in recent years.

Among the early land speculators was a successful entrepreneur from Ossining named Dr. Benjamin Brandreth, born Benjamin Daubeney in England but whose family circumstances made a name change desirable. Dr. Brandreth had found success in the patent medicine business, manufacturing pills and plasters, and harbored ideas in the mid-1800s of manufacturing a superior beer. In 1851 Brandreth purchased Township 39 (25,000 acres) from the state for fifteen cents per acre, cleared one hundred acres to grow hops, and looked forward to the brew that would result when he utilized the pure lake water of Beeches Lake, the large body of water in the township.

The experiment didn't work, and the discouraged Benjamin Brandreth was going to let the property go for taxes, but his family prevailed upon him to keep it. Summer resorts in the Adirondacks were then coming into vogue for those who could afford them. Thus, in the 1870s, the first lakeside camps were built by his son, Franklin, and the beautiful body of water was renamed Brandreth Lake.

In the early years it was a long and arduous trip by stage

110. A flag stop was required for these two vacationers at Brandreth before the depot was built. Flag stops were frequent during the early years of the Adirondack Division. Photograph by H. M. Beach. Courtesy of the Brandreth heirs.

111. The first depot at Brandreth was built by the New York Central in 1895 to accommodate the summer residents. Photograph by H. M. Beach. Courtesy of the Brandreth heirs.

112. Photographer H. M. Beach coaxed all of the teamsters at Carlson's Camp to pose with their teams, probably on a Sunday afternoon. The camp buildings from left to right are the kitchen, bunk house (barely visible), blacksmith shop, barn, and office on the far right. The tall man in the center, in front of the open bar doors, is the camp foreman, Emil Johnson. Fred Johnson was the manager of the operation. Charles A. Rumble is the well-dressed man on the far left. Photograph by H. M. Beach. Courtesy of the Brandreth heirs.

coach and buckboard for the Brandreth family to reach their Adirondack paradise. Then in 1892, when W. Seward Webb's new railroad was completed through the Adirondacks, access to their summer home became a pleasure ride, shared with many others. With coach service available, one no longer had to be affluent to enjoy a summer vacation among the beauties of the Adirondacks.

Webb's new railroad passed just to the west of Township 39, the Brandreth property, and it wasn't long before an eight-mile wagon road was built from the Brandreth lodge to a flag stop on the railroad, referred to later on the maps as Brandreth or Brandreth Lake. A small depot building was erected on the New York Central in 1895, and the Brandreth clan then arrived at their summer homes in relative comfort.

Table 7
Mac-A-Mac Railroad

Owner	Mac-A-Mac Corporation
Dates	1912–1920
Product hauled	spruce pulpwood (log length), pine and hemlock sawlogs, hemlock bark
Length	12 miles
Gauge	standard

Motive Power

No.	*Builder*	*Type*	*Builder No.*	*Year Built*	*Weight (tons)*	*Source*	*Disposition*
1	Heisler	2-truck	1253	1912	42	new	L&M Stone Co., Prospect, N.Y.
2	Alco-Schenectady	2-6-0	53845	1913		new	Bay Pond RR

Rolling stock

Flat cars (8- to 12-foot stakes)

However, access to the summer homes wasn't the only advantage afforded by the New York Central Railroad; those rails through the wilderness eventually made possible a large-scale logging operation with logging rails through Brandreth Park. The New York Central had been carrying immense quantities of logs, pulpwood, and lumber in the early years of this century; thus the logical outlet for the Brandreth timber was now available.

The Mac-A-Mac Railroad

Although previous small efforts had been made to harvest some of the Brandreth timber resources, almost the entire township still remained untouched in 1912 when the first Mac-A-Mac corporation was formed to log the immense area. This corporation was formed by three principals: John N. McDonald, a successful lumberman from Carthage and the working member of the trio; Judge Vasco P. Abbott of Gouverneur, former surrogate, St. Lawrence County; and Benjamin B. McAlpin, grandson of Benjamin Brandreth and owner of the former Hotel McAlpin in New York City.

John N. McDonald was an experienced lumberman whose father and uncle, Duncan and John McDonald, had logged for years in the Adirondack Mountains. John N. had previously lumbered at Cranberry Lake and Tug Hill and brought many experienced brothers with him to Brandreth.

113. It's time to break open the skidways, filled with logs from the early winter's cut on the steep slop above. Mac-A-Mac operation, Brandreth. Circa 1915.

114. The Dimelow's landing at Brandreth required a large bridge fill of six hundred cords of wood for the siding to reach the banking ground. It was a cold day when the landing crew began to load the flat cars. Circa 1915.

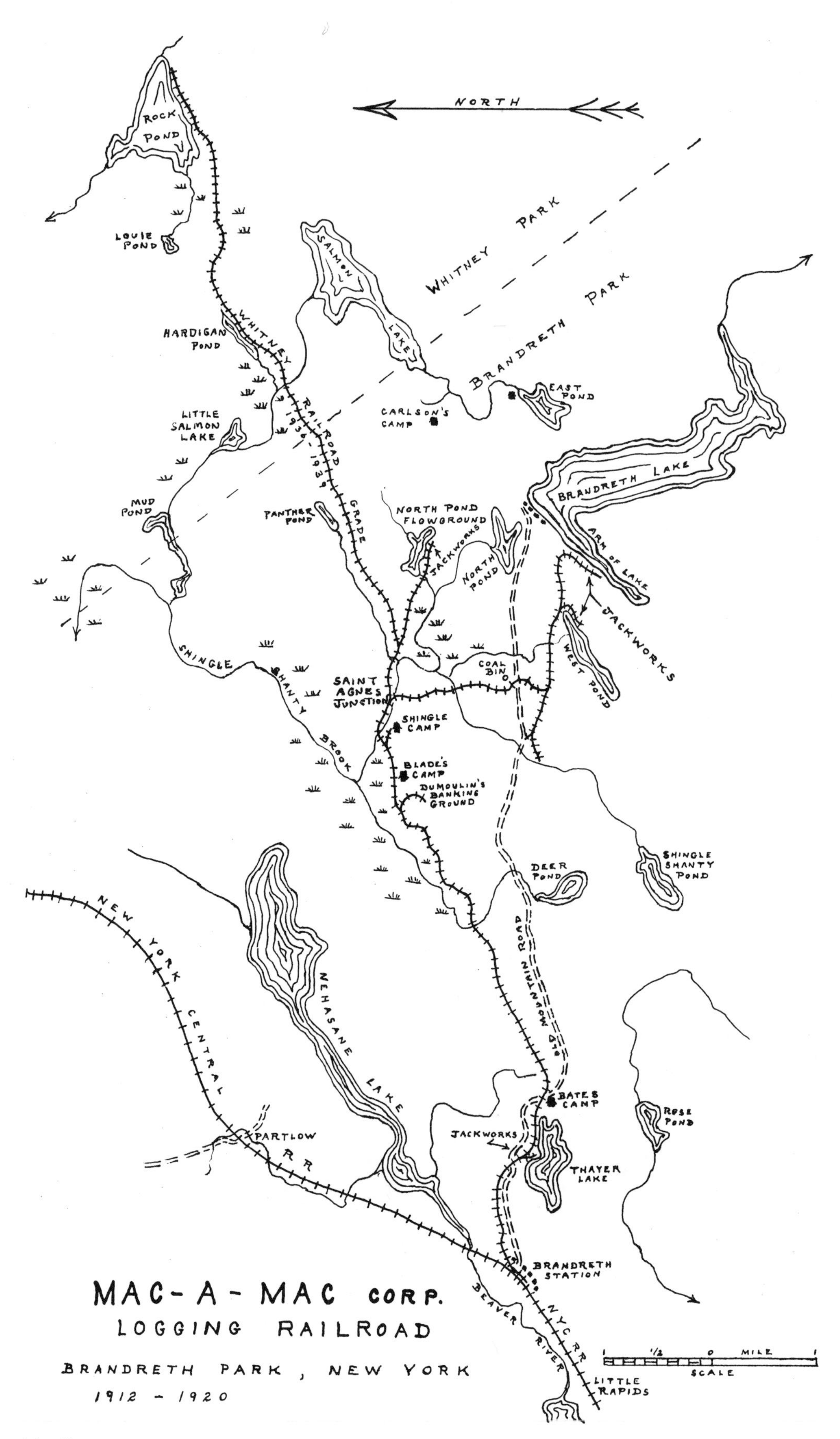
NORTH
ROCK POND
LOUIE POND
SALMON LAKE
WHITNEY PARK
BRANDRETH PARK
HARDIGAN POND
WHITNEY RAILROAD GRADE 1936-1939
EAST POND
CARLSON'S CAMP
LITTLE SALMON LAKE
MUD POND
PANTHER POND
NORTH POND FLOWGROUND
BRANDRETH LAKE
ARM OF LAKE
JACKWORKS
NORTH POND
WEST POND
SHINGLE SHANTY BROOK
COAL BIN
SAINT AGNES JUNCTION
SHINGLE CAMP
BLADE'S CAMP
DUMOULIN'S BANKING GROUND
DEER POND
SHINGLE SHANTY POND
NEW YORK CENTRAL
NEHASANE LAKE
OLD MOUNTAIN ROAD
BATES CAMP
ROSE POND
PARTLOW RR
THAYER LAKE
BRANDRETH STATION
BEAVER RIVER
NYC RR
LITTLE RAPIDS
1 1/2 0 MILE 1
SCALE
MAC-A-MAC CORP.
LOGGING RAILROAD
BRANDRETH PARK, NEW YORK
1912 - 1920

Map 7.

A contract was drawn up with the Brandreth family for the purchase of the softwood stumpage on more than twenty thousand acres of virgin timber, areas around the water bodies being excluded. Cruise reports promised five hundred thousand cords of pulpwood. Only softwood was to be harvested, all to be cut into long-length pulpwood for shipment to St. Regis Company at Deferiet. All except some of the hemlock and the white pine, that is, some of which was larger than six feet on the stump and which was to be cut into sawlogs. A sawmill was to be built at a later date. The operation would last twenty years, it was said, and would supposedly employ between five and six hundred men. According to the announcement in the Tupper Lake Free Press, the logging of the Brandreth tract was to be "one of the largest undertakings of its kind ever accomplished" in the Adirondacks.

The Mac-A-Mac Corporation financed and built a logging railroad into the Brandreth tract or "park." A switch was put on the New York Central at the Brandreth flag stop, and rails were laid easterly for nine miles, as far as Brandreth Lake, and into other water locations where the log-length pulpwood could be loaded by jackworks. Beginning in 1912, a total of about twelve miles of standard-gauge rail were eventually laid out as surveyed by James Brownell of Carthage and constructed by Closs Carlson of Wanakena, the railroad builder who had migrated to New York State with the Rich Lumber Company at the turn of the century.

Headquarters was established in 1912 at a small village built at the junction with the New York Central, a depot also established and referred to by the New York Central as Brandreth. Activity was brisk for a few years at this settlement, where the Mac-A-Mac Corporation built a general store and post office, a boardinghouse, an engine house for the two company locomotives, and seven family homes

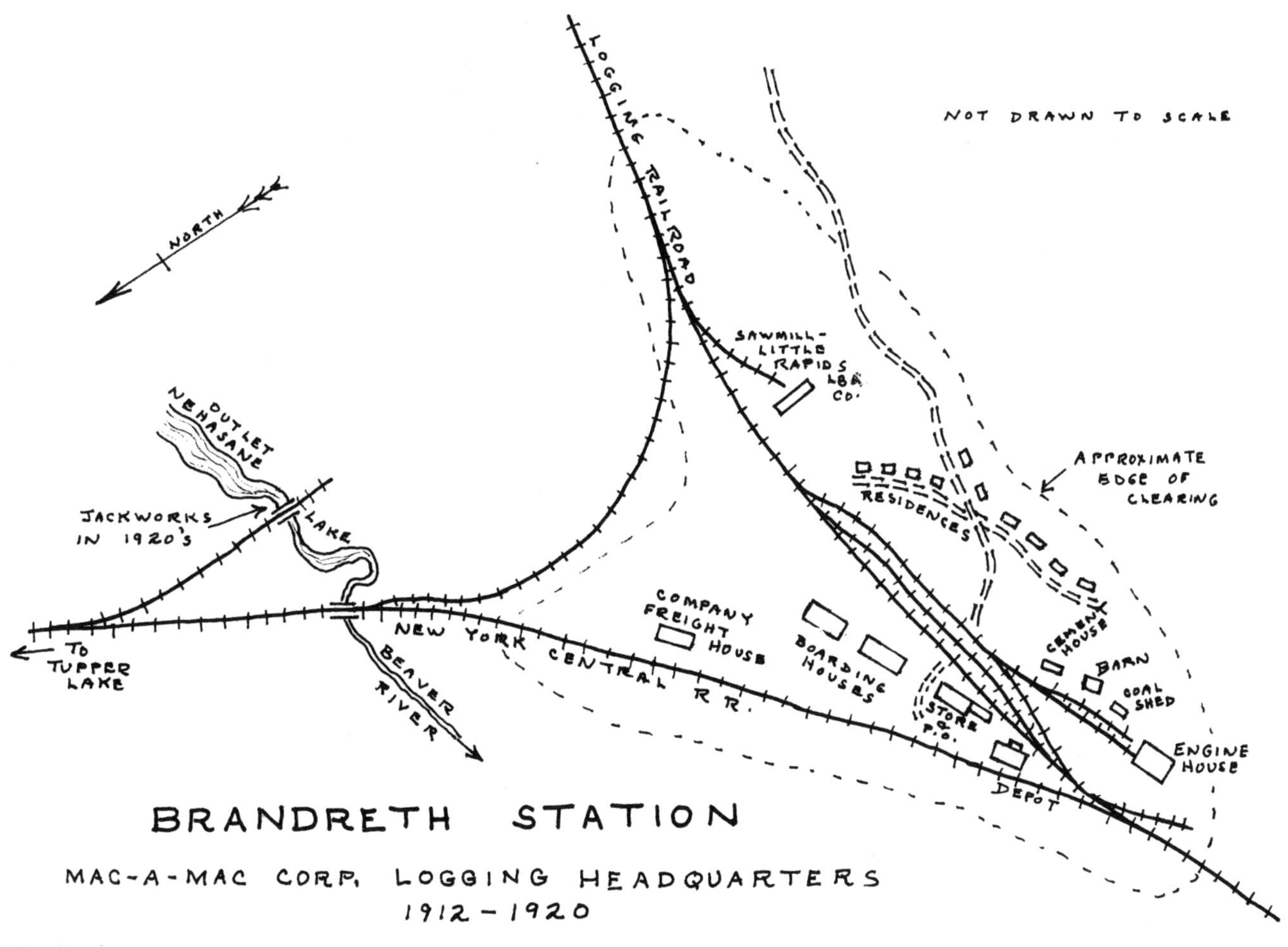

Map 8.

115. Four large sled loads of sixteen-foot spruce pulpwood logs have arrived at Carlson's Landing in 1914 for reloading on the railroad. The load on the left side is the proclaimed record load containing 298 logs, drawn by a single team. Photograph by H. M. Beach. Courtesy of the Brandreth heirs.

plus other buildings and railroad sidings. The woods camps were usually located on or near the railroad but at times also at more distant sites. Some of the camp sites are noted on the accompanying map. Camp bosses in 1917 were reported to be Leon Kelly, Fred Baxter, Fred Johnson, and a Dumoulin.

The pulpwood was cut in sixteen-foot lengths and hauled by sled during the winter either to a railside landing or, in most cases, to a lake or a "flow ground," where it was piled on the ice. Most of the wood was peeled where it had fallen, the bark removed with a spud during the peeling season, which usually ended by mid-August at the latest. Logs were then left at the stump for later skidding and sled hauling after the snow fell.

With the spring breakup of the ice, the logs landed there during the winter would be afloat and by late spring would be gathered in and enclosed within a boom at the jackworks. The company used small gasoline launches or old steam-powered boats to pull the booms of logs across the lakes or the flow grounds to the loading points. These little boats were relics, and at least one is said to be yet on the bottom of Brandreth Lake.

There were four jackworks used on the Brandreth operation, steam-powered conveyors that hauled the pulpwood logs out of the water on an endless chain and conveyed them up onto a ramp where they could be rolled onto the log cars alongside.

The Mac-A-Mac purchased two locomotives for the Brandreth job. The first acquired was a new Heisler geared locomotive, and the second one was a mogul (2-6-0), purchased new the following year. Locomotive engineer was usually George LaFountain, a huge man who weighed over three hundred pounds, with his son Ernie LaFountain or Louis Geebo as firemen. John McDonald's brother Sandy worked as brakeman on the company's railroad and at times as engineer. Elmer B. Martin also worked as engineer at Brandreth, Woods Lake, and Big Moose.

John McDonald actually had many of his family mem-

bers on the job. Brother Henry McDonald managed the company store at Brandreth Station and served as head clerk. Brothers Eugene and George were logging contractors who ran large camps for the corporation. Brothers Duncan and Allan worked for the corporation. The boardinghouse was run by Paul LaPorte and his wife. Other jobbers or contractors remembered as working at Brandreth were Marshall Wicks, Bill McCarthy, George Bushey, and Alec Deschane.

Four camps were normally operated during the cutting season, and one or more trainloads of logs were brought out daily to Brandreth Station, where the cars were picked up by

116. Ramps with endless chains were used to convey the logs up to the railroad loading deck at one of the jackworks on the Mac-A-Mac Railroad in Brandreth. A railroad car was loaded, staked, and wired each twenty minutes. Photograph by E. F. McCarthy, 1918. From the collections at College Archives, SUNY ESF.

117. The engine crew paused with the Heisler on a cold day before picking up the loaded cars parked on a camp siding. Circa 1916. Courtesy of the Brandreth heirs.

118. The Heisler locomotive with a full load of spruce pulpwood logs paused on the shore of Thayer Lake for photographer H. M. Beach. Circa 1915. Courtesy of the Brandreth heirs.

the New York Central for movement to the pulpmill at Deferiet. The operation continued for nine years, culminating in 1920, and it has been estimated that 300 to 350 thousand cords of pulpwood were removed during that period. As so often happened during that era, operations fell somewhat short of expectations.

A reported price of $750,000 was paid for the softwood stumpage, but a good profit must have been realized, as pulpwood prices rose during the First World War from $5.00 to at times as much as $30.00 per cord.

LITTLE RAPIDS LUMBER COMPANY

After the completion of the Brandreth operation in 1920, John McDonald moved his railroad equipment on north to Bay Pond, a railroad operation covered in a later chapter, but that was not the death of the little village at Brandreth depot. In 1925 a sawmill was constructed at the settlement by the newly formed Little Rapids Lumber Company, organized by some legendary figures in Adirondack lumbering history—John E. Johnston, Clarence Strife, and Harry P. Gould, president of Gould Paper Company.

Little Rapids Lumber Company was organized for the purpose of cutting the hardwood stumpage on the Brandreth tract, and headquarters were set up in the former Mac-A-Mac facilities at the depot. Logging was supervised by Clarence Strife, who utilized army surplus tanks made by Holt for yarding the logs and Linn tractors for the long haul of sled trains into the sawmill at Brandreth. The Linn tractor, made in Morris, New York, could haul eighteen to twenty loaded sleds.

White pine lumber was sawn by a night shift for sale to Plunkett Webster Company, which had an office in Utica.

119. New York Central locomotive No. 1669 couples onto a string of log cars at Brandreth, the junction of the Mac-A-Mac logging railroad with the NYC. The spruce pulpwood was shipped in sixteen-foot lengths to the St. Regis Company in Deferiet. Circa 1915. Courtesy of the Brandreth heirs.

Plunkett Webster even had company man Scott Lyng on location to do the sawing and grading of the pine.

The sawmill continued until 1931 when the depression caused financial failure of the Little Rapids Lumber Company.

Brandreth Park saw its last major logging during World War II, when George McDonald, John's brother, conducted an extensive but unprofitable spruce pulpwood operation in 1942, and Clarence Strife picked up the remaining accessible hardwood in 1943. The succeeding growth, like much of the private lands in the Adirondacks, shows some of the low quality and poor growth that has resulted from high-grading and indiscriminate cutting.

John McDonald's brother Sandy stayed on at Brandreth Station after cutting had ceased and served as superintendent for the railroad company. He was known as "the mayor of Brandreth" until his retirement move in 1947 back to Croghan. The Brandreth depot eventually burned, and vegetation quickly covered over what had once been a busy settlement in the 1920s and early 1930s.

Brandreth Park is now the property that is the longest continually owned by the same family in the Adirondacks. A family corporation was set up many years ago and the presently existing heirs, about eighty, share ownership as a tenancy in common. Land ownership of the corporation is now down to about ten thousand acres, with recreation rights on much of the former ownership. With the abandonment of the New York Central Adirondack Division, an auto road was built along an old trail in 1960 to reach the thirty-four camps on the lake that Benjamin Brandreth once thought would make a superior beer.

John McDonald has left his mark in the Adirondacks as a successful lumberman. He logged in many locations and on many ownerships, but of all his jobs, none captured the attention of the photographers of logging and woods scenes as much as his Mac-A-Mac railroad at Brandreth Park.

CHAPTER FOURTEEN

Whitney Industries Railroad

Whitney Park had its origin in 1897 when retired secretary of the navy William C. Whitney purchased sixty-eight thousand acres of virgin forestland in the Town of Long Lake, Hamilton County.

Whitney bought the properties in partnership with the renowned character Patrick C. Moynehan. "Dandy Pat" Moynehan was a self-educated Adirondack lumberman who made a fortune in a number of clever timber deals. Unsubstantiated rumors have it that he bested William C. Whitney out of large amounts of money.

When W. Seward Webb hired Gifford Pinchot, and later Henry C. Graves, to set up a forest management program on Nehasane Park in 1896, Whitney put his large tract on the same forestry project. Cutting began on the Whitney tract old-growth spruce in 1898. Under Graves's supervision, the spruce was cut to a ten-inch minimum diameter. Most of the spruce was either driven down the Raquette River or through the two Tupper Lakes to the Santa Clara Lumber Company sawmill. About 250 million board feet was cut before the harvest program came to an end in 1907.

During the same period, Pat Moynehan was also logging white pine and spruce on the adjacent Brandreth property. There was no river close enough for a river drive; the logs were loaded on the New York Central railroad at Brandreth Station.

William C. Whitney died in 1904, the partnership with Moynehan was dissolved by 1912, and the property was eventually incorporated as Whitney Industries Inc. by heir Cornelius Vanderbilt Whitney. More land was added, pushing the acreage first up to about eighty thousand and later to one hundred thousand acres. Twenty-seven years later, in 1934, the Whitney heirs decided to again harvest some mature timber on Whitney Park. Fred A. Potter was the Park superintendent at the time, and he consulted well-known

Table 8
Whitney Industries Railroad

Owner	Whitney Industries Inc.
Dates	1934–1939
Product hauled	hardwood sawlogs, spruce pulpwood
Length	13 miles
Gauge	standard

Motive Power

No.	*Builder*	*Type*	*Builder No.*	*Year Built*	*Weight (tons)*	*Cylinders Drivers*	*Source*	*Disposition*
2	Alco-Schenectady	2-6-0	53845	1913			Bay Pond RR	1941 Emporium Forestry Co.
33	Baldwin	0-6-2T	6264	1882	38	16" x 22" 38"	Bay Pond RR	unknown

forester Arthur Recknagel for advice on timber management. Potter was the grandson of Dr. Benjamin Brandreth. Negotiations were made with logging contractor John N. McDonald to harvest both the softwood and the hardwood timber and to complete the cut by 1950. The softwood was to be cut at the yearly rate of forty thousand cords and the hardwood at the rate of twenty million board feet.

McDonald was no stranger to the area; he had managed the large timber harvest operation in adjacent Brandreth Park from 1912 to 1920. As the contractor with Whitney, McDonald was to purchase all of the softwood cut and was also to yard the hardwood logs for sale to the Oval Wood Dish Company. There was a large quantity of old-growth hardwood still standing on the Park.

McDonald began the large operation on the northern portion of the Park in 1934. The pulpwood was yarded with horses and Linn tractors at Round Pond or its inlet. A dam was built at the outlet that allowed him to boom the wood across the flooded lake and float it down the Bog River to Big Tupper Lake and thence to a jackworks at the foot of the lake at Underwood. At that point, the pulpwood was loaded on New York Central rail for shipment to the St. Regis Company plant at Deferiet. The St. Regis scaler Earl Trudeau worked with the Whitney Park scaler Don Mace at the woods landings to record the measurement of pulpwood cut. The hardwood logs were cut in the winter months only.

When the south portion of the Park was next harvested, a different scheme had to be used; it was not possible to float the pulpwood down to Big Tupper. Whitney Industries chose to build a railroad into the southern part of the Park and agreed to have John McDonald build it for them. McDonald's railroad equipment used on the Bay Pond operation was apparently purchased by Whitney Industries.

Whitney Industries obtained the right from Brandreth Lake Corporation to use the former railroad bed across Brandreth Park that McDonald had built back in 1912 when cutting off that area. Seventy-five-pound Dudley rail was laid again on the old roadbed in 1935 and 1936 for about seven miles, to a point one mile short of the North Pond flow ground. A new roadbed was then built for $2^1/_3$ miles over Brandreth property to the Whitney line and then almost four miles easterly into Whitney Park to the south shore of Rock Pond. The Brandreth owners were paid a use fee of ten cents per cord for wood taken out. In order to move the loaded cars north toward Tupper Lake on the New York Central, a new junction had to be built with the New York Central north of the old Brandreth junction (see Map 7).

This was the height of the depression years, and Whitney Industries was cooperating with President Roosevelt's NRA program to attempt to put a force of about three hundred men in the woods. But it was the first year of Social Security, and it was reported to be difficult to get men to sign up. Many Russian and Polish workers were used in the construction. Woods wages were then about $50 per month plus room and board.

Oval Wood Dish Company put their own contractors to work when logging the hardwoods, two of them being Cornelius Buckley and Bert Franks. Camps were located at Little Salmon Lake, Rock Pond, Antediluvian Pond, and St.

120. John McDonald ran the well-used No. 33 Baldwin on the Whitney Park operation. This thirty-eight-ton saddletank had almost sixty years of hard logging service to its credit in Pennsylvania and the Adirondacks with at least three owners. Circa 1902. Courtesy of Franklin Brandreth.

Agnes. Oval Wood Dish paid $3 per thousand board feet for the hardwood stumpage and cut just the best. They took sixteen-foot logs only.

For this Whitney operation, John McDonald brought in the same two locomotives he had used on the Bay Pond railroad, which had closed down in 1932. There was the Alco-Schenectady mogul, which McDonald had bought new in 1913 for the Mac-A-Mac railroad, and also the little Baldwin 0-6-2 saddletank, which McDonald had picked up from the Emporium Forestry Company in 1924. The saddletank still carried the number 33 and was used in construction of the line and for switching. Loading of the hardwood logs was done with a gasoline-powered loader mounted on a flat car.

Locomotive engineers were George LaFountain, his son Ernie, and Elmer B. Martin, who had previously worked as engineer at Brandreth, Woods Lake, and Big Moose. George LaFountain had returned from a Florida retirement to operate McDonald's mogul, dropping 100 pounds in the process to weigh only 265 pounds. George operated that same locomotive on three different railroads.

This little railroad never proved very successful, however, and was only used for about three winter seasons, hardly enough to pay off the investment. For some reason only a small amount of pulpwood rolled out over the Whitney railroad, and apparently a jackworks was never established at Rock Pond. The last run of log cars was made on October 5, 1939. The rails were all pulled before the year was spent.

The railroad equipment was all purchased afterward by the Emporium Forestry Company, including 13.75 miles of rails, switches, and ties, locomotive no. 33, a snow plow, a steam log loader mounted on a flat car, and the equipment from the machine shop at Brandreth. Emporium never used this equipment, however; two years later it was all sold to a dealer, the A. A. Morrison Company, for $45,000.

CHAPTER FIFTEEN

Partlow Lake Railroad

Soon after purchasing his large Adirondack holdings and building the Mohawk and Malone Railroad, W. Seward Webb determined to set a good example of proper forest management on his vast acreage. Although he apparently had no intention of retaining all of the land he purchased at the time he was obtaining railroad rights-of-way, he did build, in 1893, the famed lodge on Lake Lila. It was situated within the forty-thousand-acre Nehasane Park that was to be owned by the family for many years. The remainder of his vast acreage was eventually sold to New York State.

To establish a forestry program on his park, Webb turned to a young European-trained forester, Gifford Pinchot. Pinchot had formulated a forest management plan for George W. Vanderbilt, Webb's brother-in-law, in 1892, and Webb hired him soon afterward on the recommendation of Vanderbilt.

Gifford Pinchot was promoting new ideas in the young profession of forestry, attempting to show that applied forestry did not primarily mean setting aside public forest reserves or the planting of trees; it meant controlling use of the ax, not stopping it. He promoted the establishment of natural reproduction, not merely the planting of seedlings.

In 1896 Pinchot began a two-year experiment on Webb's Nehasane Park, demonstrating the method of selective cutting in a spruce stand. From this project came a book, *The Adirondack Spruce*, in 1898, written by Pinchot and financed by Webb. It was the first forestry book in the United States prepared for the management of a particular forest species or type. William C. Whitney's adjoining sixty-eight-thousand-acre Park also came under the professional care of Pinchot at that time, making a total of over one hundred thousand contiguous acres in a pioneer effort of forest management.

In 1898 Gifford Pinchot went on to become the chief forester for the United States, and Henry S. Graves carried on his work, beginning in 1900. The virgin spruce timber on the entire property was removed during this period, 1896–1904, cutting to a ten-inch minimum diameter at breast height (DBH).

Table 9
Partlow Lake Railroad

Owner	W. Seward Webb
Dates	1900–c. 1905
Product hauled	spruce sawlogs
Length	3.5 miles
Gauge	standard
Motive power	New York Central locomotives

Partlow Lake Railroad

To reach the old-growth spruce timber in the area around and beyond Partlow Lake, Webb had a five-mile railroad built into a small lake known as Sylvan Lake, later called Partlow Mill Dam. The logging line began at a point north of milepost 84 on the New York Central. The junction with the NYC was known as Partlow and appeared on the New York Central timetables from about 1900 through 1905.

The logging line was built about 1900; operations lasted for five years. The locomotive and other equipment were apparently leased from the New York Central or operated by the New York Central itself. W. Seward Webb was listed as president of the railway and C. H. Burnett the superintendent.

An official station was established at Partlow, providing enough traffic so that two passenger trains and two regular freight trains stopped daily each way. This made three station stops for the New York Central within Webb's Nehasane Park, the other two being Keepawa, which was used for

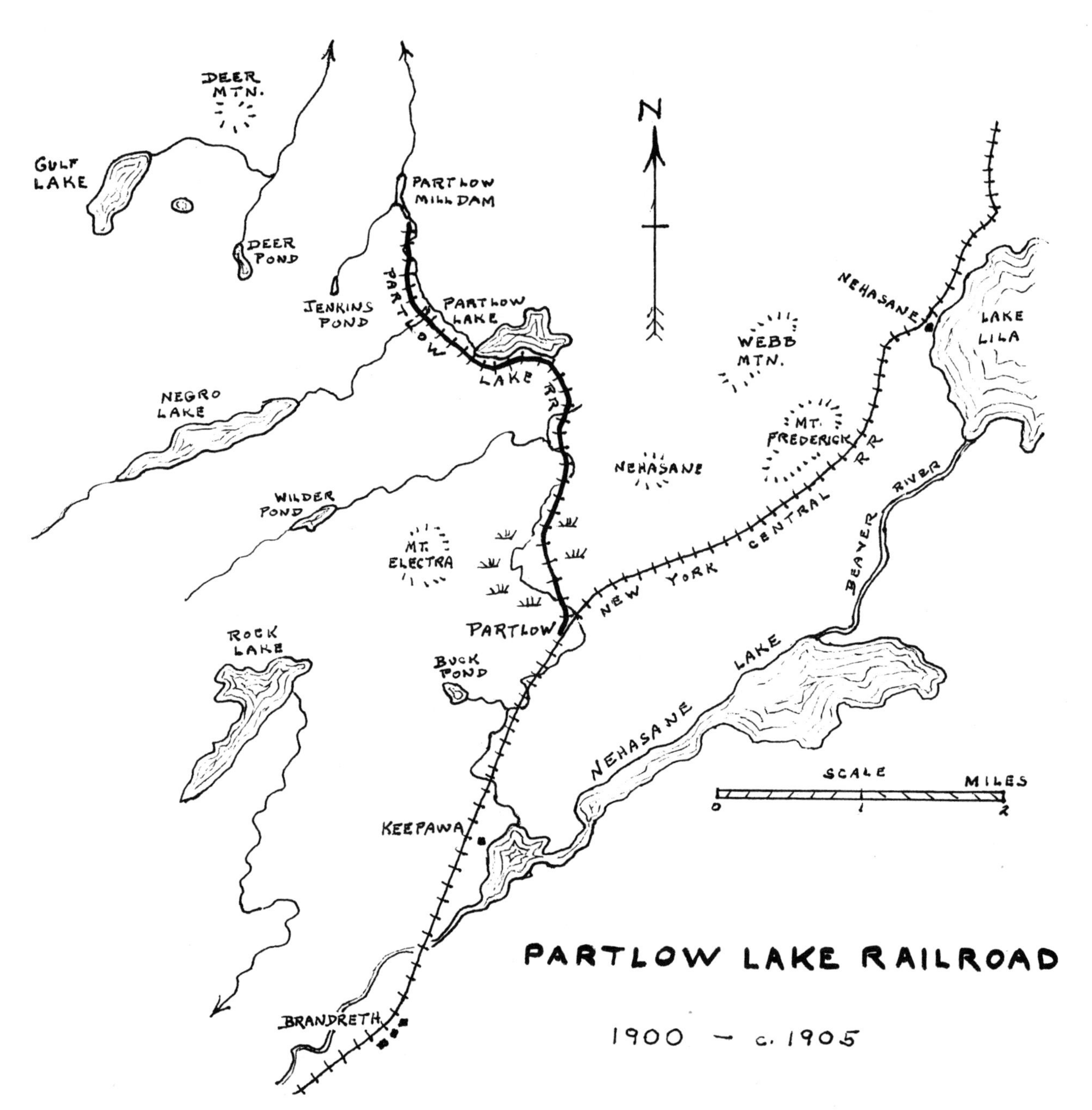

Map 9.

Webb's park personnel, and Nehasane, W. Seward Webb's private station. The Partlow station was removed from the timetables soon after 1905; logging operations had been completed.

In 1919 a new siding was established at Partlow. During the mid-1920s logs were loaded on NYC cars by John E. Johnston at the Partlow and Keepawa sidings, using an old steam log loader.

In 1914, F. A. Gaylord was employed as forester on Nehasane Park, and heavier cuts were made of softwood and hardwood timber. The eventual results were not favorable, however; their forestry practices failed to establish a thrifty young spruce forest. As land managers were to experience time and again in succeeding years, spruce cannot often be perpetuated in northern hardwood forest types after heavy cuttings. However, in the years to come, northern hardwoods would surpass to a large degree the demand for and value of red spruce timber.

CHAPTER SIXTEEN

Horse Shoe Forestry Company Railroad

From the resources of one of the wealthy Adirondack landowners came the strangest of logging railroads. It was strange in its conception and strange in the variety of products it carried. To call this little railroad a genuine logging railroad is stretching the definition just a mite. But it did carry an interesting variety of forest products. It was created as an innovative oddity by an odd individual—A. A. Low.

Abbot Augustus Low received his education in the family's successful shipping and importing business in New York City, where he built a reputation as an innovative leader and inventor. Many years of summer residence by the family at Lake Luzerne had created in Low a love for the Adirondacks, prompting him in the early 1890s to build a rustic summer retreat on Bog Lake in Herkimer County.

Table 10
Horse Shoe Forestry Company Railroad

Owner	Abbot Augustus Low
Dates	1897–1911
Product hauled	sawlogs, pulpwood, maple sap, spring water, wine, jellies
Length	approx. 14 miles
Gauge	a portion was standard and a portion 3-foot gauge
Motive Power	
Type	*Source*
0-4-4	Baldwin, c/n 5498, 11 x 16, built 1898, former elevated on Manhattan Ry.
0-4-2T	Porter, c/n 3121, 10 x 16, purchased new 1904
4-4-0	"Washington," narrow gauge
Shay, 2-truck	c/n 918, built Aug. 1904, standard gauge, sold to Porterwood Lumber Co., Porterwood, W.V., now on display at Biddeford, Me.; wgt., 37 tons
Equipment	
American log loader, flat cars	

The only access to this vacation home of Low's was a whistle stop on the New York Central's Adirondack Division, located about four miles south of the station at Sabattis. Mrs. Low named their station stop Robinwood.

Low had formed a friendship with John A. Dix, an active lumberman and later governor of New York, and through this acquaintance had been encouraged to purchase large forest tracts in the vicinity of his Bog Lake home and to harvest the timber. Beginning in 1892, Low bought several large land holdings around Bog Lake, Big Trout Pond, Lake Marion, Hitchins Pond, and Horseshoe Lake. By 1896 he had accumulated forty-five thousand acres of prime timberland and lake country. Such a large contiguous ownership in the Adirondacks was not at all unusual during this era, however, for there were a number of large nearby holdings of wealthy title holders. Low was able to amass about two-thirds of the Bog River headwaters in his domain. He set out to profit from his acreage in his own unique way.

One of his first developments was to reach an agreement with the New York Central and Hudson River Railroad in 1896 for the construction of a complete and full-service railroad station at Horseshoe Lake, about nine miles north of the Robinwood whistle stop. Low built it at his own expense, constructing a replica of the railroad depot located at Garden City, Long Island, complete with telegraph service, ticket office, and freight and express service. It took some persuasion for the U.S. government to allow the establishment of a post office at this small settlement, but when they did, Low was appointed the first postmaster. The Horseshoe depot was the most attractive station between Utica and Malone, and Low soon turned it over to the New York Central for the payment of $1.00, but he got the services he desired.

Low's electrical and mechanical talents as an inventor had been evident for many years in New York City; it was said that at the time of Low's death in 1912 only Thomas Edison had more patents registered in the U.S. Patent Office than did A. A. Low. In addition to many mundane things, his patents covered more elaborate devices such as an electric rat trap and a submarine. At his new Adirondack domain, Low had the opportunity to extend his innovative skills in many new directions. Most of these were in the field of forest products; these new activities went forth under the name of the Horse Shoe Forestry Company.

Timber harvest began also in 1896, cutting softwood logs in the vicinity of Mud Lake and decking the logs along the Bog River for river driving down to the mills at Tupper Lake the following spring. However, forty-five thousand acres is a large piece of land, and Low was already planning a diverse enterprise with a railroad as a key link.

Construction of Low's private railroad began in 1897 and was eventually extended out from both sides of the New York Central tracks for a total of at least seven and a half miles of track, possibly more. It is now difficult to locate all of the track sites. An unusual feature of Low's railroad was that he used two different gauges for a while, standard on one side of the New York Central track and narrow on the other.

It appears that the track on the east side of the New York Central was all built to standard gauge. Low purchased a standard-gauge Shay locomotive in 1904 and also had a former New York City elevated engine of standard gauge.

The five miles or more of track built on the west side of the New York Central were narrow gauge, apparently three feet between the rails. This portion was often referred to as the Wake Robin railroad and serviced the maple syrup operation as well as carrying logs and pulpwood. Low used an older American-type locomotive on these tracks. Reports indicate that the Wake Robin tracks were converted to standard gauge in 1903–4.

122. The former New York City Elevated engine posed on top of the lower dam, which was built 1903 as the first of two dams. The 250-foot dam was a combination hollow reinforced concrete structure with a fish-way and the standard gauge tracks on top. Circa 1904. Engineering News, *1908. Courtesy of R. Wettereau.*

121. River drivers on the Bog River posed at the falls where the Bog River enters Tupper Lake. Courtesy, Special Collections, Feinberg Library, SUNY at Plattsburgh.

Other equipment on the railroad included an American log loader, steam shovel, crane, and several flat cars. The maple sap was hauled from the sugar-bush collection tanks to the evaporator house in galvanized tanks mounted on flat cars. Low, who it is said seldom tied his shoelaces, would often be seen riding on a flat car seated on a chair or a bench.

The initial chore of the Horse Shoe Forestry Company railroad was the haulage of sawlogs cut north of Horseshoe Lake to the Bog River for the 1898 spring log drive. Large quantities of pulpwood were also transported. Logging was a major activity along the corridor of the New York Central at the turn of the century; more than one hundred lumberjacks would "spring out" at the nearby Wilderness Inn at Sabattis.

Low was not one to sell his timber in log form if he could make a more valuable product. In 1897 construction began on a sawmill on the Bog River near Hitchins Pond, a band mill complete with box and barrel shops. The complex of

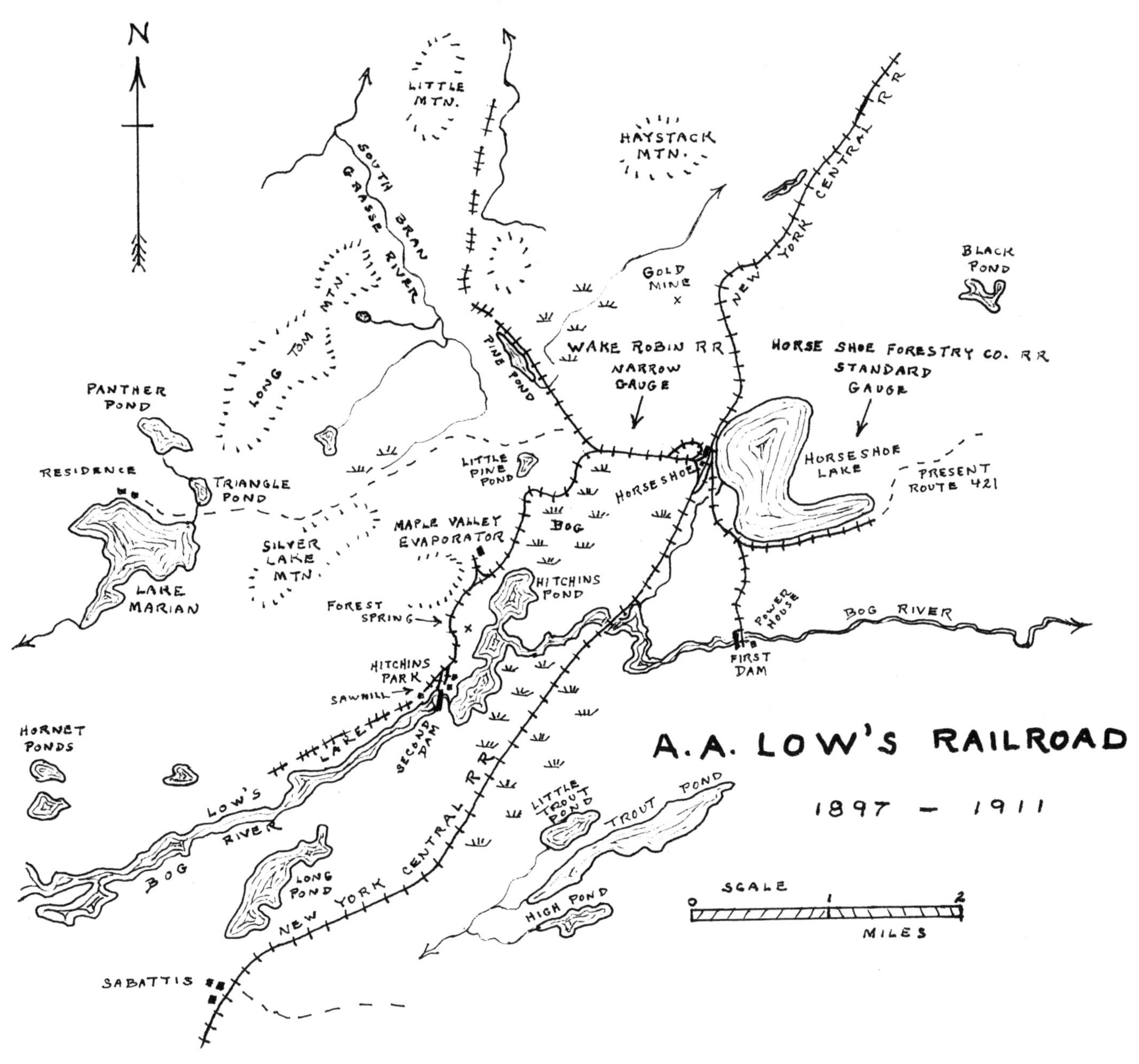

Map 10.

buildings at the mill site included a three-story boarding house for employees. Mrs. Low, who seemed to be the one naming the various sites, called this location Hitchins Park.

The next product idea to come from Low's innovative mind was spring water, bottled in square returnable bottles patented by Low and shipped to New York City in returnable shipping cases made by the Horse Shoe Forestry Company. The water came from two large springs in a ravine located halfway between Hitchins Park and Horseshoe depot and about twenty feet below the rail elevation. It was bottled and packed at the spring and conveyed up to the railside loading platform by a steam-powered conveyor. The spring water project, labeled Virgin Forest Springs on the bottle, lasted from 1899 until the market waned in about 1905.

One of the most interesting products hauled by the Horse Shoe Forestry Company railroad was maple sap. The first of three sugar houses was built in 1898 and began operation the next spring. The ten thousand taps that were made that spring were said to yield four thousand gallons of syrup. Some of the sap collection was done by horse-drawn tank sleds and loaded by hand from buckets. But where possible a collection system of pipes and troughs was set up to convey

123. The Washington, a 4-4-0, was equipped with an ornately capped stack, fluted domes, and an oil lamp. The crew was filling a flat car-mounted tank with maple sap from one of the large collection tanks that were set up railside on log crib works. The scene was on the narrow gauge Wake Robin operation in 1901. From Adirondack Sugar Bushes, *Horseshoe.*

124. A. A. Low would occasionally ride the rails throughout his Park, perched upon a bench built in the center of a flat car. Circa 1905. From the Collections of the St. Lawrence County Historical Association.

the sap to collecting tanks located railside. From there it was gravity-loaded into large sap tanks on railroad flats. Three separate sugar bushes were established on the property, each with a steam-heated evaporator house. Records show that twenty thousand gallons of syrup were made in the peak year of 1907.

The maple syrup operation was a sophisticated and quite successful venture using an evaporation system devised by Low and his managers. His maple products won prizes at the fairs in Pennsylvania and Vermont. It's been claimed that Low invented the first apparatus to granulate maple sugar. Low provided the maple sugar cakes that were profitably peddled by newsboys on the New York Central's Adirondack Division. Other commodities produced were Horse Shoe Grist Mill and Staff of Life cereals, wine (elderberry, wild cherry, and grape), and wild raspberry and wild blueberry jam. There was a two-hundred-acre potato farm.

Among the products produced by the Horse Shoe Forestry Company was a most unusual one—"natural wood fuel." In 1900 Low was granted a patent for a type of wood fuel that he made by sawing tree limbs into wafers, removing the center, and stringing them on rope to dry. Packaged in burlap bags, they were sent to the city and marketed under the name "physic coal." It was a thoughtful effort by Low to find a market for logging debris, but it had little sales appeal and soon failed.

The mineral content of the water used by the three locomotives and the other steam boilers was continually causing a build-up of scale in the boiler tubes, and to attempt to remedy the problem, Low started a "boiler compound" factory at the Maple Valley site. The product's effectiveness is questionable in view of the fact that the family's long-time superintendent, Armand Vaillancourt, referred to it as a "witches' brew."

A. A. Low was not one to bypass any innovative use of the natural resources at hand, and in 1903 he constructed a dam on the Bog River, backing the water up into Hitchins Pond. In addition to stabilizing the stream flow for the annual log drives, the dam was provided with turbines to drive two electric generators. The 250-foot dam had a fishway surmounting its 30-foot height and had railroad tracks along the top.

Within two years of this installation, the sawmill and all of the buildings in the Park, including the depot, were electrified. It soon became obvious that the water thus impounded was insufficient for year-round demands, and in 1907 a smaller second or upper dam one hundred feet long was completed on the same river and above Hitchins Pond, also with a power station. The Bog River above this dam was raised almost six feet, and a huge flow was created for eleven miles above. That's when Low's big troubles began.

A number of discouragements began in 1907. It was a dry

125. The narrow-gauge track leading to one of Low's magnificent sugar houses made a sharp curve on the downgrade. Circa 1905. From Adirondack Sugar Bushes, *Horseshoe.*

year and forest fires were again spawned, one burning through a portion of Low's land. His discouragement intensified because he had already suffered fire losses in 1899 and 1903. The New York Central had paid Low $21,500 for losses suffered by fires caused by locomotives in those two years. Now the dreaded losses continued on land he had been thoughtfully nurturing.

Then in July 1908 Low received a letter from the New York State Forest, Fish, and Game Commission notifying him that his upper dam was causing flooding of state-owned lands, demanding that the flooding cease and specifying damages to be paid. Low ignored the notice—he was sure the flooding was confined to his land ownership.

Then in September of that same year, another forest inferno swept through the southern portion of Low's ownership, across the Mud Lake and Big Trout Pond areas, completely destroying the hamlet of Sabattis, then called Long Lake West. Total damage was assessed at $335,000. The state pursued its demand for damages on flooded land; Low refused to acknowledge the validity of the claim but paid the fine under protest. It wasn't until 1950, prodded by the persistence of Low's son, that the Supreme Court of New York ruled that the state had not owned the flooded land and that the fine must be returned to the heirs with interest.

The forest fires and the demands by the state were a discouraging mix of setbacks for Low. Thousands of acres of timberland were in ashes, undermining support for his timber industry and maple syrup production. The anticipated lawsuit against the New York Central and Hudson River Railroad, whose locomotives had ignited the conflagration, brought no consolation or hope that he might be able to continue the activity of the Horse Shoe Forestry Company. All of Low's bold and innovative enterprises quickly faded. His enthusiasm died.

Within a matter of a month's time, the railroad equipment was sold, and most of the buildings were taken down again to reduce the hazard of further fires. By the summer of 1911 all was quiet on the Horse Shoe. The sawmill equipment and some of the railroad equipment was sold to the Emporium Lumber Company and used at their new Conifer location. Clyde Sykes of Emporium had a high regard for Low and his family and willingly helped in the disposal.

In October 1910, Low made out his will at the age of sixty-six, and he died a year later. Much of the property remained in family ownership until the mid-1970s. Now all that remains in the Adirondacks to memorialize the inventive and spirited genius of A. A. Low are the two dams, still functioning.

CHAPTER SEVENTEEN

The Emporium Forestry Company and the Grasse River Railroad

For more than a century, the manufacture of hardwood lumber has been a mainstay of the lumber industry in the northeast. And of all the operators, probably none has ever equaled the heights attained by William Sykes and his Emporium Lumber Company. There were five major company band mills, and there was the largest railroad logging activity in the northeastern United States, extended over a period of sixty-six years and two long generations. In excess of 1.1 billion board feet of hardwood timber went through the Emporium saws. William Sykes and his family carved a niche in history to which this brief account could not give proper credit.

William L. Sykes's origins were in north central Pennsylvania, in the forests of hemlock and white pine and northern hardwoods. From 1883 until 1918 Sykes and partners operated a number of sawmills in the Keystone State, principally the three large hardwood band mills at Keating Summit, Galeton, and Austin. For most of that period, the mills operated as the Emporium Lumber Company behind the drive and initiative of Sykes and his hard-working associates. They created a colorful history of building a lumber empire out of the virgin hardwood forests of Pennsylvania and New York, a task that could never be duplicated in today's world.

Thirty-six years of intense logging in Pennsylvania, most of it with company-owned railroads, had depleted almost all of the merchantable timber on company-owned timberlands and operations in that state. It came to an end in 1918 when the last of the mills in Pennsylvania, the one at Galeton, was sold. A brief account of that Pennsylvania portion of the Emporium era is told in the book entitled *Rails in the North Woods: Histories of Ten Adirondack Short Lines*, published by North Country Books.

Many years before the Pennsylvania timber well had gone dry, the Emporium officials were seeking new frontiers. Reports filtered in regarding large acreages of virgin hardwood timber in the Adirondack Mountains that were awaiting a buyer. Rough cruises were eventually made that indicated an average of six thousand board feet of hardwood per acre over the varied terrain of low mountains and swamp ground. And then the land purchases began, some as early as 1905. These would prove to be only the first of many wise land purchases made by the Sykes family over the next thirty-eight years, with the company eventually accumulating almost 125,000 acres of prime hardwood timber.

The initial timberland purchase was made in the northern edge of the Adirondacks, eighteen thousand acres in the Town of Clare, in the year 1905. The former owner, Canton Lumber Company, had been logging off the white pine, but the company went into receivership in 1892, and Emporium eventually acquired ninety thousand acres in the area at $15 to $20 per acre. About 22 percent of the purchased area was virgin timber, but the hardwood timber on the remainder was untouched. The Emporium management had originally planned to build their New York sawmill at Canton, and thus the reason for acquisition of land in the far north. Company logging railroads would have the potential of hauling in a westerly direction to junction with the New York Central for transport of logs to Canton.

Further reconnaissance regarding the opportunities for additional timberland purchases in the future indicated that Canton was not the best location for a sawmill. Land purchases would have to progress in an easterly direction, and the obvious area for a sawmill would be in the upper watershed of the Grasse River. A remote site was then selected in

Table 11
Emporium Forestry Company and Grasse River Railroad

Dates:

Sykes and Caflisch (Pa.)	1883–1892
Emporium Lumber Company (Pa.)	1892–1918
Emporium Forestry Company (N.Y.)	1911–1949
Grasse River Railroad	1915–1957
Heywood-Wakefield Company	1949–1957
Product hauled	sawlogs, pulpwood, lumber, passengers, misc. supplies
Length	approx. 80 miles total
Gauge	standard

Locomotive Roster for New York State Operations

Grasse River RR Locomotives[a]

No.	*Builder*	*Type*	*Builder No.*	*Year Built*	*Weight (tons)*	*Cylinders Drivers*	*Source*	*Disposition*
33	Baldwin	0-6-0ST rebuilt 0-6-2ST	6264	1882	38	16 "x 22" 38"	1900, Miles Co. Used at Pa. 1913 to Conifer	1926 to Bay Pond RR
38	Baldwin	0-4-4T Forney	12599	1892	20	14" x 16" 42"	1908 from Post & Henderson Co. Used at Pa.	1923 to Tupper Lake Chem. Co.
42	Brooks	0-6-0	999	1883		17" x 24" 48"	1911 from BR&P RR. Used in Pa.	1918 to Pittsburgh steelmill
43	Brooks	0-6-0	946	1883		17" x 24" 48"	1911 from BR&P RR. Used in Pa.	1918 to Pittsburgh steelmill
60[b]	Schenectady	2-6-0	3902	1892	60	19" x 24" 64"	1919 from NYC	1950, scrapped at Conifer.
61	Schenectady	2-6-0	3897	1892	60	19" x 24" 64"	1923 from Rochester Iron & Metal Co.	1926 to Hudson Ship Bldg. & Repair Co.
63	Schenectady	4-6-0	4553	1897	70	20" x 26" 57"	1923 from Potato Creek RR	
68	Schenectady	2-6-0	3910	1892	60	19" x 24" 64"	1923 from Pa. Wood & Iron Co.	1951, scrapped at Conifer
71	Schenectady	2-6-0	3913	1892	60	19" x 24" 64"	1923 from Pa. Wood & Iron Co.	1949, scrapped at Conifer
3	Baldwin	0-4-4T Forney	13026	1892			from Mid-tolkian Country Club	converted to gasoline, scrapped at Conifer
2	Alco-Schenectady	2-6-0	53845	1913			1941 from Mac-A-Mac Corp. no. 2	1951, scrapped at Conifer
1	GE-Erie	B-B Diesel	30783	1950	44		bought new by Heywood-Wakefield	1959 to East Branch & Lincoln RR

(continued on page 108)

Table 11 (*cont.*)
Emporium Forestry Company and Grasse River Railroad

Emporium Forestry Company Logging Locomotives

No.	Builder	Type	Builder No.	Year Built	Weight (tons)	Cylinders Drivers	Source	Disposition
34	Lima	Shay 2-truck	235	1889	30	11" x 10" 29"	1905, C.H.&.W.A. Rexford. Used at Pa. 1912 to Conifer	1934, scrapped at Conifer
35	Climax	Climax 2-truck	560	1905	50		new, 1905. Used at Pa. 1912 to Conifer	1946 to Dexter Sulfate P&P Co. Scrapped 1950
36	Lima	Shay 2-truck	874	1904	37	—	Jerseyfield Lbr. Co.	Scrapped at Conifer
37	Lima	Shay 3-truck	756	1903	50	12" x 12" 32"	Campbell & Hagenbuck. Used at Pa. 1971 to Conifer	1943, scrapped at Conifer
39	Lima	Shay 3-truck	1548	1905	65	12" x 15" 36"	1910 from Chapman Coal Co. Used at Pa. 1917 to Conifer	1934, scrapped at Conifer
40[c]	Lima	Shay 3-truck	687	1902	65	$14^1/_4$" x 12" 36"	1910 from Lackawanna Lbr. Co. Used at Pa. 1917 to Conifer	1950 to Elk River Coal & Lbr. Co. Scrapped 1962
44	Climax	2-truck	948	1909	30		1927 from Mt. Hope Coal & Coke Co.	1949 to S. New York Ry., no. 44
51	Lima	Shay 3-truck	2758	1914	75	12" x 15" 36"	1930 from Jerseyfield Lbr. Co. no. 1	1951, scrapped at Conifer
52[d]	Lima	Shay 3-truck	974	1905	75	14" x 12" 35"	1930 from Jerseyfield Lbr. Co. no. 2	1951, scrapped at Conifer
81[d]	Lima	Shay 2-truck	2881	1916	60	12" x 12" 36"	1934 from Basic Refractories Co.	1942, scrapped at Conifer
82[d]	Lima	Shay 2-truck	2755	1914	50	11" x 12" 32"	1934 from Basic Refractories Co.	1942, scrapped at Conifer

the Adirondack wilderness for a new Emporium location, at a spot along Dead Creek, a small tributary of the Raquette River. To this site, about ten miles west of the Tupper Lake village, which became known as Conifer, the Emporium Lumber Company's fortunes were to gradually shift.

The Conifer site was not entirely wilderness at that time. The small firm of George A. McCoy and Son had just set up a small circular sawmill at the location, preparatory to logging a nearby thirty-five-thousand-acre tract belonging to the International Paper Company. A wagon road had been cut out to Childwold for hauling out the lumber. A few log buildings had also been erected when Emporium purchased the entire location in 1910.

A contractor named A. S. Heltman was sent in to begin constructing the many village residences that would be needed, and an Emporium crew began construction of a

Table 11 (*cont.*)
Emporium Forestry Company and Grasse River Railroad

Motor Railcars

No.	*Description*	*Source*	*Disposition*
0	Evans Rail Bus	Arlington & Fairfax	
1	name: Will Roll Some; built from old Pullman car powered with Sterling engine; engine built 1906; 8 cyl., 100 HP	shop-built by Roy Sykes	
2	name: Rolliam; converted from 1905 Thomas Flyer; 4 cyl., 50 HP	shop-built by Roy Sykes in 1916	
4	motorized track car converted from Baldwin No. 3, removed boiler, powered with Sterling gasoline engine	shop-built	scrapped
11	Name: Jumping Goose; converted from 1931 White bus	shop-built by Roy Sykes	1956 to Rail City Museum
12	Rail interurban car	from Lancaster, Oxford and Southern RR	1961 to Strasburgh RR

Rolling Stock

150 +/-	log cars, shop-built
	flat cars, 80,000 lb. capac., steel underframe
	flat cars, 80,000 lb. capac., steel channels center sills
10	boxcars, 60,000 lb. capac.
4	cabooses
2	dump cars, center dump, shop-built
4	dump cars, side dump, shop-built
	snow plow, shop-built
	tool car
	auto railer
	motor car (track maintenance)
	tank car with 30,000 lb. steam pump and 5,000 gal. tank, shop-built 1916
3	passenger cars (coaches)
4	Barnhart log loaders
	Marion steam shovel

[a] Some of the Grasse River Railroad locomotives were at times used on the logging operations.
[b] Converted to oil burner.
[c] Company's best and largest Shay.
[d] Never used.

large band sawmill suitable for the Emporium scheme. The Conifer site then was like a frontier town, and life was at first not easy for the owners and associates from Pennsylvania who had been accustomed to the convenience of modern living.

A sawmill worthy of Sykes's grand intentions was needed, and the opportunity for purchase presented itself when A. A. Low became discouraged and shut down his forest products operation at Horseshoe. The entire lot of sawmill machinery was purchased for $15,000 and moved to Conifer. The head rig was a Clark Brothers seven-foot band saw and the carriage a twelve-foot Clark with a ten-inch shotgun steam feed. A

126. *The Conifer sawmill when first built, probably 1911. The log structures built by George McCoy still stood behind the initial crew sent in by Emporium. Camp buildings, from the left: horse barn, root cellar in center, blacksmith shop behind on hill, bunkhouse in rear on right, lounge and dining room, and cook shack. Note the corduroy road in the foreground, made of logs and covered with a little dirt. Courtesy of the Sykes family.*

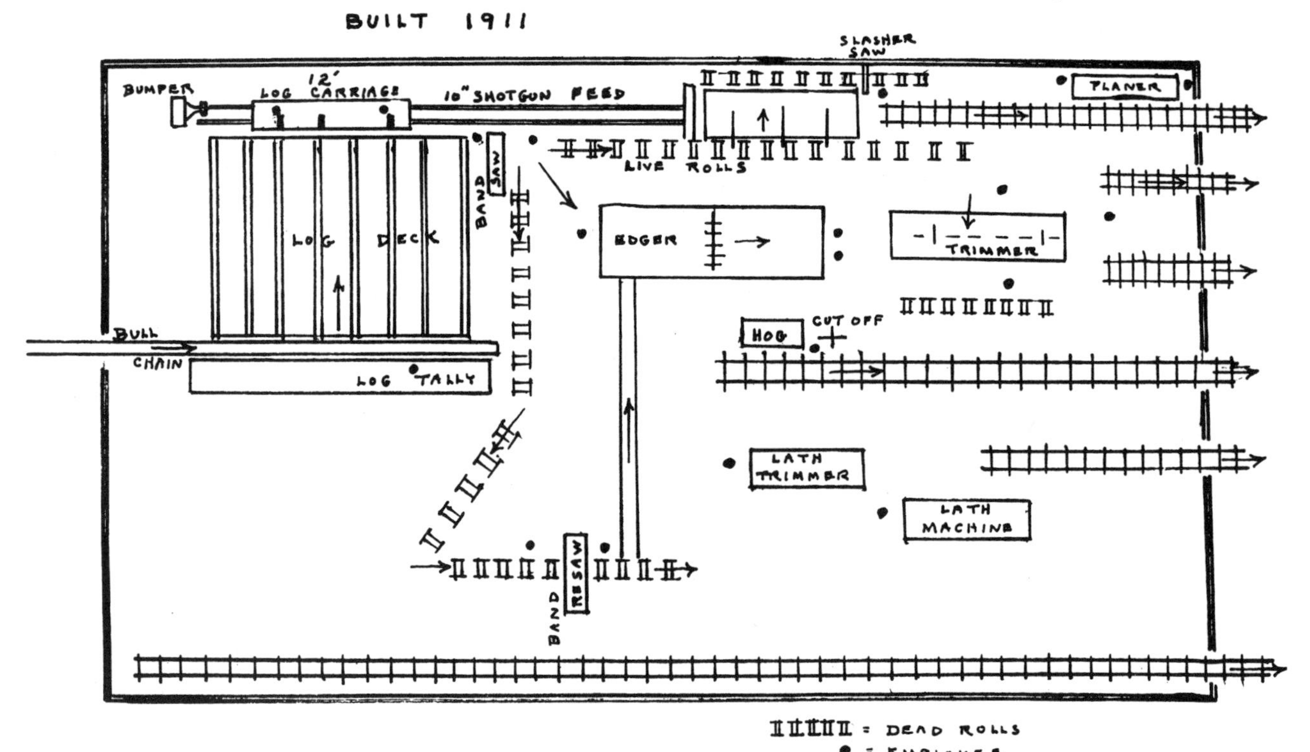

127. *Machinery layout.*

four-saw edger, two-saw trimmer with additional jump saw, and a Mershon band resaw for low-grade material were the other major machines in the sawmill. During the many years that the Emporium operated this mill, three different carriages were installed, all of them made by Clark Brothers of Olean. The last carriage had steam set works and air dogs, the steam supplied by a telescopic pipe installed alongside the steam carriage feed.

The sawmill began operation in October 1911. For thirty-seven and a half years this mill operated steadily under Emporium ownership at a production rate of 8–10 million board feet a year, a total probably of close to 350 million. The first sawyer was Fred Woods, followed by other capable sawyers such as Jud Butler and Ambrose Edwards. Edwards spent a career of fifty-one years in an Emporium sawmill, starting in the company's Galeton sawmill. Clinton Fuller was superintendent, and Paul Thomas supervised the yard. This sawmill, as it turned out, was not a particularly efficient mill.

The village of Conifer grew to be a company town of sizable proportions, one of the last of such villages built at the end of an era. At its peak there were over a hundred dwellings, a boardinghouse, hotel, store, post office, railroad depot, school, church, doctor's office, pool hall, soda fountain, and assorted buildings. Plus, of course, the large sawmill, which employed over a hundred workers. The Emporium owned all of the buildings except for the schoolhouse, and the homes were rented to the married employees for a rental fee varying from $5 to $18 per month.

To handle the New York operations, William Sykes and associates formed a new corporation in 1912, the Emporium Forestry Company. Control was still primarily in the same hands, but now Sykes's three sons had progressed to the

128. A string of empty log cars line the log pond behind the Emporium sawmill at Conifer. The planning mill is on the far right. Photograph by H. M. Beach.

129. The Clark log carriage at the Emporium sawmill in Conifer with the two riders, the dogger and the setter, standing ready. From the collections at College Archives, SUNY ESF.

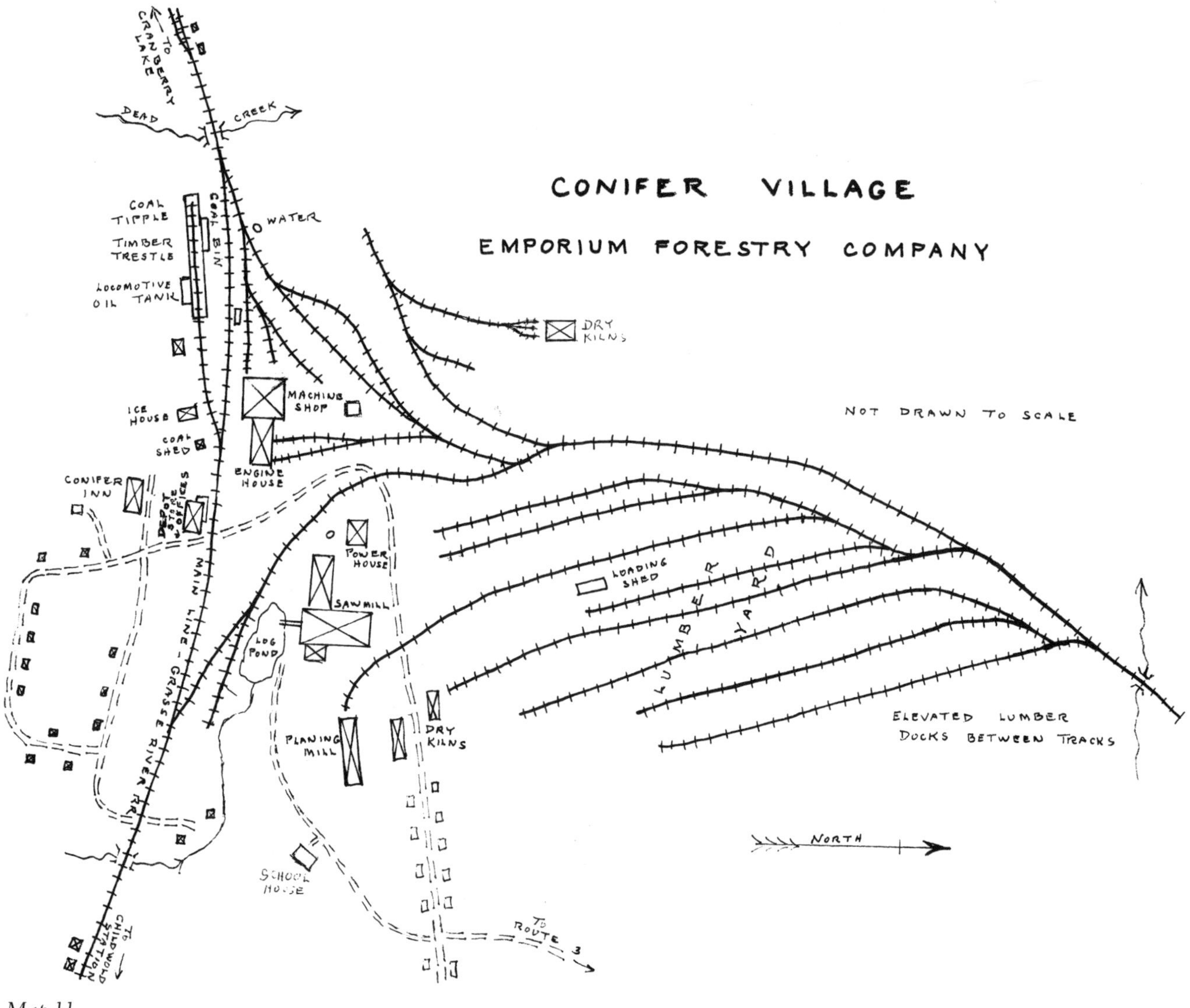

Map 11.

130. The Emporium Forestry Company store and office when newly constructed alongside the Grasse River Railroad tracks. Circa 1912.

Table 12

Hourly Wages of Emporium Sawmill Employees in 1915

Sawyer	35 cents
Dogger	21 cents
Setter	25 cents
Tail sawyer	23 cents
Edgerman	25 cents
Tail edgerman	19 cents
Scalers	20 cents

Source: Data from C. W. H. Douglass, "Report on Lumbering Trip to the Emporium Forestry Company," Cornell student paper, 1916.
Notes: Total cost of lumber production from standing tree to lumber loaded on cars was $16.00 per MBF. Average freight rate: softwood, $2–5 per MBF; hardwood, $5–8 per MBF.

W. L. SYKES 1882 SYKES & CAFLISCH 1883 EMPORIUM LUMBER CO. (PENN'A) 1892

G. W. SYKES, PRESIDENT & GEN. SUPT.
E. J. JONES, VICE PRESIDENT
W. CLYDE SYKES, TREASURER
A. L. OWEN, SECRETARY
ROY O. SYKES, VICE PRESIDENT & SALES MANAGER

MILL AT CONIFER, N. Y.

Emporium Forestry Company

INCORPORATED 1912. (N. Y.)

MANUFACTURERS OF LUMBER

Planing Mills, Dry Kilns, Flooring Plant.

GENERAL OFFICE CONIFER

BRANCH OFFICES 5635 GRAND CENTRAL TERMINAL NEW YORK CITY

161 DEVONSHIRE STREET BOSTON, MASS.

131. Letterhead.

Incorporated under the Laws of the State of New York.

No. 31 310 Shares

Emporium Forestry Company

CAPITAL STOCK, $500,000.00

This Certifies that W. T. Turner, is the owner of Three Hundred ten Shares of the Capital Stock of Emporium Forestry Company, full paid and non-assessable, transferable only on the books of the Corporation by the holder hereof in person or by Attorney upon surrender of this Certificate properly endorsed.

In Witness Whereof, the said Corporation has caused this Certificate to be signed by its duly authorized officers and to be sealed with the Seal of the Corporation this 21st day of May A.D. 1914

W. T. Turner, Treasurer. W. L. Sykes, President.

SHARES $50.00 EACH

132. Stock certificate.

point of assuming executive leadership. The slate of officers of the new corporation read as follows: George W. Sykes, president and general superintendent; E. J. Jones, vice president and general counsel; Roy O. Sykes, vice president and sales manager; W. Clyde Sykes, treasurer; and Arthur L. Owen, secretary. Frank P. Sykes, a cousin of W. L. Sykes, was the first woods superintendent for the Emporium Forestry Company.

William L. Sykes and William S. Caflisch, the company patriarchs, were still on the scene, however, guiding and directing their new venture. Caflisch was a gruff individual who was all business and often bickered with Sykes. He was a popular man, though, and would jump right in and work with the men, although he is reputed to have used a little physical force on occasions when a man was seen slackening up.

Sykes, in contrast, was one who would take moments off from work and talk at length with an employee if the latter showed an interest in the company activities and in advancing. At times Sykes could be quite long-winded, but he sure loved to talk about trees and forestry.

133. The principal officials of the Emporium, including the founders, struck a pose at the Conifer mill in 1916. Left to right: Arthur L. Owen, J. C. Shumberger, William L. Sykes, Joshua Sykes (standing, father of W. L.), William C. Caflisch, Evan J. Jones, Frank P. Sykes, William L. Turner (sitting in front), and William S. Walker (behind).

134. Locomotive No. 43, a small Brooks rod engine, departed from its usual Grasse River Railroad duties to haul a log train along the North Tram. Circa 1920.

Logging Rails

Based on their extensive railroading experience in Pennsylvania, the Emporium management built and operated a first-class logging railroad, the finest ever to operate in New York or New England. The quantity of raw forest products moved and the longevity of the pike attest to its supremacy. The primary product moved by the Emporium Forestry Company railroad over the thirty-six years of operation was hardwood sawlogs, of course, and the equipment was geared to that product.

135. Shay No. 51 of the Emporium was a three-truck locomotive built in 1914. Purchased from Jerseyfield Lumber Company in 1930, it was used by Emporium and Heywood-Wakefield Company until scrapped in 1951.

The first rails laid were those that created a link with the outside world, a connection from the new sawmill village of Conifer to a junction with the New York Central. In May 1911 the company sought proposals to build this line, which was about a mile in length plus sidings. Claude Carlson, a railroad builder who had migrated north into the Adirondacks with the Rich Lumber Company in 1903, made a proposal to build the line at a cost of $3,120 per mile. But he didn't get the job; a Frank Greco bid $2,560 per mile in June 1911. Greco did the construction, although the company was quite unhappy with his slow progress. The junction with the New York Central became known as Childwold station.

The first logging rails were put down in 1913, extending south along Dead Creek. In that year, too, the rail line was pushed through to Cranberry Lake, a sixteen-mile line that was to become the sturdy main line of the Grasse River Rail-

Railroad Construction Proposal

When making his bid on May 23, 1911, to construct the two miles of railroad, Claude Carlson broke down his proposal as follows:

Clearing right-of-way	$240 per mile
Making ties	560 per mile
Grading	1440 per mile
Laying iron	400 per mile
Surfacing	480 per mile
	$3,120 per mile

Grading: 14 feet wide
Ties: made on location from beech, birch and maple
Trees cut for clearing made into sawlogs at $1 per MBF
Right-of-way brushed 60 feet wide
Bark peeled from hemlock trees to be cut and piled for $2 per cord

Prices include bridges and culverts and the switches and wye needed at Childwold station

Source: Letter in Emporium Forestry Company Records, courtesy of The Adirondack Museum, Blue Mountain Lake, N.Y.

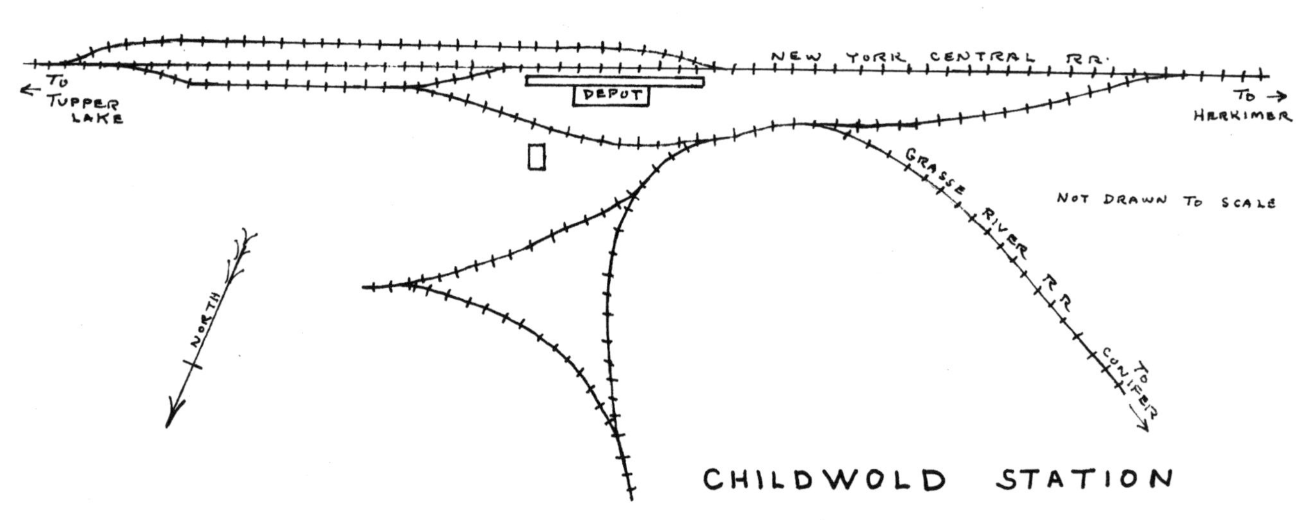

Map 12.

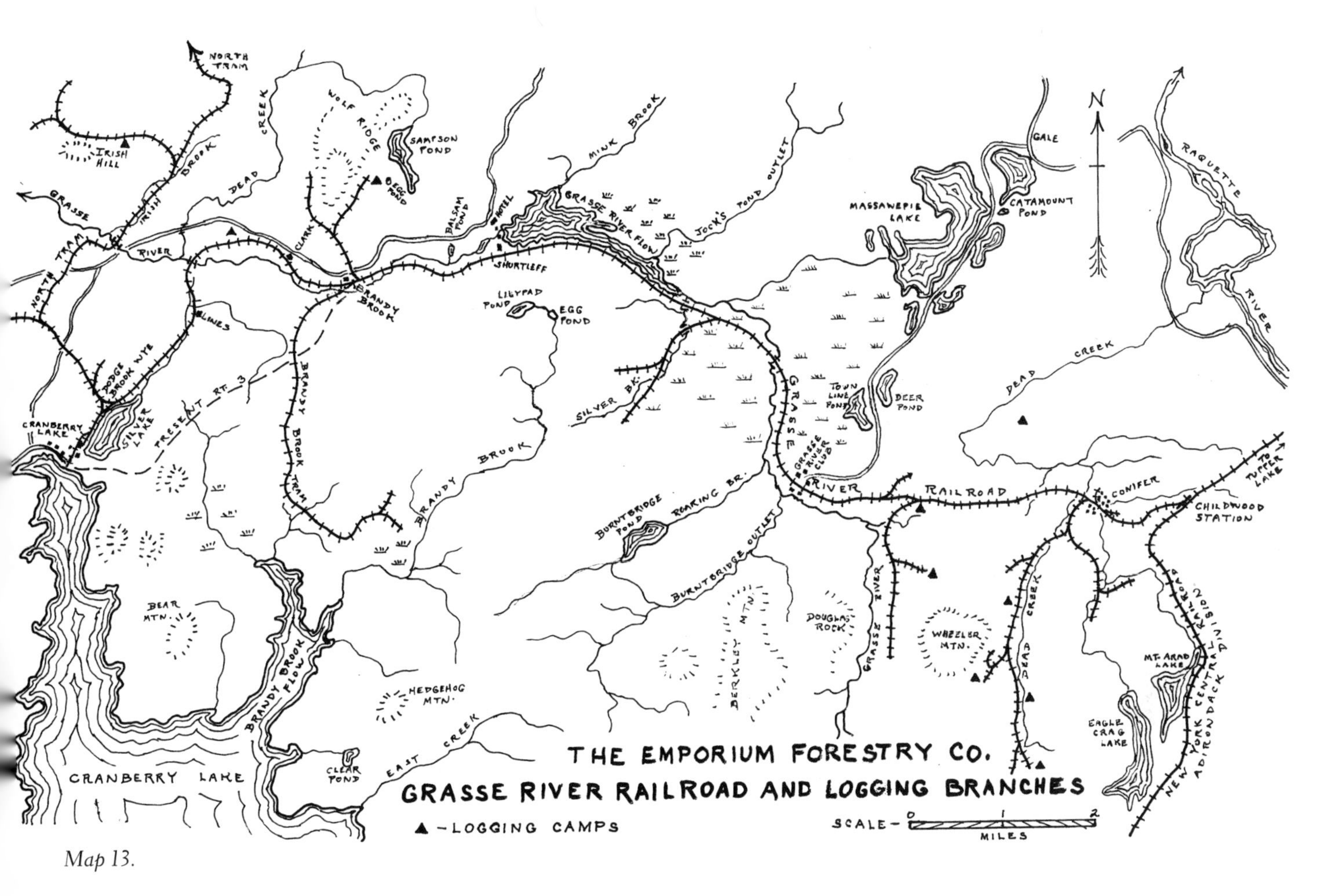

Map 13.

road. Cranberry Lake was a beautiful and somewhat inaccessible Adirondack lake that a few fortunate ones enjoyed as a summer resort.

More railroad equipment began to arrive from Pennsylvania when two of the Emporium's three sawmills in that state shut down in 1913. A first-class railroad was in order for the Adirondacks.

From a point just east of Cranberry Lake, at a junction that became known as the Dodge Brook Wye, the company built a main trunk line in a northerly direction to a point way up near the North Branch Camp of the Stillwater Club in the center of St. Lawrence County. Extending over thirty miles, it was to sprout many branch lines and support them for almost thirty years of hard service. It became known as the "North Tram." A single-wire telephone line ran the length of the tram.

The branch lines were temporary, of course, and many shortcuts were taken in construction techniques. Ties and rails might be laid right on undisturbed ground surface by the crew of Italian immigrants, dumping fill on top where depressions occurred and tamping down the gravel. Hollows would be filled with whole logs or trees, using species of little value, or even slabwood waste from the sawmill. If the rail wouldn't seat properly, a block was pounded under the tie.

When State Highway Route 3 was built at a later date, a crossing was created on the North Tram. The cost of building an overhead railroad crossing was prohibitive, and a lot of legal delays could have occurred. To prevent this, the Emporium agreed to have a stop sign erected that faced the railroad. All trains had to stop before proceeding.

The Emporium Forestry Company employed an interesting mix of rod and geared locomotives, with a large stable of each. As with their previous railroading experience, it was customary to use the rod engines on the trunk lines for faster transport and the geared engines (Shays and Climaxes) on the cruder-built branch lines where the logs were loaded. However, it was not unusual also to find a Shay hauling a long string of loaded log cars down the North Tram to the sawmill. From the most northerly point at the far end of the North Tram, the distance down to the Dodge Brook Wye was about thirty winding miles. For the slower moving Shay, it meant a

136. The sharp curve on the Emporium's North Tram was known as Pine Curve, for an obvious reason. Courtesy of the Sykes family.

six-hour trip home from the far end to reach Cranberry Lake. If the string of cars was going on to the Conifer mill, there was an additional hour haul behind a rod engine. If the Climax was being used on the North Tram, the time would be a bit shorter, because it operated faster than the Shay.

The log cars used by the Emporium were the short buggies, twenty-one feet in length, long enough for one tier of logs. They had a wood-beam skeleton frame with rails mounted on the top of the twelve-foot bunks for the Barnhart log loader to travel on when transferring from car to car. The car weighed about six tons and had a carrying capacity of twenty tons. The log buggies were made in the company shop at a cost of $250 to $350 each.

A trip-chain device was used on the log cars to safely release the binding chains holding the logs. By means of a special bar, the worker could release the logs while standing at the end of the car. The mechanism was devised by the company, probably by Frank Sykes, but was patented by William Caflisch. The brakes were hand set, using a bar known as a "Jim Crow."

Couplers on the logging equipment were the link-and-pin type, but there would usually be a mixture of link-and-pin and knuckle couplers on the Grasse River Railroad portion of the operation with the movement of the company box cars and "foreign" equipment brought in. A special draw bar was used to connect the two styles of couplers. The locomotives had a dual coupler fixture, such as the bullnose seen on the front of Climax No. 5. This was a four-pocket drawhead that would accept a link at various heights and also a removable automatic knuckle coupler that was held in the drawhead slots by a vertical pin.

Loading of the log cars was normally done with one of the four steam Barnhart loaders on the company roster. The Barnhart would transfer itself along the top of the log cars by means of a cable hooked about three cars ahead and pulled along with a patented slack pulley. After transferring, the operator would load the car behind, which he had just vacated. Archie Pollard was remembered as an expert operator who could place the tongs where needed with gentleness and yet with unmatched speed. Ed Ressler was also a good operator but not remembered as a likable man. When loading the cars with logs taken out of Cranberry Lake, one of the first things he would do would be to drop a log in the water, thoroughly wetting down the tong hooker.

The Emporium also used a gasoline-powered Raymond loader at Sabattis, likewise operated on a log car. Most of the train loads were made up of about twenty-two cars, each car with five to seven thousand board feet of logs. The train was usually loaded by 1 PM and sent on its way.

The locomotive roster of the Emporium Forestry Company and the Grasse River Railroad totaled twenty-two engines over the many years in Conifer. The Emporium Lumber Company in Pennsylvania had owned fourteen locomotives, and nine of these were brought up to New York. Of the twenty-two locomotives used in New York, eleven were rod engines of three different makes—Schenectady, Brooks, and Baldwin. There were eight of the favored Shays, although three were never used because they were in bad condition when purchased, and two Climax engines. The final locomotive was a small forty-four-ton diesel locomotive purchased by the succeeding owner, Heywood-Wakefield Company.

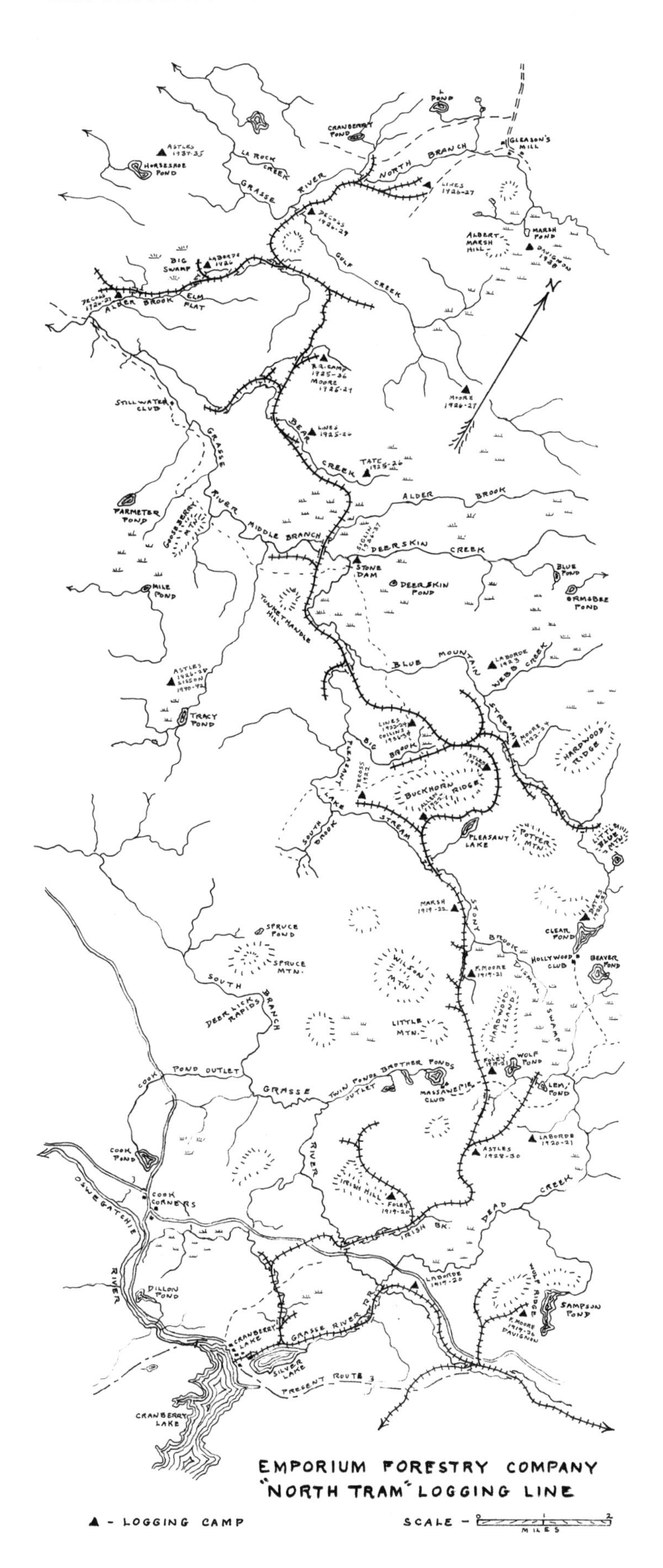
EMPORIUM FORESTRY COMPANY
"NORTH TRAM" LOGGING LINE
▲ - LOGGING CAMP
SCALE
MILES
CRANBERRY POND
GLEASON'S MILL
HORSESHOE POND
LA ROCK CREEK
GRASSE RIVER
NORTH BRANCH
ALBERT MARSH HILL
MARSH POND
GOLF CREEK
BIG SWAMP
ALDER BROOK
ELM FLAT
STILLWATER CLUB
BEAR CREEK
ALDER BROOK
PARMETER POND
MIDDLE BRANCH
DEERSKIN CREEK
STONE DAM
DEERSKIN POND
BLUE POND
ORMSBEE POND
MILE POND
TURKEY HANDLE HILL
BLUE MOUNTAIN STREAM
WEBB CREEK
TRACY POND
BIG BROOK
HARDWOOD RIDGE
BUCKHORN RIDGE
PLEASANT LAKE STREAM
SOUTH BROOK
PLEASANT LAKE
POTTER MTN.
STONY BROOK
CLEAR POND
HOLLYWOOD CLUB
BEAVER POND
SPRUCE POND
SPRUCE MTN.
WILSON MTN
SOUTH BRANCH
DEER LICK RAPIDS
LITTLE MTN.
SMALL SWAMP
HARDWOOD ISLAND
POND OUTLET
GRASSE
TWIN PONDS
BROTHER PONDS
OUTLET
MASSAWEPIE CLUB
WOLF POND
LEM POND
COOK POND
RIVER
IRISH HILL
DEAD CREEK
COOK CORNERS
IRISH BK.
OSWEGATCHIE RIVER
DILLON POND
GRASSE RIVER R.R.
SAMPSON POND
CRANBERRY LAKE
SILVER LAKE
PRESENT ROUTE 3

Map 14.

137. The temperature was 38 degrees that morning as the train backed into a siding, ready for work. From the collections at College Archives, SUNY ESF.

138. William L. Sykes pointed out the virtues of a large yellow birch log to his visitors while Mr. Flint, the former chancellor of Syracuse University (in white), assists with a peavey. Shay No. 39 waits patiently.

139. The log cars used by the Emporium were the short, one-tier skeleton cars with loader rails fastened on the bunks. The train crew paused on the newly built trestle at Conifer that formed the leg of the wye between the main line and a stub track at the mill. Note the hose on the Barnhart loader, used to draw water from a brook. Courtesy of the Sykes family.

140. Climax No. 35 was equipped with a four-pocket drawhead to accommodate a variety of coupler circumstances. While in the Conifer yard, the removable knuckle coupler was in place for shifting a boxcar. Circa 1920. Photograph by John Stock.

The Conifer mill was now in full swing, keeping about seventy-five employees busy under the direction of mill superintendent Clinton Fuller. Included among the duties of yard superintendent Paul Thomas were the expanding dry kiln facilities that were so necessary even then in the trade. George Sykes designed the first Conifer kiln, a railroad kiln that held six cars with a capacity of ninety thousand feet.

Timberland purchases were continuing in the area around Cranberry Lake and in the vast region to the north of it. Although buying activities continued until 1943, the main land purchases were made by the year 1920. Large tracts on two sides of Cranberry Lake were purchased from International Paper Company after they had logged them, in some cases twice for the softwood pulpwood. In time the ownership was to total about 190 to 195 square miles.

Grasse River Railroad

When the Emporium Forestry Company originally built the company railroad westerly from Conifer to Cranberry Lake, it had to obtain rights-of-way across other land ownerships and in doing so had to make agreements with the land owners. With what was termed a private contract, the Emporium agreed at that time to haul out certain forest products from these properties and even to haul workers and supplies to

141. The Barnhart loader was busy loading a string of empties behind Shay No. 51 on a cold winter day.

143. A consist of long-length pulpwood was about to be moved out of the Cranberry Lake yard, nudged along by No. 33 Baldwin. Photograph by McClure. Courtesy of Jeanne Reynolds.

142. No. 68 was one of the two second-hand moguls purchased by Emporium in 1923 for use both on the Grasse River Railroad and for logging activity. In this logging scene of 1930, Ike Regan is the crewman on the log car and William Dilcher is one of the men on the ground. Courtesy of Jeanne Reynolds.

Table 13
Approximate Cost of a New Locomotive in 1916

Equipment	*Weight (tons)*	*Cost ($)*
Rod locomotive	45–50	7,000–8,000
Shay	95	12,000
Climax	65	9,500

Source: Data from C. W. H. Douglass, "Report on Lumbering Trip to the Emporium Forestry Company," Cornell student paper, 1916.

woods operations of other companies as well as to the Emporium logging camps. Local residents were able to grab a free lift out to the Childwold depot, a gift of a cigar helping to please the train crew.

The New York Public Service Commission took a dim view of this practice and deemed that a common carrier service was being provided by Emporium. To meet the requirements of the law, this sixteen-mile portion of the railroad property from Childwold Station to Cranberry Lake had to be formed into a separate corporation and operated as such. Accordingly, the Grasse River Railroad was formed on January 9, 1915, although it did not begin operations as such until May 1916.

Once the Grasse River was a separate corporation, it became necessary that ownership of the equipment and necessary real estate be transferred over to it. Four locomotives, Nos. 33, 38, 42, and 43, were "sold" to Grasse River ownership, as well as three coaches; a caboose; ten boxcars of sixty-thousand-pound capacity; about twenty-four rebuilt flat cars; the company's Marion steam shovel, which was built in 1912; two gasoline-powered railroad cars; and other miscellaneous rolling stock. Also transferred was the 16.87 miles of track from Childwold to Cranberry Lake depot plus land and buildings. Other locomotives were added at later dates. The normal procedure was to maintain three rod locomotives in operating condition to service the Grasse River line.

The sixteen-mile railroad line to Cranberry Lake had been completed in 1913 at a cost of $162,800. Fifty- to sev-

144. A carload of hardwood ready for dumping at the mill pond unloading ramp. Note the tilt of the track. From the collections at College Archives, SUNY ESF.

145. The switch points needed cleaning out after the snowstorm as Grasse River Railroad No. 2 prepared to make a run to Childwold station in December 1949. Shay No. 40 sat on the next track. Photograph by Philip R. Hastings.

enty-five-pound rail was used on the main line whereas the branches were built with forty-, fifty-six-, and sixty-pound rail. Ties were laid with a spacing of sixteen per thirty-foot length of rail. On the Grasse River main line the maximum grade was 3 percent and the maximum curve about twelve degrees. The officers were the same principals of the Emporium Forestry Company, and the only salaried official was Superintendent Roy Sykes, youngest son of William Sykes. Roy received a salary of $2,400 for his part-time duties with the Grasse River Railroad.

The principal revenue products on the Grasse River

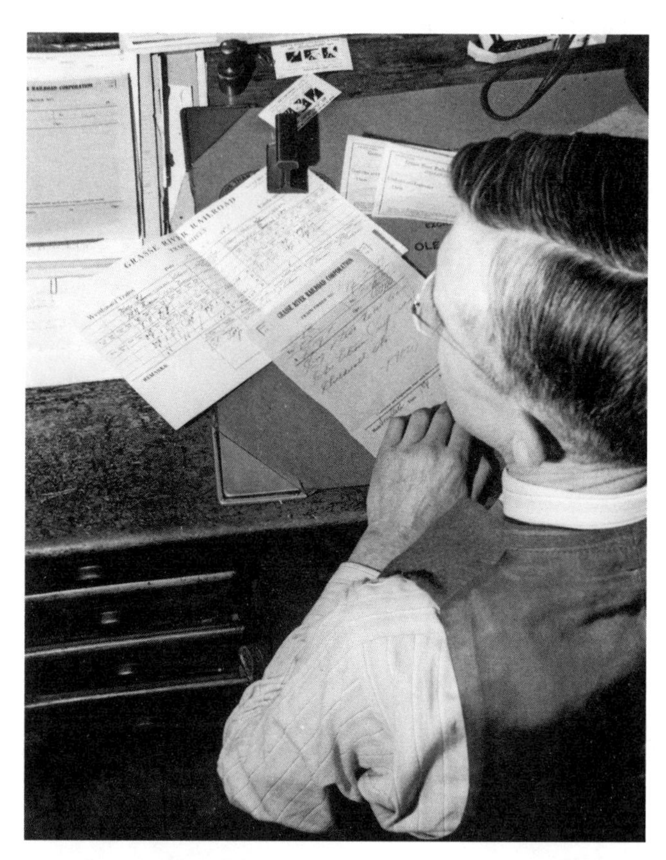

146. There was freight to move from Conifer to Childwold and the Grasse River Railroad dispatcher issued an order for No. 2 to run an extra on December 23, 1949. Photograph by Philip R. Hastings.

147. The winter storm had let up, but the temperature remained brisk as Grasse River No. 2 prepared for the run to Childwold, about to take on water at Conifer. This mogul, built in 1913, had previously seen logging activity at Brandreth. Photograph by Philip R. Hastings.

were lumber and raw forest products, supplies for the villages of Conifer and Cranberry Lake as well as intermediate clubs and camps, passengers, and the U.S. Mail. Freight items such as pulpwood from International Paper Company operations or construction material for the new dam or the new state highway were sporadic at best. The result was many lean years for the Grasse River Railroad, with its high operating expenses and sizable deficits. The year 1928, for example, had a deficit of $25,755 with expenses 140 percent of revenues. There were thirteen employees that year.

Passenger rates in the early years were five cents per mile, and baggage brought a flat charge of fifty cents up to 150 pounds. By 1922 there were two trains daily in each direction for the single combination passenger and baggage car, with a travel time of one hour from Childwold to Cranberry Lake, unless the train happened to make one of the convenient meal stops at the Conifer Inn. The transport of a corpse required a first-class ticket with a minimum of one dollar for the deceased plus a person to accompany the box.

GRASSE RIVER RAILROAD CORPORATION

Time table No. 38 — Subject to change without notice — Effective 12.01 a. m. June 26, 1927

EASTERN STANDARD TIME

FROM CRANBERRY LAKE (READ DOWN) — TO CRANBERRY LAKE (READ UP)

9	7	25	5	3	1		Daily Except Sunday		2	4	6	26	8	10
P. M.	P. M.	P. M.	A. M.	A. M.	A. M.				A. M.	A. M.	P. M.	P. M.	P. M.	P. M.
6.30			11.15	9.00		Lv.	Cranberry Lake	Ar.	9.00	11.05			6.02	
6.45			11.30	9.15		F	Clark's	F	8.45	10.50			5.47	
6.48			11.33	9.18		F	Brandy Brook	F	8.42	10.47			5.44	
6.53			11.38	9.23		F	Shurtleff's	F	8.37	10.42			5.39	
7.05			11.50	9.35		F	Grasse River Club	F	8.25	10.30			5.27	
7.15			12·00	9.45		Ar.	Conifer	Lv.	8.15	10.20			5.17	
7.20	4.45	2.05	12·08	9.45	7.20	Lv.	Conifer	Ar.	8.15	10.20	12·35	2.45	5.17	7.55
7.30	4.55	2.15	12·18	9.55	7.30	Ar.	Childwold	Lv.	8.07	10.10	12.25	2.35	5.07	†7.45
P. M.	P. M.	P. M.	P. M.	A. M.	A. M.		F: Flag stop all trains		A. M.	A. M	P. M.	P. M.	P. M.	P. M.

NEW YORK CENTRAL RAILROAD

9	7	25	5	3	1				2	4	6	26	8	10
7.45	5.03	2.20	12.23	10.09	7.35	Lv.	Childwold	Ar.	8.05	10.09	12.23	2.32	5.03	
	5.17	2.36	12.38		7.50	Ar.	Tupper Lake Jct.	Lv.	7.50	9.56		2.17		
	6.35	3.45	2.00			"	Saranac Lake	"	6.50	9.07		12.55		
	7.05		2.25			"	Lake Placid	"	6.30	8.45		12.35		
	7.34		2.55			"	Malone	"		7.35		12.05		
	10.15					"	Montreal	"				9.20		
11.30		6.15		1.15	11.15	"	Utica	"	3.45		8.30	10.30	1.30	
2.45		8.00		2.31	1.05	"	Syracuse	"	12.50		5.35	8.30	11.30	
5.31		9.02				"	Lyons	"	9.20			7.18	10.31	
4.45		9.55		4.05	2.54	"	Rochester	"	11.04		3.30	6.22	9.39	
6.45		11.40		5.30	4.30	"	Buffalo	"	9.35		1.50	4.40	8.00	
2.20		8.30		3.30	2.47	"	Albany	"	11.03		5.45	7.10	11.12	
10.55		6.40		10.50	8.45	"	Boston	"			10.00		3.15	
6.10		11.45		6.50	6.30	"	New York	"	7.10		11.20		7.43	
A. M.	P. M.	P. M.	P. M.	P. M.	P. M.				P. M.	A. M.	P. M.	A. M.	A. M.	

Light face figures denote A. M. time — Heavy face figures denote P. M. time

†No. 10 will wait at Childwold until 8.15 for N. Y. C. No. 15 ONLY when message has been received at Childwold that there are passengers for this line. Time of N. Y. C. trains shown for information only. For detailed information consult N. Y. C. time tables.

GENERAL INFORMATION

The timetable herein is subject to change without notice

It shows the time trains should arrive at and depart from the several stations and connect with other trains, but their departure, arrival, or connection at the time stated is not guaranteed.

The cash Rate of Fare published in passenger tariff on file with the Public Service Commission will in all cases prevail.

All baggage checked must be rechecked at Childwold Station. This applies to baggage to and from the New York Central Railroad.

A timetable will be furnished upon request by addressing agent at Cranberry Lake, N. Y. or Conifer, N. Y., and they will be found at the bureaus of information at the following stations. Buffalo, Rochester, Lyons, Syracuse, New York City, Albany, Utica, and many other junction points.

148. Grasse River Railroad timetable.

Realizing the need of passenger revenue to support the costly little Grasse River Railroad, the Sykes family actively promoted tourism at Cranberry Lake. Prior to World War I, the summer residents had been able to arrive at Cranberry Lake via the New York Central to Benson Mines and then the Rich Lumber Company railroad into Wanakena at the lower end of the lake. However, this service had ceased in 1914, when the Rich brothers shut down their Wanakena sawmill and discontinued their railroad.

The situation changed, however, in the post–World War I era. At that point it became convenient, as the summer season arrived, for the affluent to arrive at Childwold station on the New York Central's Montreal Express and make connections for Cranberry Lake. Arriving at Childwold at 5:30 AM, the departing passengers would rub their sleepy eyes and search with affectionate longing for that little Grasse River locomotive and her single coach. But her frequent late arrivals would make it an eager wait for these folks not accustomed to such early awakenings. Anticipation of the coming season at the home on the lake would turn impatience into the joyful realization that they were about to have that unique ride on the Grasse River, a joy they had often thought about at their winter homes in the city.

Boarding the Grasse River coach, the excited vacationers would depart Childwold and then usually stop for about an hour at Conifer, where the passengers were given an opportunity to enjoy a breakfast at the Conifer Inn next to the depot. The sumptuous serving of cereal, bacon and eggs, and thick pancakes was provided to local workers for fifty cents, but, of course, for traveling folks the price was a dollar. Many suspected a hold-up game, but after all, this was part of the summer experience, something to dream about next winter.

The early morning jaunt from Conifer on to Cranberry Lake was truly an experience to delight any youngster from an urban environment. Squeals of delight filled the little coach as deer or heron were spotted in the marshes along the Grasse River Flow or whenever the engineer had to whistle a deer off the track. The conductor was known to have once stopped the train so that a lady on board could identify some wildflowers. On arrival at the long-awaited destination, the coach would be left at the edge of the lake while the engine went off to perform a few switching duties. While awaiting

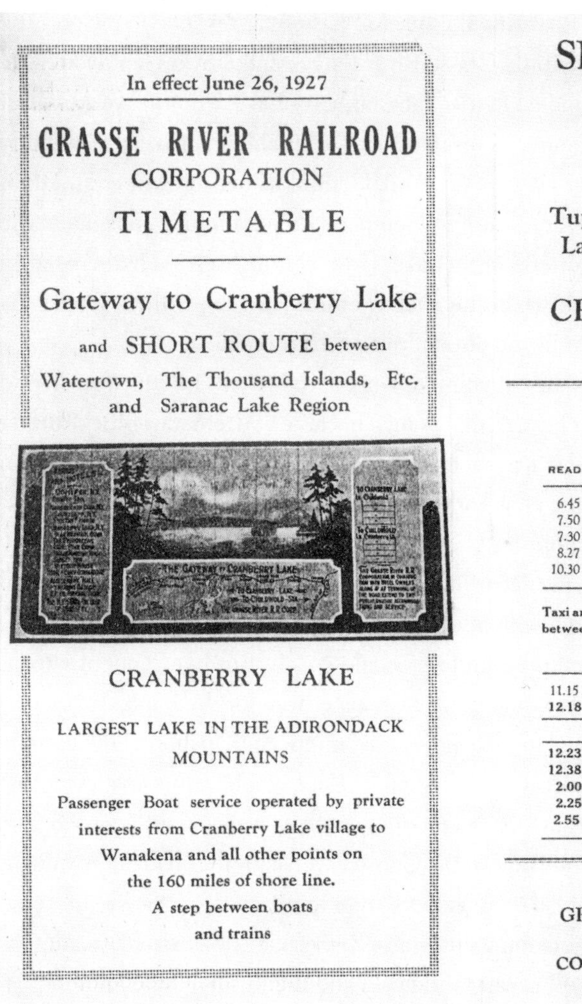

In effect June 26, 1927

GRASSE RIVER RAILROAD
CORPORATION

TIMETABLE

Gateway to Cranberry Lake

and SHORT ROUTE between

Watertown, The Thousand Islands, Etc.
and Saranac Lake Region

CRANBERRY LAKE

LARGEST LAKE IN THE ADIRONDACK MOUNTAINS

Passenger Boat service operated by private interests from Cranberry Lake village to Wanakena and all other points on the 160 miles of shore line.
A step between boats and trains

SHORT ROUTE

BETWEEN

Watertown, Carthage,
The Thousand Islands

AND

Tupper Lake, Saranac Lake,
Lake Placid and Malone

VIA

CRANBERRY LAKE
GATEWAY

Subject to Change Without Notice
Daily Except Sunday

NEW YORK CENTRAL

READ DOWN				READ UP
6.45 A. M.	Lv.	Clayton	Ar.	6.35 P. M.
7.50 A. M.	"	Philadelphia, N. Y.	"	5.19 P. M.
7.30 A. M.	"	Watertown	"	5.35 P. M.
8.27 A. M.	"	Carthage	"	4.40 P. M.
10.30 A. M.	Ar.	Newton Falls	Lv.	2.55 P. M.

Taxi and Bus Service operated by private interests between Newton Falls and Cranberry Lake (11 miles)

GRASSE RIVER RAILROAD

11.15 A. M.	Lv.	Cranberry Lake	Ar.	11.05 A. M.
12.18 P. M.	Ar.	Childwold	Lv.	10.10 A. M.

N. Y. C. R. R.

12.23 P. M.	Lv.	Childwold	Ar.	10.09 A. M.
12.38 P. M.	Ar.	Tupper Lake Jct.	Lv.	9.56 A. M.
2.00 P. M.	"	Saranac Lake	"	9.07 A. M.
2.25 P. M.	"	Lake Placid	"	8.45 A. M.
2.55 P. M.	"	Malone	"	7.35 A. M.

For further information address
GRASSE RIVER RAILROAD
CORPORATION
CONIFER . . NEW YORK

149. Grasse River Railroad timetable.

the arrival of the boat to take them out to their camps, the vacationers could consider some purchases at the general store, gaze at the men lounging on the steps, or just enjoy the smell of sawdust.

Passenger service on the Grasse River Railroad continued somewhat profitably up until the state highway was hard-surfaced in 1926 and 1927.

Cranberry Lake Sawmill

Lumber markets were strong during the World War I, and the Emporium Forestry Company deemed it advisable to build a second Adirondack sawmill. In 1917 a large double band mill was built on the northwest shore of Silver Lake, on the edge of Cranberry Lake village, and it began operations in September of that year. Having two band head-saws, the original intention was to saw a considerable amount of softwood on one side of the mill. Cranberry Lake village was expected to experience an industrial and tourism boom.

This mill was said to be the best that Sykes ever built, but as it turned out, it also proved to be the most expensive to operate. Arthur Owen was made superintendent, to run this facility with the latest in machinery such as a twin band resaw and jump saw trimmer. The sawyer was Leo D. Axtle, the mill foreman was Gene Lothiam, and the yard superintendent was Irving Hazelet. Production was almost double that of Conifer, and during its ten years of operation, the mill turned out at least two hundred million feet of lumber, often operating two shifts. At times there would be twelve million feet on sticks in the yard. In contrast to the Conifer mill, the Cranberry Lake mill had no dry kiln or flooring mill.

Although the mill did saw some spruce, pine, and hemlock, hardwood predominated. Many spruce logs were shipped out as "fiddlebutts," high-quality spruce logs used

Wages of the Train Crew, 1916

The train crew on the Grasse River Railroad consisted of six men working a twelve-hour schedule. Any overtime was still paid at the straight rate.

Wages per hour in 1916:

Engineer	25 cents
Fireman	20 cents
Conductor	24 cents

The logging train also required a three-man crew to operate the Barnhart loader at the following hourly wages:

Loader engineer	29 cents
Tong men (two)	22 cents

Loading costs for the company (wages and fuel) ran from 10 to 20 cents per thousand board feet of logs.

Source: Data from C. W. H. Douglass, "Report on Lumbering Trip to the Emporium Forestry Company," Cornell student paper, 1916.

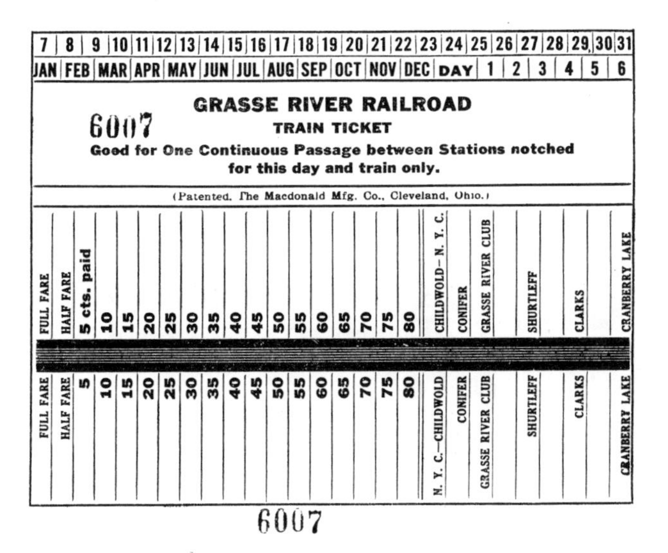
7 | 8 | 9 | 10 | 11 | 12 | 13 | 14 | 15 | 16 | 17 | 18 | 19 | 20 | 21 | 22 | 23 | 24 | 25 | 26 | 27 | 28 | 29 | 30 | 31
JAN | FEB | MAR | APR | MAY | JUN | JUL | AUG | SEP | OCT | NOV | DEC | DAY | 1 | 2 | 3 | 4 | 5 | 6

GRASSE RIVER RAILROAD
6007
TRAIN TICKET
Good for One Continuous Passage between Stations notched for this day and train only.
(Patented. The Macdonald Mfg. Co., Cleveland, Ohio.)

FULL FARE | HALF FARE | 5 cts. paid | 10 | 15 | 20 | 25 | 30 | 35 | 40 | 45 | 50 | 55 | 60 | 65 | 70 | 75 | 80 | CHILDWOLD—N. Y. C. | CONIFER | GRASSE RIVER CLUB | SHURTLEFF | CLARKS | CRANBERRY LAKE

FULL FARE | HALF FARE | 5 | 10 | 15 | 20 | 25 | 30 | 35 | 40 | 45 | 50 | 55 | 60 | 65 | 70 | 75 | 80 | N. Y. C.—CHILDWOLD | CONIFER | GRASSE RIVER CLUB | SHURTLEFF | CLARKS | CRANBERRY LAKE

6007

150. Grasse River Railroad ticket.

for the manufacture of piano sounding boards, violins, and other musical instruments, and a great deal of the softwood was put into pulpwood. A large part of the log supply came from various timber tracts purchased from International Paper around Cranberry Lake.

In the following year, 1918, their last Pennsylvania mill at Galeton closed its doors, and the company's manufacturing efforts were then solely within the Adirondacks. At the close of the Keating Summit, Pennsylvania, mill, the general office had been moved to Union Station in Utica, and then in 1926 to Conifer, where it remained. The Woods Department headquarters was located at Cranberry Lake. Loron Silliman spent many years as the Emporium's office manager.

At one point in the early years, the company considered continuing the railroad westerly from Cranberry Lake to Newton Falls, where it could connect with the Carthage and Adirondack branch of the New York Central. It had been encouraged to do so by the Watertown Chamber of Commerce and the Northern New York Development League in order to shorten the freight hauls from Canada. But the idea never received serious consideration.

The Railroad Shop

With a railroad operation as large as the Emporium's, it is not surprising that it had a first-rate locomotive repair shop. With Jesse Blesh as master mechanic, the shop crew could rebuild any of the locomotives in the large roster. Roy Zenger succeeded Blesh in later years. They built their own log cars, wood-frame railroad trucks, and other equipment.

The shop required a force of twelve workers—boiler maker, head machinist, and ten helpers. Wages in 1915 reflected top pay for a skilled mechanic, thirty-three cents per hour for the boiler maker and the machinist and nineteen cents hourly for the helpers. The only workers' labor union at Conifer was the trainmen's union, to which the railroad operators belonged.

The Emporium shop crew shared the feeling of the operating crew that the Shay locomotive was preferred over the other type of geared locomotive, the Climax. The working parts and the gears of the Shay were all accessible along the right side of the locomotive and maintenance was much easier. The Climax was a faster locomotive, however.

Of the eight Shays included in the New York roster, the mechanics and the engineers agreed that the ninety-two-ton No. 40 was the best worker. It was also the largest locomotive ever owned by Emporium. In 1926, after twenty-four years of service, the Emporium shop crew installed in No. 40 a new 225-pound pressure boiler purchased from the Lima Loco-

151. Grasse River No. 68, a Schenectady-built mogul, waited for track repair in the Conifer yard in September 1949. Photograph by Philip R. Hastings.

motive Works. In 1948 the tires (rims) on the drive wheels needed to be turned to eliminate the rough spots. However, the shop crew instead installed trucks that were taken off of Shay No. 52. These trucks were made of steel castings and had thirty-five-inch wheels instead of thirty-six, as originally found on this Shay, which then slowed down big No. 40 a little. Shay No. 52 was in poor operating condition when purchased from Jerseyfield Lumber Company and never did get repaired and used on the Emporium.

In 1920 Shay No. 37 was given new tires at the Conifer shop, the new metal tires or rims having been purchased from a firm in Germany.

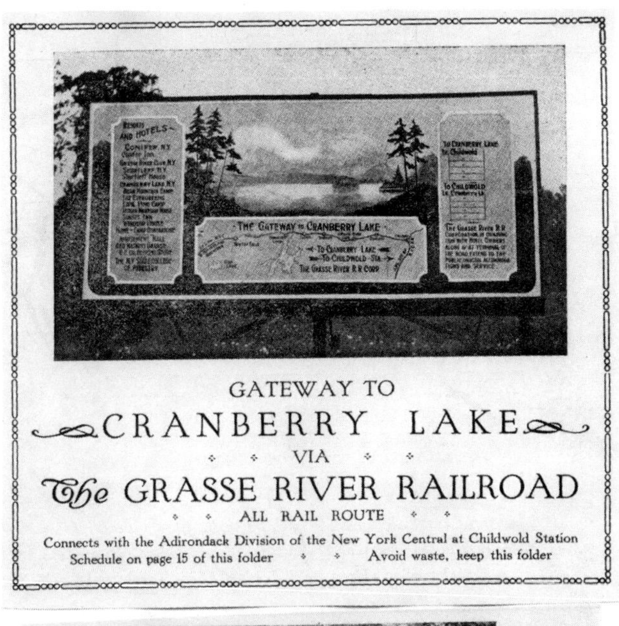

152. William L. Sykes made an effort to boost the revenues on the Grasse River Railroad by promoting tourism for Cranberry Lake.

William L. Sykes was a man who loved railroad machinery and felt an overpowering need to collect and save railroad machinery and spare parts. As one might suspect, this was an irritating habit that created quite a junk pile near the railroad shop. Once when he was on a trip to Utica, his sons took the opportunity to clean house a little and sold a considerable amount to a junk dealer. Quickly loading it on flat cars, they sent it on its way south before their father returned. But Father, returning home on the New York Central, spotted this good-looking collection of spare parts loaded on flat cars sitting on a siding and immediately wired the junk man to purchase the needed parts, unaware that they were originally his. At least, that's the way the story is told.

It was a common practice for the Emporium to rent out a locomotive to any of the smaller logging railroads in the Adirondacks for emergency needs. If a logger only had one or two locomotives, which was true of many of them, a breakdown could leave the jobber in a tough position. Sykes once offered the use of his fifty-five-ton Climax to the Glenfield and Western Railroad over on Tug Hill during a time of urgent need. The Climax was offered at a rental price of $20 per day plus transportation or a purchase price of $8,000. Only the offer of rental was accepted.

The inventive ingenuity of William Sykes was particularly inherited by his youngest son, Roy. Roy spent many hours in the shop along with the mechanics, building various types of motor cars for service on the Grasse River Railroad. It was costly to run steam on the sixteen-mile run with hardly more than a couple of bags of mail aboard, and cheaper operating rolling stock was always being experimented with.

One of the first motor cars was Roy's 1905 Thomas Flyer four-cylinder automobile, which was rebuilt in 1916 with railroad wheels and a chain drive and called the No. 2 "Rolliam." About the same time, there was No. 1, the "Will Roll Some," an old Pullman car powered with a hundred-horsepower eight-cylinder Sterling marine engine that was mounted on posts resting on the truck pedestals. Roy spoke proudly of this innovation, which cut down on the car vibration. The car carried both passengers and mail for about three years, until the auto road was built.

No. 4 motor car came forth when one of the Baldwin locomotives, No. 3, had the boiler removed and a Sterling gasoline engine installed. A few still remember No. 11, an old White bus that was rebuilt with double trucks and chain drive on the rear truck. It was called "Black Maria" by some, "Jumping Goose" by others. No. 12 was an old sixty-foot rail interurban car that had at least two different engines installed. It was large but economical and ran for ten years carrying freight and as many as sixty passengers. Among Roy Sykes's achievements was a kerosene carburetor for the cars that he designed and patented in 1914.

Small speeders were used for the logging boss to run up to the camp, powered by a Fairbanks engine and a belt drive.

It's to the credit of both the shop crew and the operating crews that no major railroading accidents ever occurred on

153. Grasse River Railroad's rail motor car No. 11 was waiting to meet the southbound New York Central train No. 2 at Childwold depot on December 23, 1949. Photograph by Philip R. Hastings.

either the logging railroad or the Grasse River Railroad—quite unusual for a logging railroad. There were, of course, the minor mishaps and snafus that kept the men always alert, such as the occasion in 1918 when three loaded cars got away from conductor Fraser. Fraser had neglected to align the safety switch properly, and the cars jumped the track and were wrecked. Fraser was demoted for negligence.

Link-and-pin couplers caused the occasional loss of a finger or hand when the brakeman was not totally alert to the train movement. In 1920, for example, Dr. Cohen had to amputate the little finger of a workman in the Conifer yard after it became crushed between couplers. Occasionally a fatal accident had to be reported to the authorities, such as the incident in 1927 when a wood cutter not in the employ of Emporium tried to jump aboard the caboose while the train was running at about ten miles an hour. He missed his footing on the step, fell, and rolled under the train, succumbing shortly afterward.

Fires were a constant threat to any sawmill industry, both forest fires and mill fires. When New York State adopted the 1916 regulations requiring the railroads to use only oil-burning locomotives during the fire season, the Emporium engine shop was well equipped to convert the locomotives from coal burning to oil, but it wasn't a cheap alteration. The motive power on the Grasse River Railroad was switched over to fuel oil, but the Shays remained coal burners. The cost to equip Shay No. 37, for example, with a burner and tank was quoted as $876 but apparently was never authorized as an acceptable expense.

The Emporium had only one serious forest fire in the New York operation, quite a contrast to the constant forest fires they had experienced in Pennsylvania. In 1941 the Grasse River flow and the Grasse River Club were swept by a fire, which put quite a scare into the inhabitants of Conifer, as they, at one stage, could view the flames down the right-of-way.

Fire extinguishers were placed in the locomotives, and a

154. Driver Rob Mills loads the morning mail at Childwold, bound for Conifer. Photograph by Philip R. Hastings.

155. The Cranberry Lake sawmill, with Silver Lake on the left side of the log-dumping track and the log pond on the right. Note the tilt of the dumping track. Circa 1920.

patrol of one or more men followed each train at a suitable distance to look for fires when coal burners were allowed to operate. When extra manpower was available, the company might put a crew to work cleaning up the combustible material along the rights-of-way, usually clearing fifty feet on each side of the track. Special care was given along the main line

156. An aerial view of Cranberry Lake village in the early 1920s shows the Emporium sawmill (lower center), with the elevated lumberyard docks visible above the mill. A boom of pulpwood can be seen in Cranberry Lake, about to be loaded onto the overhead conveyor for movement over to the lumberyard, where it will be loaded on cars. Circa 1922. Photograph by D. P. Church.

of the Grasse River Railroad because of the importance of creating good appearances for the general public and the state commission members.

The company built a five-thousand-gallon tank car with a steam pump and three lines of hoses, which was kept at Conifer. Also, filled water barrels were placed at the crests of long track grades.

The only serious mill fire under the Emporium ownership was the burning of the planing mill in 1920. This fire of unknown origin was first reported about twenty minutes after midnight and burned all day with a heavy loss. Another fire in the 1920s destroyed the engine house.

Logging Activities

Almost all of the logs sawn by the Emporium Forestry Company for the first thirty years were harvested on the company's own timberland. At its peak, the Emporium acreage was close to 150,000 acres of hardwood timber, almost entirely within St. Lawrence County. Initial cuttings were in the old-growth hardwood near Conifer and westward toward Cranberry Lake. The Brandy Brook Tram extended for five miles down to the Brandy Brook Flow vicinity and was used in the period 1913–17. The North Tram was built later, extending northward for a total of thirty miles or more, in addition to many miles of branch lines.

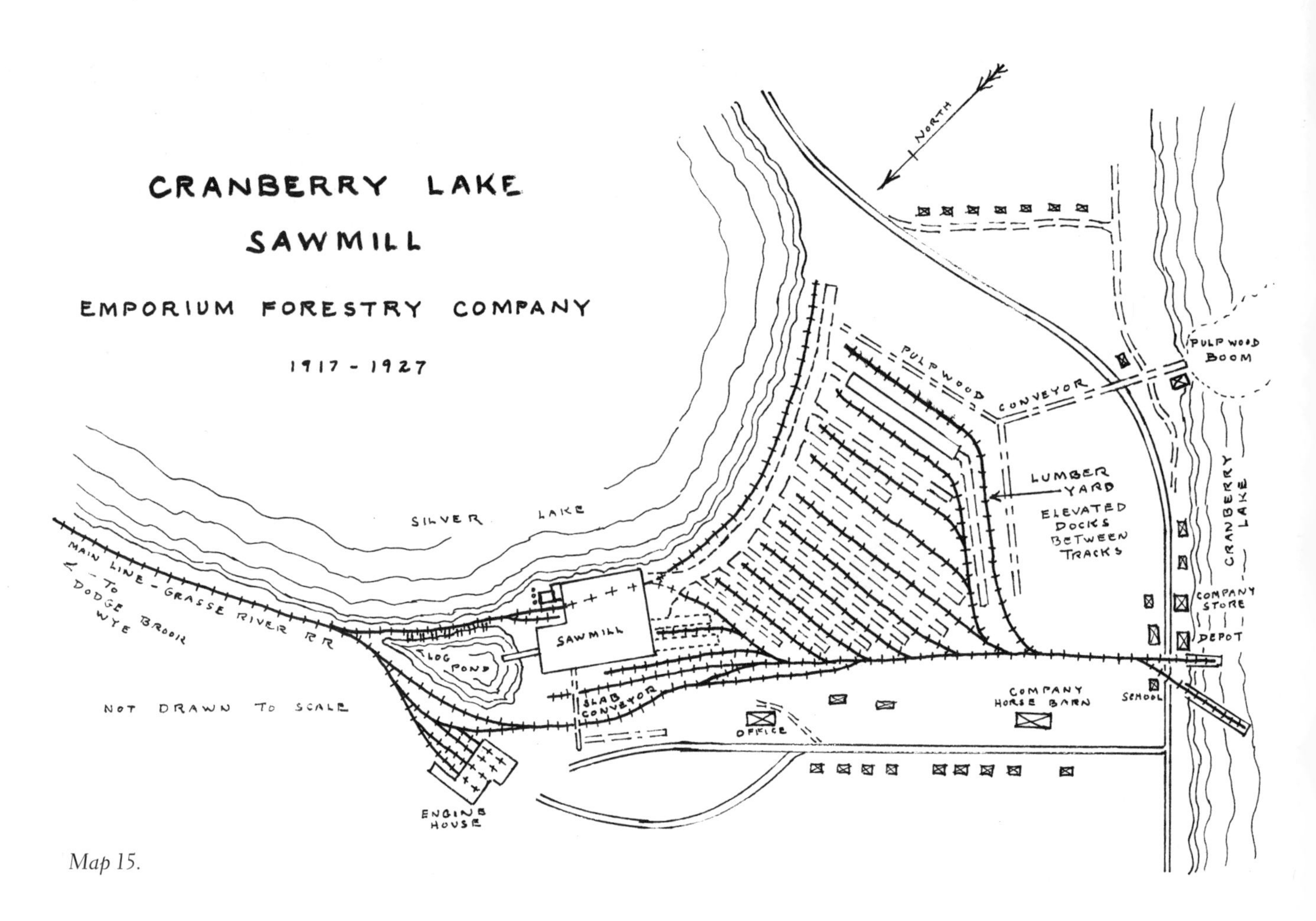

Map 15.

The first woods superintendent on the New York operation was Frank Sykes, who was followed later by his son, Chester W. Sykes. William N. Kellogg was with the company for a year or two, until 1923, when he moved away from the Adirondacks. Later woods superintendents were Victor Noelk, who came up from Pennsylvania with the Emporium, and Bill Prince.

Frank Sykes was the man who developed and patented the portable log slide described in the chapter on logging. The slide was commonly used on the Emporium operations, as well as by other Adirondack loggers.

Harry C. VanHorn, a graduate forester, was the company forester and logging engineer from 1920 until his tragic death in 1934, when dynamite caps exploded while he was yanking open a drawer that had become stuck. John W. Stock and Robert Windsor were company foresters at a later date. Cranberry Lake was the headquarters for the Woods Department from 1916 into the 1940s and was also the center of the railroad operations.

Although the Emporium sawed only hardwood and a limited amount of white pine, the spruce and fir pulpwood on many of the tracts was logged for the International Paper Company. Hemlock had only a limited market; in fact, when the North Tram was being built up to the Clare Tract, many hemlock trees were dumped into the hollows for fill.

During the early years at Conifer, the logging was done by both contract jobbers and company crews, but in the late 1920s and early 1930s, the use of company camps was discontinued. The contractor or jobber would agree on a price to harvest the wood and to place it next to the tram rails for loading by company train crews. The jobber would build his own logging camp at a site approved by the company. Local jobbers included: Ovid DeCoss, Oliva Proulx, Charles Satterlee, North Siglin, Adolphie Cusson, John Davignon, and A. M. LaPlatney of Harrisville, who had the first tractor in the woods in the 1920s.

If the softwood pulpwood was to be cut, it was handled either by the Emporium's jobber or a contractor working for

157. The Emporium shop crew was busy with repairs on Shay No. 51 in 1949. From left: Rob Mills, master mechanic Ray Zenger, and Vic Noelk, Jr. Shay No. 51 was the smaller of the two Shays that came from Jerseyfield Lumber Company in 1930. Note the bullnose coupler on the front. An automatic knuckle coupler was fastened in place with a pin, and there were four pocket drawheads for link-and-pin coupling. Photograph by Philip R. Hastings.

158. Shay No. 40 was proclaimed to be the largest and best engine that the Emporium owned among the roster of twenty-two locomotives in New York. It was operated by the company for forty years in Pennsylvania and New York before being sold to a lumber company in West Virginia. Shown above performing switching duties in the Conifer yard. From the collection of Pat McKenney.

the International Paper Company on their lands around Cranberry Lake. Contractor Harte LaFountain was cutting pulpwood for International Paper in the 1910s and 1920s and towing the wood down the lake with a boat known as *Old Ironclad.* The four-foot wood was sluiced through the dam at the foot of the lake and sent on down the Oswegatchie River. The Grasse River Railroad also delivered pulpwood to the lake at that time; it was dumped in and driven down to the Newton Falls Paper Company.

A few years later, four-foot pulpwood was being railroaded in the other direction by the Grasse River Railroad. Contractor John E. Johnston was towing wood down the lake in the early 1930s with the same towboat, jacking the wood out and loading it into open-top rack cars at Cranberry Lake. Grasse River moguls No. 60 and 68 were active in

159. Railroad top officials enjoyed a friendly fraternity. From left: Sir Henry Thornton, president of Canadian National Railways; William L. Sykes, president of the little Grasse River Railroad; and Patrick E. Crowley, president of the New York Central Railroad. Photograph taken at Syracuse University, June 1926. Courtesy of Robert Groman.

BANGOR and AROOSTOOK RAILROAD COMPANY
1931-32 No. 581
PASS MR. W. L. SYKES
PRESIDENT
GRASSE RIVER RAILROAD CORPORATION
OVER ALL LINES
UNTIL DECEMBER 31ST 1931 UNLESS OTHERWISE ORDERED AND SUBJECT TO CONDITIONS ON BACK
COUNTERSIGNED BY
PRESIDENT

SOUTHERN NEW YORK RAILWAY INCORPORATED
1932 No. A 132
PASS W. L. Sykes
ACCOUNT President-Grasse River RR Corp.
UNTIL DECEMBER 31ST UNLESS OTHERWISE ORDERED AND SUBJECT TO CONDITIONS ON BACK
VALID WHEN COUNTERSIGNED BY A. J. STRATTON OR C. W. COLLINS
COUNTERSIGNED
PRESIDENT

1932 No. A 8549
Lehigh Valley Railroad
Pass W. L. Sykes
President
Grasse River R.R.Corp.
BETWEEN ALL STATIONS UNTIL DECEMBER 31, 1932
SUBJECT TO CONDITIONS ON BACK
VALID WHEN COUNTERSIGNED BY H. R. GERMAN OR W. H. SHARP
COUNTERSIGNED
PRESIDENT

GRASSE RIVER RAILROAD CORPORATION
1931-32 № 3
PASS W. L. Sykes
ACCOUNT President.
BETWEEN ALL STATIONS
PRESIDENT

SALT LAKE & UTAH RAILROAD COMPANY
D. P. ABERCROMBIE, RECEIVER
1931 No. E 366
PASS - - Mr. W. L. Sykes - -
President,
Grasse River Railroad Corp.
BETWEEN ALL STATIONS
UNTIL DECEMBER 31, 1931, UNLESS OTHERWISE ORDERED AND SUBJECT TO CONDITIONS ON BACK
VALID WHEN COUNTERSIGNED BY MYSELF OR V. MESSINGER
COUNTERSIGNED BY
GENERAL MANAGER

BUFFALO & SUSQUEHANNA RAILROAD CORPORATION
1931 No. 711
PASS *** Mr. W. L. Sykes ***
President
ACCOUNT Grasse River Railroad Corporation
OVER ENTIRE SYSTEM
UNTIL DECEMBER 31ST, 1931, UNLESS OTHERWISE ORDERED AND SUBJECT TO CONDITIONS ON BACK
VALID WHEN COUNTERSIGNED BY MYSELF OR BY E. J. URTEL
COUNTERSIGNED BY
A. M. Darlow
VICE-PRES. & GEN'L MGR.

CLINCHFIELD RAILROAD COMPANY
"CLINCHFIELD ROUTE"
1932 No. A 1039
PASS Mr. W. L. Sykes - - - - - -
ACCOUNT President,
Grasse River Railroad Corporation.
BETWEEN ALL STATIONS
UNTIL DECEMBER 31, 1932, UNLESS OTHERWISE ORDERED OR RESTRICTED BELOW AND SUBJECT TO CONDITIONS ON BACK
VALID WHEN COUNTERSIGNED BY O. E. KISER OR E. M. VOGEL
COUNTERSIGNED
GENERAL MANAGER

ERIE RAILROAD COMPANY
1931-32 S23073
PASS Mr. W. L. Sykes ----
good for 1932
ACCOUNT President
Grasse River Railroad Corporation
OVER ALL LINES
UNTIL DECEMBER 31, 1931
PRESIDENT

160. In exchange for a free pass on his sixteen-mile Grasse River Railroad, William Sykes received free passes for most of the nation's major railroads.

161. The Jumping Goose (No. 11) was shop-built by Roy Sykes from a 1931 White bus. Rob Mills is at the controls.

162. Motor railcar No. 1, sometimes called "Will Roll Some," was a rail interurban car and was later sent to the Strasburgh Railroad.

hauling the pulpwood to Childwold for transfer to the New York Central, which then hauled it to Piercefield.

The Emporium had purchased some large timber tracts from the International Paper Company on the east and south sides of Cranberry Lake and in the 1920s decided to cut the hardwood logs in that area. A long sled haul would have been required, so the company chose instead to tow the logs down the lake on rafts. Hardwood logs do not float well, and thus rafts were made by laying four or five large spruce logs, fifty feet long, parallel in the water and fastening the front ends at intervals along another sixteen-foot log laid across the ends. The hardwood logs were then placed crossways on the raft by a Barnhart loader, which was placed on a float where it could reach the logs piled on the shore. The thirty-five-foot tugboat *Wallace* would be seen pulling about ten of these rafts up the lake from the Sucker Brook area,

163. An engine house fire in the early 1920s left little but metal scrap. A Grasse River engine and combine await at the depot (right center).

165. It was a bright winter day as the team of horses dragged a string of logs down the log slide on an Emporium Forestry Company logging operation. From the collections at College Archives, SUNY ESF.

164. The portable log slide that was built and patented by Frank Sykes, in use by Emporium. A horse would walk along the side of the slide pulling a string of attached logs.

166. Proulx's logging camp on Sucker Brook contracting for Emporium Forestry Company in 1918. The side walls of the bunk houses were made from logs fastened vertically. Photograph by E. F. McCarthy. From the collections at College Archives, SUNY ESF.

Chair Rock Flow, or possibly West Flow. One trip would carry fifty thousand board feet, enough to fill a train, loading the cars with another Barnhart parked on a trestle built out into the lake at Cranberry Lake village. This operation was one of the many jobs ably supervised by Henry Cope, an excellent woodsman who drove logs on the Connecticut River and on Pennsylvania's Susquehanna River before coming to the Adirondacks to work for Sykes. George Sykes called him the most widely experienced woodsman he had ever known.

Hardwood logs from the Edgar Tract, east of Cranberry Lake, were hauled to the Cranberry Lake mill in the 1920s with Linn tractors. The company had four of them, each capable of hauling six or seven large sleds of logs. The Linn was one of the earliest types of woods tractors, equipped with

Table 14
Emporium Forestry Company Logging Costs

Average Logging Costs by Company Crews in 1922

Product	*Work*	*Cost*	*Standard*
Sawlogs: hardwood and softwood	short skid to track side	$78	MBF
	longer haul on sleds to tram	$9–10	MBF
Pulpwood: spruce	peeled, cut 16-foot lengths, piled at tram	$6	cord
Hemlock bark	piled at tram road[a]	$4.50	cord
Estimated expense for loading and hauling to mill		$5[b]	MBF

Average Jobbing Costs in 1926–1927

Product	*Cost*	*Standard*
Hardwood sawlogs	$16.50–22.00	MBF
Softwood sawlogs	$14.50–22.00	MBF
Peeled pulpwood	$8.00–11.00	cord

Source: Emporium Forestry Company Records, Adirondack Museum, Blue Mountain Lake, N.Y.
Note: MBF: thousand board feet.
[a] An extra $.50 was charged for loading.
[b] Comprised 50 percent for load and haul expenses and 50 percent for tram road overhead.

crawler tracks on the rear and a set of runners in the front for steering. Results were excellent in deep snow. In the early 1930s the Linn tractors were hauling hardwood logs up to the Grasse River Club station, where they were loaded on the railroad.

Lumber Products and Sales

At the heart of success with this type of business, there has to be a good sales organization. The Emporium Forestry Company and its older brother, the Emporium Lumber Company, were noted for high lumber prices, usually $4 to $5 per thousand above the market, possible because the company had a reputation for supplying high grades and quality service.

Sales officers were maintained at the industrial hubs of Utica, Buffalo, New York City, and Boston, at first under the managing direction of Clyde Sykes. Roy Sykes took over as sales manager when Clyde succeeded Turner as treasurer. William Erhart, the ambitious young nephew of William S. Walker, who had managed the company's Austin, Pennsylvania, sawmill until shortly before its demise, came back after a short span to continue his service to the company. Will ran the Boston sales office from 1919 until about 1932.

Industrial and residential flooring was always a major item in the company's hardwood sales, and Erhart was instrumental in boosting this market. Large quantities of industrial hardwood flooring went to the textile mills scattered throughout New England from Rhode Island to Maine. Since the textile plants used humidifiers, the flooring was air dried to 20 percent moisture content and end matched only. The Conifer mill had a large flooring mill with nine employees that also specialized in finished kiln-dried flooring for bowling alleys, a large market at the time.

Another large-market item during this era, in addition to furniture lumber, was hard maple heel stock in thicknesses of 8/4", 9/4", and 10/4". Much maple also went into billiard cue stock, using 5/4" white or sap wood, and also into hammer boards for drop forging plants, which specified that the $2^1/_2$" maple be straight grained.

The market for piano action stock was then very active, demanding white maple (sapwood of hard maple) in 1 to $1^1/_2$" thickness for the moving parts of the piano, although the manufacturers did use red birch (heartwood of yellow birch) for the hammers and basswood for the keys. The Emporium would end dry the lumber for piano action stock, standing the lumber on end along the edge of the lumber

CONTRACTS WITH INDEPENDENT JOBBERS

Contracts with jobbers included many agreements stating which party would incur which expenses. For example:

CONTRACT WITH JOHN DAVIGNON OF TUPPER LAKE, JUNE 1930

Sawlogs, hardwood and white pine: $21.50 per MBF; to be skidded to within 75 feet of tram road
Pulpwood, spruce and fir: $10 per cord; cut, peeled, and delivered onto cars from sleds
Contractor to insure the employees
Contractor to build and conduct the camps, lumber supplied free
Job to be completed by April 1931

CONTRACT WITH A. M. LAPLATNEY, OCTOBER 1940

Sawlogs to be cut and delivered from Brody tract to banking ground along railroad track at Cranberry Lake, placed from 8' to 75' from track: $9 per MBF
If metal grab left in log, company to charge $25 per log
Company to let jobber use company log slide

Source: Emporium Forestry Company Records, courtesy of The Adirondack Museum, Blue Mountain Lake, N.Y.

167. The Barnhart loader sat on a separate float while loading logs on a raft in the foreground on Cranberry Lake. The raft was constructed of softwood logs. Circa 1916. From collection of Pat McKenney.

168. The Emporium's Barnhart loader had been loading log cars at the Cranberry Lake "store dock." The empty floats that carried the hardwood logs down the lake can be seen in the foreground. Courtesy of Jeanne Reynolds.

169. Eight sleds filled with hardwood logs are in tow behind the Emporium's Linn tractor. Circa 1920. From collection of Pat McKenney.

docks and leaning them against special pegs to minimize surface contact. At times they stood the lumber up in some pulpwood-rack railroad cars that the International Paper Company stored at Conifer. The lumber air dried much faster, of course, with all of the surfaces exposed to the air currents (discounting the opinion then held by some that the water ran down the board by gravity), and, more important, the maple would dry with a good uniform white color.

Large, clear timbers were sawn from yellow birch and occasionally maple for such uses as turnings for columns by a New York City firm. The sizes could be anywhere from 4" x 4" up to 12" x 12", but the timbers had to be clear and thus from the outer part of the logs. Sawyer Ambrose Edwards re-

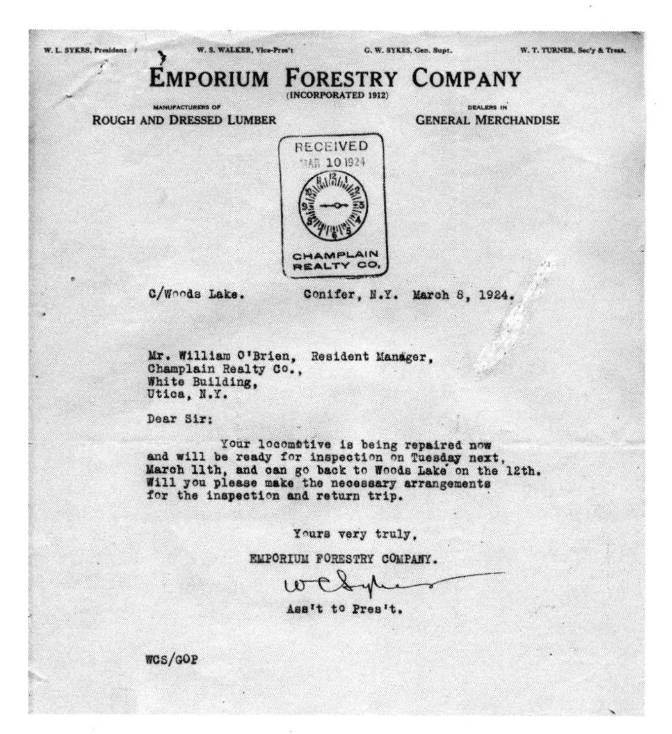

W. L. SYKES, President W. S. WALKER, Vice-Pres't G. W. SYKES, Gen. Supt. W. T. TURNER, Sec'y & Treas.

EMPORIUM FORESTRY COMPANY
(INCORPORATED 1912)

MANUFACTURERS OF ROUGH AND DRESSED LUMBER DEALERS IN GENERAL MERCHANDISE

RECEIVED
MAR 10 1924
CHAMPLAIN REALTY CO.

C/Woods Lake. Conifer, N.Y. March 8, 1924.

Mr. William O'Brien, Resident Manager,
Champlain Realty Co.,
White Building,
Utica, N.Y.

Dear Sir:

Your locomotive is being repaired now and will be ready for inspection on Tuesday next, March 11th, and can go back to Woods Lake on the 12th. Will you please make the necessary arrangements for the inspection and return trip.

Yours very truly,
EMPORIUM FORESTRY COMPANY.
W C Sykes
Ass't to Pres't.

WCS/GOP

170. The Emporium occasionally made locomotive repairs for other Adirondack railroad loggers at the Conifer shop.

171. The lumber scaler at Emporium's Cranberry Lake sawmill displays some of the top grade "white maple," standing on end for drying. Circa 1920.

called making many clear 12" x 12" pieces of birch, which shows the excellent quality of timber the Emporium had as they logged the old-growth Adirondack hardwoods.

Some pine was sawn at Conifer, and although never a major item, many patterns of finished pine were made at the planing mill there. Pine was not a major species on the Emporium timberlands, but quality was good and large quantities of thirty-six-inch-wide pine boards dropped off the band saw.

The New York State operations were never as profitable for Sykes and associates as had been the three large Pennsylvania sawmills. Railroad operations were more expensive in the Adirondacks, and the Grasse River Railroad never proved to be much of a money maker. Lumber manufacturing costs were greater, and the cost of maintaining a village the size of Conifer was stupendous, justified only by the fact that in no other way could one hundred employees become available.

George Sykes, oldest son of W. L., acted as general manager of the Emporium Forestry Company and proved to be quite capable, admitted even by those who might have disliked him. He was outspoken, but his word was good. William Sykes's original partner, William Caflisch, had been the general manager in Pennsylvania and also at the time of the move to New York. However, Caflisch died in April 1917, a death surrounded by tragic circumstances when most of his immediate family was killed in a railroad accident en route to his funeral. George Sykes stepped into his shoes.

As the depression years came on, the lumber markets subsided, and the company found itself laboring with heavy debt. The debt load had increased from $1.4 to $1.9 million between 1912 and 1926. The profitable Pennsylvania mills had run out of timber, and the Cranberry Lake mill had proven to be very costly to gear up into production.

A heroic policy of retrenchment and sale of land was therefore deemed to be the only way to achieve needed debt reduction. The costly operating sawmill at Cranberry Lake had to go; it produced its last board on April 1, 1927. From 1929 to 1945, many thousands of acres of timberland were sold with large sales to Tri-River Power Corporation and the State of New York. The last major land sale during the era of debt reduction was the one to Draper Corporation in 1945. The sale was a neat package of almost seventy-two thousand acres, which included some areas still not cut over for hardwood. Another step taken to improve the financial position of the company was the leasing of the Conifer sawmill to one of the major lumber customers, Heywood-Wakefield Company, in 1945.

Most of the accessible timber had been picked over when, in 1941, death overtook the hardwood king, William L. Sykes. Railroad logging up on the North Tram ceased about this time, and some of the rail began to be lifted. Railroad construction had been continuing up until about 1937, with a total of at least eighty miles of track laid since the onset of construction in 1911.

Thirty years of railroad logging came to an end on the Adirondack properties. No longer was the roar of steam heard across the tamarack swamps or over the knolls of birch and maple. No longer did the Emporium Forestry Company woods camps vibrate with activity during the winter months. No longer did the loggers come up Cranberry Lake in a motorboat from camps on the Edgar Tract, eager for a Saturday night at the Columbian Inn, where they would carouse all night and then sleep it off out on the grass.

Despite the objections of some of the corporation members, the Emporium split apart its assets of the manufacturing facilities from the raw material source. Ever since the wheels began to turn in 1911 at the Conifer mill, the company had supplied almost all of the logs from their own lands. Now logs had to be procured outside for delivery by trucks or via the New York Central Railroad from more distant points.

Costs were rising, and the Sykeses were losing their enthusiasm to continue. After two full generations as a hardwood dynasty, there were none in the third generation with a desire to keep alive the Emporium name. Internal disagreements clouded future direction.

The Grasse River Railroad, for many years a drain on corporation profits, had to go next, and the Adirondacks lost a source of many fond memories. The rail on the remainder of the North Tram road was taken up soon after Draper Corporation bought the property in 1945, and in 1948 the rail on the Grasse River line was taken up from Cranberry Lake village east to a point about one mile west of Conifer village. This left an imposing roadbed of precisely two miles for the Grasse River Railroad from Conifer to Childwold Station plus sidings. There were still about eight locomotives sitting around, mostly unused, but even those in operating condition didn't run steam much longer under the Emporium ownership. During the World War II years, the Conifer mill had been sawing lumber under a manufacturing agreement with Heywood-Wakefield Company, and when, in 1949, this large furniture manufacturer from Gardner, Massachusetts, offered to buy the mill, the Sykeses were unanimous in their desire to "get out of it."

In April of that year the deeds were signed, conveying all of Conifer village, including the sawmill, and the remaining two-mile vestige of the railroad, but only about seventy acres of land in vicinity of the village. After sixty-seven years of continual operation at seven different sawmill locations, the dynasty built by the hardwood king, William L. Sykes, had splintered and come into new hands. The two New York sawmills had manufactured approximately 450 million board feet of lumber during the Emporium's thirty seven and a half years of operation.

Heywood-Wakefield Ownership

Heywood-Wakefield rebuilt the sawmill into a unit much more efficient than the former Emporium mill, capable of sawing twelve million feet per year. New equipment was installed such as an eight-foot band mill, Prescott carriage, Prescott band re-saw, and a jump saw trimmer built by Paul Thomas. The outmoded overhead lumber docks were removed and a new sorting system built.

The future seemed to have a rosy glow for the inhabitants of Conifer, with mill employment going up to 125 to 150. Many of the old Emporium supervisory personnel were kept on, including Loren Silliman as superintendent, Paul Thomas as mill manager, and Victor Noelk in charge of log procurement. Roy Sykes came with them to handle lumber sales. But log costs had now risen excessively, and it wasn't "like it was" during Emporium's better times.

The Grasse River Railroad (what was left of it) had a face-lift, too. The remaining steam locomotives were either sold or scrapped, and in 1950 a new little forty-five-ton General Electric diesel locomotive was purchased. This little engine performed admirably for the next few years as carloads of lumber were taken down to the junction at Childwold Station and flat cars loaded with logs were brought back to Conifer, when the New York Central delivered them at the Childwold siding. In 1953 fifty million feet of logs and lumber moved in and out of the operating area with the aid of their rail-mounted diesel cranes. Included were 725 outbound cars of lumber.

Heywood-Wakefield Company enjoyed ownership of their little railroad more than any commercial enterprise ever has. Proudly proclaimed as the nation's smallest 100-percent dieselized railroad, anything was permitted when visiting company officials from Gardiner, Massachusetts, wanted to toy with their real live railroad. A caboose was dolled up, fancy toilet and all, and named *Perry's Pride* as a joke on company treasurer Henry Perry. The railroad's annual reports, issued by President Richard Greenwood, were a tongue-in-cheek masterpiece of proclaiming what he

172. *The Conifer sawmill during the later Heywood-Wakefield Company years. Circa 1955. Photograph by Paul S. Davis.*

173. *The Grasse River Railroad No. 1 was the only diesel locomotive used by any Adirondack lumber operator. The forty-four-ton GE-Erie B-B was purchased new in 1950 by the Heywood-Wakefield Company, successor to the Emporium Forestry Company at Conifer. In 1959 the little diesel was shipped to the East Branch and Lincoln Railroad in Lincoln, New Hampshire.*

174. *The Grasse River Railroad was proclaimed by its owner to be the nation's smallest 100-percent dieselized railroad in the 1950s. Shown here was the complete motive roster for the two-mile railroad between Childwold station and Conifer. Larry Nicklaw is the engineer moving freight to Conifer, including some logs that had arrived over the NYC. The diesel never saw any service on actual logging operations. Photograph by Paul S. Davis.*

175. *Conductor Bill Tarbox waves from Perry's Pride, the small caboose dolled up as a fancy plaything for the company brass during the Heywood-Wakefield ownership of the Grasse River line. Circa 1954.*

deemed to be the shortcomings of the Internal Revenue Service and the Washington administration in general. They are now a collector's item.

For eight and a half years all went fairly well, if one can ignore the figures in red, but in November 1957 there came the dreaded cry that Sykes heard only once—FIRE! Although only part of the buildings burned, it was the death blow for both the industry and the nation's happiest little railroad. Heywood-Wakefield later set up a circular sawmill to saw out the remaining log pile.

The village of charred remains and homes of forlorn residents was purchased the following year by Charles Ruderman for salvage activity. There was some hope by the locals that activities might begin again, but that was in vain. The ghostly mess left by the fire remained almost untouched for many years afterward to remind them of the once-better days.

The homes have now been purchased by their occupants and in many cases have been given proper maintenance. Workers traveling daily to Tupper Lake or elsewhere for the day's labor include many of the next generation who know little or nothing of the Emporium Forestry Company. What had once been a proud village of a hundred dwellings with a company employing almost three hundred workers is now an isolated little bedroom community, out of sight from travelers along Route 3.

William L. Sykes and associates have left their mark, both in the woods and in the homes of the public. Better than 1,100,000,000 board feet of hardwood passed through its saws in the Pennsylvania and New York sawmills, most of it prime timber grade. There's no doubt that a good quantity of this wood still graces the bedrooms and living rooms of homes throughout the nation. But as with all mortal men, the hardwood king is gone, and so is his Emporium.

GRASSE RIVER RAILROAD CORPORATION
GENERAL OFFICES, CONIFER, N. Y.

The President's Letter

To the Stockholders:

About six months late, but not as late as we would have liked under the circumstances, I respectfully submit the Annual Report of your Railroad for 1957. Our attempts to forego the expense and trouble of telling stockholders that their Railroad floundered in financial quicksand for the 9th straight year was thwarted by a few over-zealous stockholders who stubbornly continue to associate railroading with profit. Our past experiences certainly should not have given them any such impression!

In the light of panic, shock and disillusionment in many business circles because of the continued slow-down of orders for goods and services, it must be comforting to stockholders to note that the Grasse River officers and directors are calm, cool, collected and unconcerned. Our experience with a business that consistently loses money year after year enables us to judge present conditions objectively and to check off the 1957 accomplishment as just another fly in the ointment.

We massed a perfect score of zero for incoming carloads in 1957. This contrasts with 6 the year before and 32 in 1955. Outgoing carloads, which are always more important, were 461 in 1957 as against 594 in 1956. We cannot explain this but you will be relieved to know that an employee more loyal to management is now doing the tabulation of incoming and outgoing cars. It is much easier to get new men from time to time to count carloads than increase the actual volume.

A quick review will reveal that the overall revenue decreased 31% with an operating expense decrease of only 7%, bringing about a decrease in net income of 98%. The important symbols in this gobbledygook are the percent signs. Or to say it in the vernacular, something went haywire again.

The greatest achievement of the year was the reduced contribution to the Department of Internal Revenue. For the third successive year we have continued to trim down our contribution toward keeping government bureaucrats "gainfully" employed. Our taxes dropped from $7,340 in 1955 to $7,182 in 1956 . . . and down to a handsome $2,295 last year. Such a tax gain ought to compensate for anything else that happened last year!

Every professional annual report should mention earnings. We have just mentioned them which should classify this report as official and in keeping with accepted accounting practices.

The downward "ski-jump" pattern of operations, illustrated on the front cover of this report, should be apparent at a glance. In no instance, during 1957, has any figure gone up. Total revenues, operating expenses, Federal income tax, operation ratio, net income, earnings per share and deficit per share have dropped to new lows. Some of these lower figures work to our advantage; others, like earnings and dividends, are sufficiently unimportant to have any true significance.

The elaborate and costly efforts made in 1955 in re-conditioning our passenger seating facilities (at the request of the ICC) have paid off handsomely. Passenger revenue in 1955 was at a new low. After the coach seats were cleaned and repaired, the 1956 revenue increased almost 4 times. And in 1957, passenger revenue went up another 33⅓%! Translating these complicated percentages into dollars and cents, the 1955 passenger traffic income was 60¢; in 1956, it was $1.50 (excluding a conscience contribution of $1.00 from Canada) and in 1957, paying passengers dropped $2.00 into the coin box! Such an experience with passenger traffic makes us wonder, quite frankly, why other railroads cannot increase passenger revenue as the Grasse River Railroad has done so consistently for three years.

Our experience in operating your property in a progressively declining pattern steels us against such newspaper headlines like: "President Asks Congress to Help Railroads." To railroad men inexperienced in unprofitable year-end results, such headlines ignite, perhaps, a spark of hope. We at Grasse River, however, already know how to lose money regularly without getting into astronomical figures that resemble the national debt or the unbalanced budget.

Nor have the rumbles about mergers caused us to perk up our ears to television antennae proportions. In all fairness to the many stockholders with huge investments in this link in America's vast transportation network, it must be reported that we have not been approached by adjacent railroads regarding our possible interest in a merger. Should such a proposition present itself, your capable management will exercise all of its many "horse-trading" skills, many of which haven't been practiced for years.

There is no substitute for good management except, perhaps, an inexhaustible supply of money. With no printing plants or U. S. Mints located on our right of way, we have had to rely wholly on railroading know-how and a fair share of luck. In both these categories we look for considerable improvement. In the meantime, your management is leaving no stone unturned in its efforts to give you the same kind of progressive, efficient and cool-headed leadership that has long showered your Railroad with managerial distinction.

[illegible signature]
President

176. *The Heywood-Wakefield Company made great sport out of their ownership of the Grasse River Railroad, as evidenced by the cover of the 1957 annual report and the tongue-in-cheek letter to the stockholders.*

C H A P T E R   E I G H T E E N

Paul Smith's Electric Railway

Paul Smith came to New York from Vermont in 1850 with the shrewd business ability that was to shape him as an Adirondack legend. While employed on the Erie Canal boats he often indulged in his favorite pastime of hunting and eventually guided in the unbroken wilderness at Loon Lake. In 1852 he applied his business acumen to the opening and running of Hunter's Home, an inn for men only, near Loon Lake. A barrel of whiskey was provided in the living room with a community dipper attached; a drink cost four cents.

Regaled by Paul Smith's storytelling and his ability as a woodsman, the professional men making up his guest list implored Smith to build a finer hotel. Among other desired comforts, they wanted their wives to be able to meet the unusual Paul Smith.

Thus in 1858 Paul Smith moved down to Lower Saint Regis Lake and began what was to be one of the most renowned Adirondack hotels. His hotel enterprise lasted for seventy-two years and was never known as anything but "Paul Smith's." As the facilities grew, eventually encompassing five hundred rooms, the guest list grew to include a fashionable clientele. New York City papers included regular social-note columns from Paul Smith's.

The travel arrangements to Paul Smith's were tedious in those early years. Even after the Delaware and Hudson Railroad completed a branch to Point of Rocks at Ausable Forks in 1868, it was still a bumpy forty-two-mile journey to Paul Smith's by horse and buggy or by stage. But before the century was out, other railroads were to make entry into the northern Adirondacks. In 1886 two different railroads were built that considerably shortened the stage ride to two locations for Paul Smith's guests, some of whom were definitely not accustomed to long, rough stage rides.

In the 1880s the Chateaugay Railroad was building its narrow-gauge line from Plattsburgh south to Saranac Lake. Loon Lake was reached in November 1886, and from there the Paul Smith stage had only a fourteen-mile journey. Then the distance was cut in half a year later when the Chateaugay

Table 15
Paul Smith's Electric Railway

Dates	1906–c. 1932
Product hauled	hotel guests, softwood logs, pulpwood
Length	6.5 miles
Gauge	standard

Motive Power

Builder	*Type*	*Builder No.*	*Date Built*	*Source*
Alco-Cooke	0-6-0	25241	1902	ex-NYC #3723
NYC, W. Albany shop	4-4-0		1886	ex-NYC shop
J. G. Brill	electric combine car		1907	new, scrapped 1984

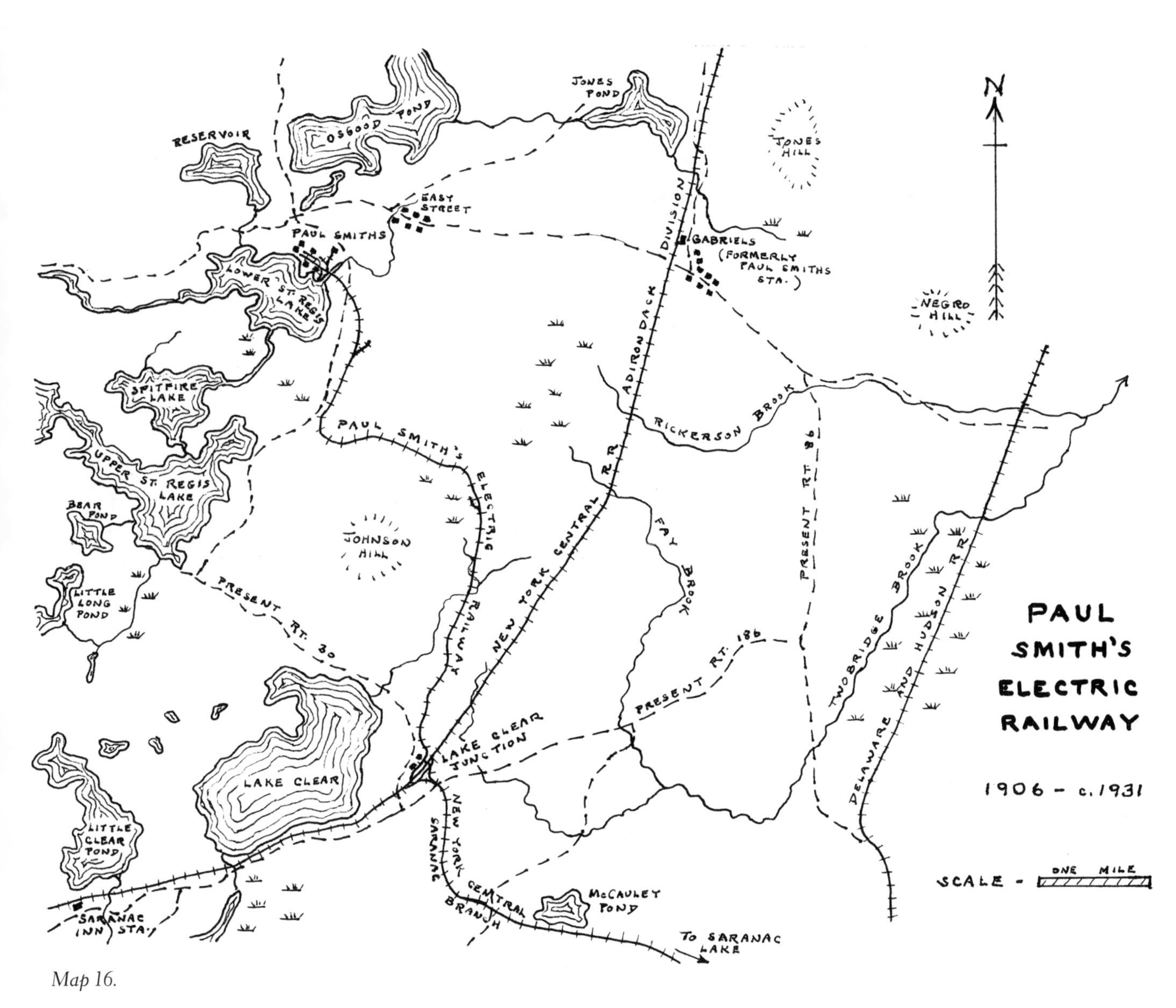

Map 16.

was completed to Saranac Lake and a depot established at Bloomingdale, referred to then as Paul Smith's Station.

At the same time another railroad pierced the Adirondack wilderness. John Hurd's Northern Adirondack Railroad reached Brandon in 1886, a point eight miles west of Paul Smith's Hotel in the opposite direction from Bloomingdale. This depot also became known as Paul Smith's Station.

It was not many years before a third railroad came through to open the north country even more for hotel tourism. W. Seward Webb's Mohawk and Malone Railroad was built southward from Malone in 1892, and a station was established. Also known initially as Paul Smith's Station, though later changed to Gabriels, this depot was only four miles from the famous and expensive hotel of Paul Smith's.

At that point, however, with three railroads, each with a depot known as Paul Smith's Station, the transportation of guests by the hotel became somewhat complex. The hotel stage had to meet fourteen different trains each day on the three railroads, requiring the constant use of two or three stages. Paul Smith's solution to simplifying this complication was to build his own railroad to connect with the most likely depot, the Mohawk and Malone (New York Central in 1893).

Paul Smith's Railroad

It was not really a logging railroad; this was certainly not its purpose for existence. But since Paul Smith's railroad did haul logs and pulpwood during its unique existence, especially in the early years, it is included here.

Surveys were completed in 1904 by M. J. Corbett. The logical location to connect with a mainline railroad would have been a four-mile connection with the New York Central at Gabriels, but the grade up Easy Street Hill was too much. Thus a 6.5-mile line was built over relatively flat and swampy terrain to the next depot traveling southerly on the New York Central—Lake Clear Junction. Stage connections at the other depots, Bloomingdale and Brandon, were also broken. From that point on, guests en route to Paul Smith's would arrive only on the New York Central and Hudson River Railroad and depart at Lake Clear.

Construction of the Paul Smith railroad was completed by 1906; regular service began in August of that year. The line was laid with sixty-pound rail, somewhat lighter than that used on the New York Central. Cost of the railroad, including rolling stock, has been reported at $75,000. Five round trips were made daily to connect with the New York Central passenger runs. Travel time from Paul Smith's to Lake Clear Junction was thirty minutes.

As Paul Smith's electric railroad was being built in 1905, there arose public interest and planning to construct an extensive network of electric railroads from Lake Champlain into the heart of the Adirondacks. The Ticonderoga Union Terminal Railroad Company was formed in 1905 to organize construction from Ticonderoga over to a connection with Paul Smith's new electric railroad. A principal freight item, it was proclaimed, would be large quantities of pulpwood in transit to the Ticonderoga paper mill. The construction, which was announced to begin in September 1905, never happened.

Motive power on Paul Smith's railroad was initially two former NYC steam locomotives, used primarily in construction of the line. But Paul Smith had other plans from the very beginning. He had already been generating electric power for the hotel at his nearby Keese's Mills plant since 1898 and thus wanted to electrify his new railroad. Only eleven days after the new railroad opened on August 20, 1906, inquiry was made at General Electric for advice concerning a trolley line. Soon after, Paul Smith ordered an electric interurban combine car from the J. G. Brill company of Philadelphia. Both the motor and the electrical equipment were supplied by General Electric. Cost of the car was $25,000.

When the new electric car was delivered, probably in 1907, it exhibited some serious malfunctions that required one or more trips back to the General Electric factory in Schenectady. The car was said to have been an experimental first of its kind.

The car drew 5200 volts AC from the overhead wire and converted this internally to 600 volts DC. This power current in turn powered the electric motors mounted on the trucks. The malfunction occurred in the conversion to direct current. Initially a mercury air rectifier was employed to make the conversion, but operations fell short of expectation. General Electric replaced this rectifier with a motor generator in about 1908, and it served well for the remaining twenty-two years of the car's operation.

177. The photographer captured all three waiting at the Lake Clear Junction depot. From left: Saranac Lake Local with engine No. 942, the New York Central Montreal Express with engine No. 2004, and Paul Smith's Electric Railroad combine. Circa 1910. Courtesy of The Adirondack Museum.

178. Paul Smith's electric combine was built by the J. G. Brill Company in 1907. Courtesy of Paul Smith's College.

The car was equipped, when new, with four trolley poles, one pole on each end and one on each side. However, the power line was strung over along the side of the track, said to be the only electric operation in the state to do so. And as the car was never turned around, the pole on the off side was eventually removed.

Paul Smith was renowned for his shrewdness and ability to tactfully extract the maximum from his guests with high prices. There was reportedly an occasion when the train was actually held up and the passengers robbed on the way from the hotel. When told about it, Smith supposedly laughed and replied "fool of a highway man, holding up passengers after they've left here. What did he expect to find on them?"

Fare remained fifty cents one way for the full six-and-a-half-mile trip. Travel time was twenty-five to thirty minutes at the average speed of thirteen miles per hour. Speed had to be kept down because of the rough roadbed, not because of lack of power. Private railroad cars of the wealthy who stayed at Paul Smith's or who had camps nearby were brought in and put to rest on one of the two sidings that ended at the hotel store.

One of the wealthy patrons who customarily had her private car towed in was informed that in the event of a train wreck the last car often received the most damage. With that warning in mind she purchased an old baggage car, which she arranged to have attached behind her private car for the six-mile haul to Paul Smith's.

179. Pine and spruce logs decked along Paul Smith's Railroad. Courtesy of Paul Smith's College.

180. It's wintertime and the hotel is closed, the ideal time to haul logs with the electric car. Courtesy of Paul Smith's College.

Logging Activities

There was not a large area of timber that was close enough to be serviced by Paul Smith's railroad, but a few short log-loading spurs were built in the vicinity of Johnson Hill. The maximum spur capacity was seven log cars.

A longer spur was built out to a point in the lake near the hotel, and in the winter tracks were actually laid out on the ice. Log cars were backed out onto the ice and the logs unloaded there to await the spring thaw. In the spring they were boomed together and floated to a sawmill at another location on the lake.

The sawmill existed on the bay between Kellogg Point and the mouth of the Barnum Pond outlet. It's been reported that a plan was proposed to extend the track over to the mill; it was never built. Neither was the proposed wye ever built to turn the electric car around.

Pulpwood was also taken out by rail in the early years of operation. With the limited area for timber harvest and the primary focus on passenger transport, one can assume that there was never a large volume of log movement on Paul Smith's unique railroad.

The railroad ceased to operate about 1930, possibly as late as 1932. The connection to the New York Central was broken in 1929. Any traffic over the line after that probably used the small gasoline-powered jitney, which carried freight and mail only. The electric car remained at the location after abandonment, slowly deteriorating for many years on the Paul Smith's main line until it finally was demolished in 1984.

Paul Smith died in 1912, after which his two sons carried on the business. But times were changing by the time the hotel burned down in 1930. No effort was attempted after that to carry on the grand showmanship of one of the Adirondacks' greatest showmen. The entire estate was used to found Paul Smith's College, specializing in hotel management and forestry.

CHAPTER NINETEEN

Kinsley Lumber Company Railroad

NOT MUCH IS KNOWN about the small railroad operation built by the Kinsley Lumber Company in the Town of Franklin. The company had been formed in about 1896 or possibly earlier by partners Frank G. Smith, Arthur G. Leonard, and John O'Rourke; Smith served as president.

A sawmill and office were built in Onchiota, a small village at the north end of Rainbow Lake and situated in the narrow space between the New York Central and the Delaware and Hudson Railroads. Large tracts of timberland were obtained in the area of Debar Mountain, north of the Loon Lake mountains. Much of the land was purchased from the Patton Company and also from Henry Patton.

Soon after the company was incorporated, a logging railroad was built westerly from the New York Central Railroad into the Debar Mountain area and thence northerly to Debar Pond, about seven miles in total length. The railroad junctioned with the NYC at a point one and a half miles north of Inman (or Loon Lake Station). About a mile and a half westerly from the NYC, the railroad passed through a small charcoal kiln settlement known as Tekene. The right-of-way extended about three miles from the NYC before it entered into the company's timber tract.

Nothing is known about the Kinsley Lumber Company locomotive other than it was a small saddletank with the number 2 painted on it. Very little information has been passed down on this operation.

Table 16
Kinsley Lumber Company Railroad

Dates	c. 1896–1905
Product hauled	sawlogs
Length	7 miles
Gauge	standard, portion rebuilt narrow at one point in time
Motive power	saddletank locomotive, identification unknown

The railroad had only a short period of operation under the Kinsley Lumber Company ownership; the Onchiota sawmill was reportedly destroyed by fire after a brief existence. In August 1899 the company sold the twelve-thousand-acre Debar Mountain timberland for $61,718 to a man who had been purchasing large amounts of Adirondack property, William Rockefeller. Included in the sale was one logging camp outfit, all of the logs that had been cut, the railroad equipment, and locomotive No. 2. Also included were the rails and ties on the first three miles of right-of-way before reaching the conveyed land.

Upon the release of news that William Rockefeller had purchased the Debar Mountain tract, his Malone attorney, John P. Kellas, was deluged with letters from other landowners also wishing to sell their Adirondack property. One party who owned a few acres adjoining Rockefeller's "preserve" in Santa Clara added that if Rockefeller did not purchase his property, he expected to open a saloon in close proximity to the Rockefeller land. The stratagem failed; no other purchases were made.

Rockefeller had purchased the Debar Mountain tract strictly as a business investment, and plans for continued timber harvest and railroad operation were put into effect. In the winter of 1899–1900 about three hundred men were put to work in the logging camps on the property. Fifteen thousands cords of pulpwood were cut and yarded that winter, ready for rail transport. In late January 1900 manager N. W. Tarbell brought the locomotive over to the Malone railroad shops for a complete overhaul. A new paint job was applied and the locomotive lettered with the name of Rockefeller's son, William G. Rockefeller, on the side. Charles W. Flana-

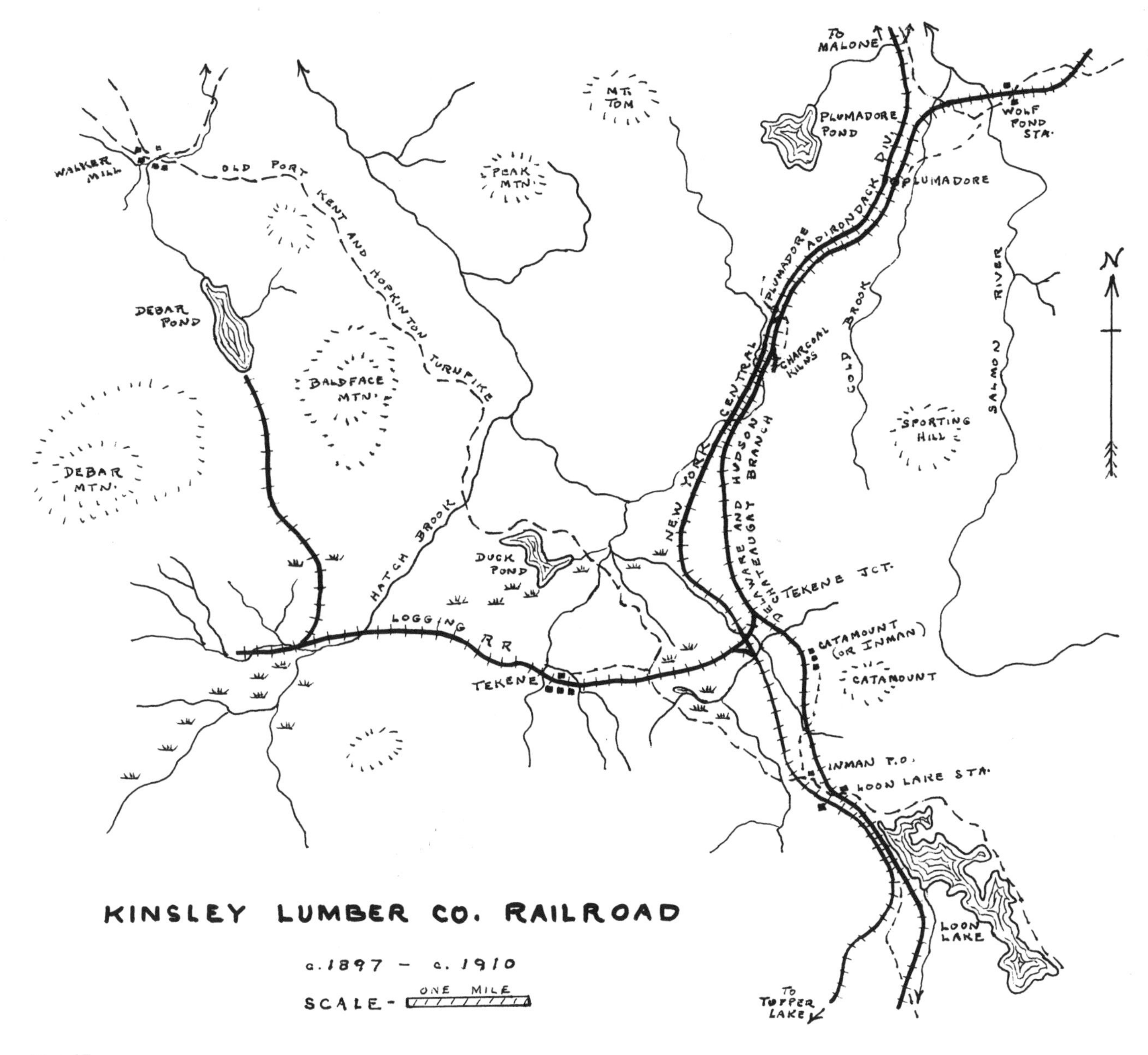

Map 17.

gan was engaged as engineer. A contract was made with the Brooklyn Cooperage Company to accept the hardwood logs at the Tupper Lake mill, to be railroaded there on the New York Central. However, there was a roadblock.

The three-mile stretch over which the Kinsley railroad passed between the New York Central and the Kinsley timberland was owned by the Chateaugay Ore and Iron Company, which was controlled by the Delaware and Hudson Railroad. Shortly before 1903, the D&H had narrowed the track on the Kinsley road to a three-foot gauge, at least as far out as the charcoal kiln settlement of Tekene. A high overpass was built in 1903 over the New York Central tracks in order for the Tekene line to reach the Chateaugay Branch of the D&H on the east side of the New York Central. Until 1903 the Chateaugay Branch itself had been narrow gauge.

But this presented a problem for William Rockefeller's loggers. The track further in on the Rockefeller property was still standard gauge, as well as his railroad equipment. And he wanted to run his log cars onto the New York Central without reloading, of course. The rails and ties on the right-of-way belonged to Rockefeller but the land belonged to the D&H. A confrontation was brewing.

On Friday, September 17, 1903, a gang of laborers in the employ of William Rockefeller moved in to widen the rails on the logging line to standard width. Before that night was over, word of the action reached D&H officials at Plattsburgh; they immediately moved into action. At one o'clock in the morning, a train pulled out of Plattsburgh with D&H attorney Thomas B. Cotter on board, headed for Port Henry. The fifty-five mile run was made in a little more than an hour.

Upon arrival at Port Henry, Cotter awakened Judge Chester B. McLaughlin of the Appellate Division of New York City, who spent summers at Port Henry, and they were able to procure an injunction that restrained Rockefeller from altering the track gauge. Another special train was on hand that then sped on to Tekene junction with the Delaware and Hudson party of Attorney Thomas Cotter, chief engineer James McMartin, and about eighty section men.

Upon arrival, Judge McLaughlin's order was served upon Rockefeller's agent, who then had no choice but to desist. The large section crew, aided by another gang of one hundred men from Saranac Lake, immediately went to work; by the end of that afternoon they had restored all of the track to narrow gauge.

The reason behind Delaware and Hudson's opposition to standard-gauging is confusing and lost in history. In 1902–3 the D&H standard-gauged the entire length of the former three-foot-gauge Chateaugay Railroad, from Plattsburgh to Lake Placid. That task was completed in September 1903, the same month the D&H opposed standard-gauging the logging line at Tekene junction.

Be that as it may, the Delaware and Hudson had what they considered good reason for retaining control of the logging right-of-way over their land and may have even later obtained an option for the right-of-way on the Rockefeller property. They had covetous eyes for westward expansion.

John Hurd's Northern Adirondack Railroad from Moira south to Tupper Lake had gone bankrupt in 1895 and had been operating since then as the New York and Ottawa Railroad. But the NY&O was on a shaky financial foundation, and both the New York Central and the Delaware and Hudson were vying for control. The D&H had ideas of extending Kinsley's logging railroad beyond Debar Pond, eventually to reach westward to St. Regis Falls for a connection with the former Northern Adirondack and the markets in the western Adirondacks. But they lost out; the New York Central took control of the former Northern Adirondack in 1906.

The Brooklyn Cooperage Company harvested the remaining timber on the Rockefeller tract, probably using their own railroad equipment. They purchased the right-of-way over the existing roadbed, although it is not known how the problem of track gauge was settled. The track was pulled up in about 1917, and the William Rockefeller heirs sold the property in 1923 to Reynolds Brothers of Malone, although Brooklyn Cooperage still owned the right-of-way.

In 1926 the Brooklyn Cooperage Company gave a five-year lease to the Gould Paper Company and John Johnston to use the entire seven-mile roadbed as a log truck road for logging the property beyond Debar Mountain.

SECTION TWO

Logging Railroads That Connected with the Northern Adirondack Railroad

THIS FIFTY-TWO MILE RAILROAD, originally the Northern Adirondack Railroad (later New York Central and Hudson River Railroad, Ottawa Division), became the trunk line from which five different companies built logging railroad operations.

CHAPTER TWENTY

NORTHERN ADIRONDACK RAILROAD

JOHN HURD, the visionary builder of railroads and creator of mills and villages where none existed, the shrewd and reckless speculator of timberlands, left a big mark in the northern Adirondacks. It was back in 1882 that "Uncle John Hurd," a Connecticut industrialist and son-in-law of P. T. Barnum, made his appearance in northern New York and basked in a few short years of glory. He built the second railroad to penetrate beyond the then existing Blue Line into the Adirondacks and built what might be called the first logging railroad in the Adirondacks.

In the early 1880s, lumber speculators from afar, especially Michigan, began to invest in Adirondack timberlands, most of which were far removed from any network of transportation as it then existed. The locals considered the purchase of thirty-one thousand acres in the towns of Brandon and Brighton by John Torrent and Patrick Ducey of

Table 17
Northern Adirondack Railroad

Dates	1883–1895, Northern Adirondack Railroad 1895–1904, New York & Ottawa Railroad 1904–1937, New York Central & Hudson River Railroad, Ottawa Division
Product hauled	softwood logs, lumber, misc. forest products, passengers
Length	52 miles
Gauge	standard

Motive Power[a]

No.	*Type*	*Builder*	*Builder. No.*	*Year Built*	*Name*	*Source*	*Disposition*
1	4-4-0	Schenectady	1648	1882		NYCHR no. 156	NY&0
2	4-4-0	Schenectady	1431	1881		NYCHR no. 148	NY&0
3	4-4-0	Rhode Island	1373	1884	Santa Clara	new	NY&0
4	4-4-0	Rhode Island	1308	1884	Black Mountain	new	NY&0
5	4-4-0	Rhode Island	1578	1885	Tupper Lake	new	NY&0
6	4-4-0	Rhode Island	1378	1884	Bay Pond	new	NY&0
7	4-6-0	Rhode Island	2780	1892	Mountaineer	new	NY&0

Rolling Stock (as of 1889)
3 passenger cars
4 baggage, mail, and express cars
64 boxcars
100 platform (log) cars

Note: In 1889, all equipment was leased except the first two locomotives and forty-five log cars.
[a]Applies only to John Hurd's Northern Adirondack Railroad.

182. The second St. Regis Falls depot, built in 1918, served until the termination of the railroad in 1937. The building then served as a town hall until it was torn down in 1964. Courtesy of the St. Regis Falls Historical Society.

Muskegon, Michigan, to be absurd; it was very impractical to think that a railroad would ever be built to that remote site. This was followed in 1882 by the purchase of sixty thousand acres of land in Townships 10, 11, 14, and 17 by John Hurd, allied with Charles Hotchkiss, also of Bridgeport, Connecticut, and lumberman Peter MacFarlane of Michigan. Included in their purchases was an old water-powered sawmill in St. Regis Falls, which they converted to steam power, capable eventually of sawing sixty thousand board feet in an eleven-hour day. By the fall of 1883, Hurd, Hotchkiss, and MacFarlane were in the lumber manufacturing business in a large way and had logging camps in operation.

John Hurd knew, of course, that the success of his new lumber venture would depend on a railroad connection; thus, within a year he began construction of a railroad between Moira and St. Regis Falls. Moira was a station on the Ogdensburg and Lake Champlain Railroad, later part of the Rutland Railroad. Building under the rights of a charter he had obtained entitled the Northern Adirondack Extension Company, he had rails completed over the twelve miles to St. Regis Falls by 1883, providing an outlet for the lumber then being produced at their mill in that village. Within a short time, the population of St. Regis Falls village increased fivefold, and real estate values quadrupled.

John Hurd, however, was entertaining great visions of an expanded railroad and forest-products empire. Therefore, with little delay he had his crew pushing southward into the large expanse of isolated timber holdings. Hofer Meakham was in charge of the survey crew, succeeded later by Will LaFountain. Construction supervisor M. Callahan of Canaan, Connecticut, had a hardy crew of Italian laborers toiling under the watchful eyes of construction bosses Frank and Ed Berkeley. Pay for the laborers was $1.25 per day less a charge of $6 or $7 per month for board and room.

The construction crew was laying down the fifty-six-pound rail at a rate of about five to six hundred feet per day with minimum concern about fine grading or the elimination of curves. One of the first locomotive engineers in the line, William "Old Bill" Cudhea, once recalled how a particular trouble spot was treated. At Bay Pond the railroad was laid over a narrow neck of the pond that linked the two principal bays. It was only a short stretch across the neck of water, but the construction crew was bogged down there much of the summer. Hundreds of large pine trees, some four feet through, were dumped into the swampy area and immediately sank out of sight. Eventually, a track was put down and a locomotive gingerly taken across. The tracks sank behind them and, as Cudhea recalled, they were sure the engine and everything else were going under. It didn't happen, but the lake crossing was an expensive one, reportedly costing as much as $200,000 to make the fill.

About six miles beyond St. Regis Falls, the new railroad reached a small settlement of crude dwellings then known as Humphreys. Hurd and partners were soon to expand the site

181. A gathering of nicely attired patrons awaited the arrival of the southbound passenger train at the original St. Regis Falls depot. The date was after the New York Central had taken over Hurd's Northern Adirondack in 1904. The locomotive was No. 789, the Tupper Lake, a 4-4-0 originally purchased by John Hurd in 1885.

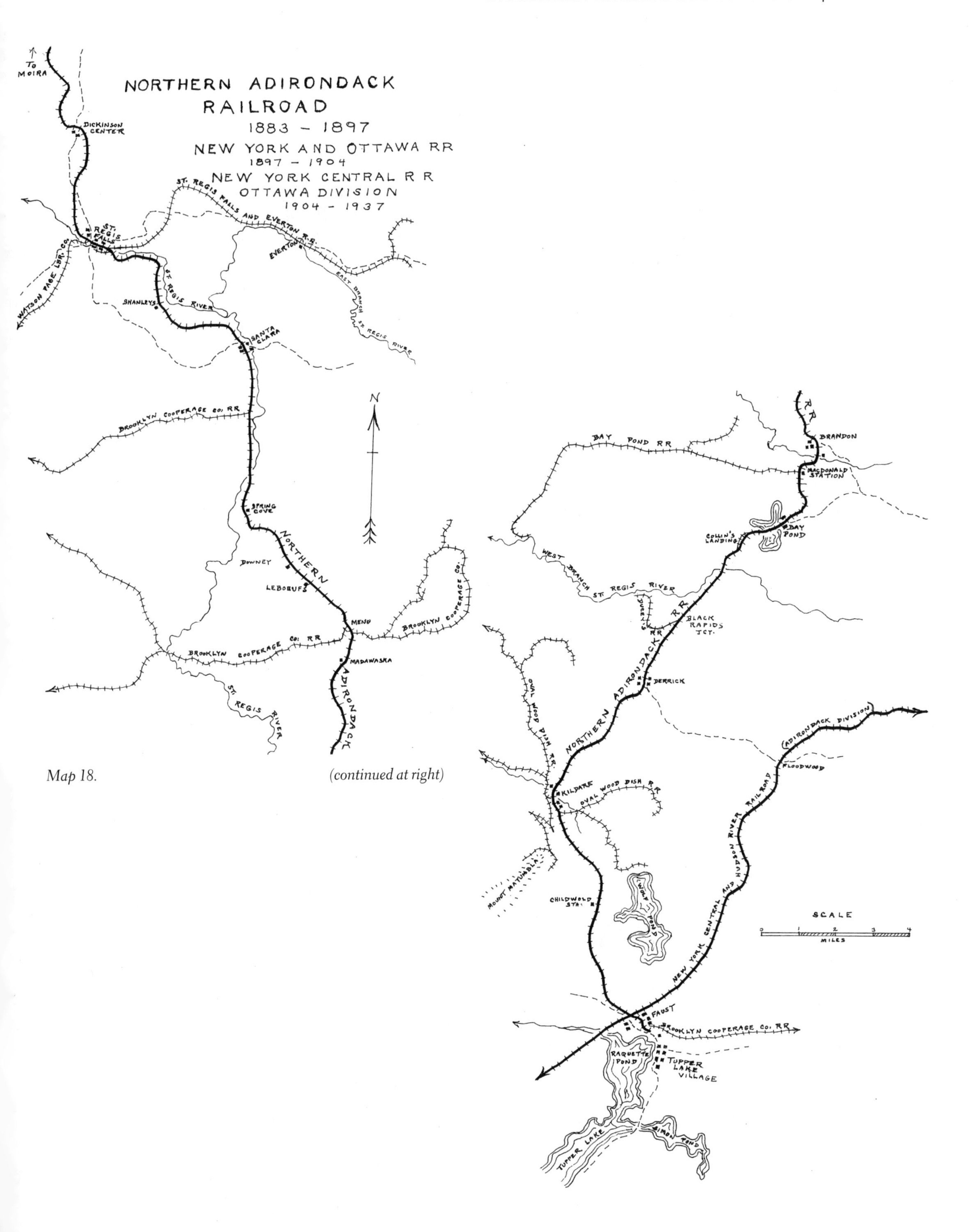

Map 18.

(continued at right)

183. The track-laying crew of the New York and Ottawa Railroad displays all the tools needed for new construction. Courtesy of the St. Regis Falls Historical Society.

184. Locomotive No. 7 of the Northern Adirondack Railroad was a thirty-ton 4–6–0 built by Rhode Island Locomotive Works in 1892. It was named the Mountaineer.

into a thriving location for lumber manufacture as well as railroading headquarters. The village became known as Santa Clara, which, according to some sources, was named by Hurd in honor of his wife. Other records, however, show that Mrs. Hurd's name was Amy, not Clara. Before long, rails were laid to a camp known as LeBoeufs, probably named for Alfred LeBoeuf, who came from Michigan with MacFarlane to work as woods superintendent on the St. Regis River. It was at LeBoeufs that Hurd's railroad crossed over the Blue Line into the Adirondack Park.

By 1884 Hurd's crew had progressed twenty miles beyond St. Regis Falls. John Hurd was not an easy man to work

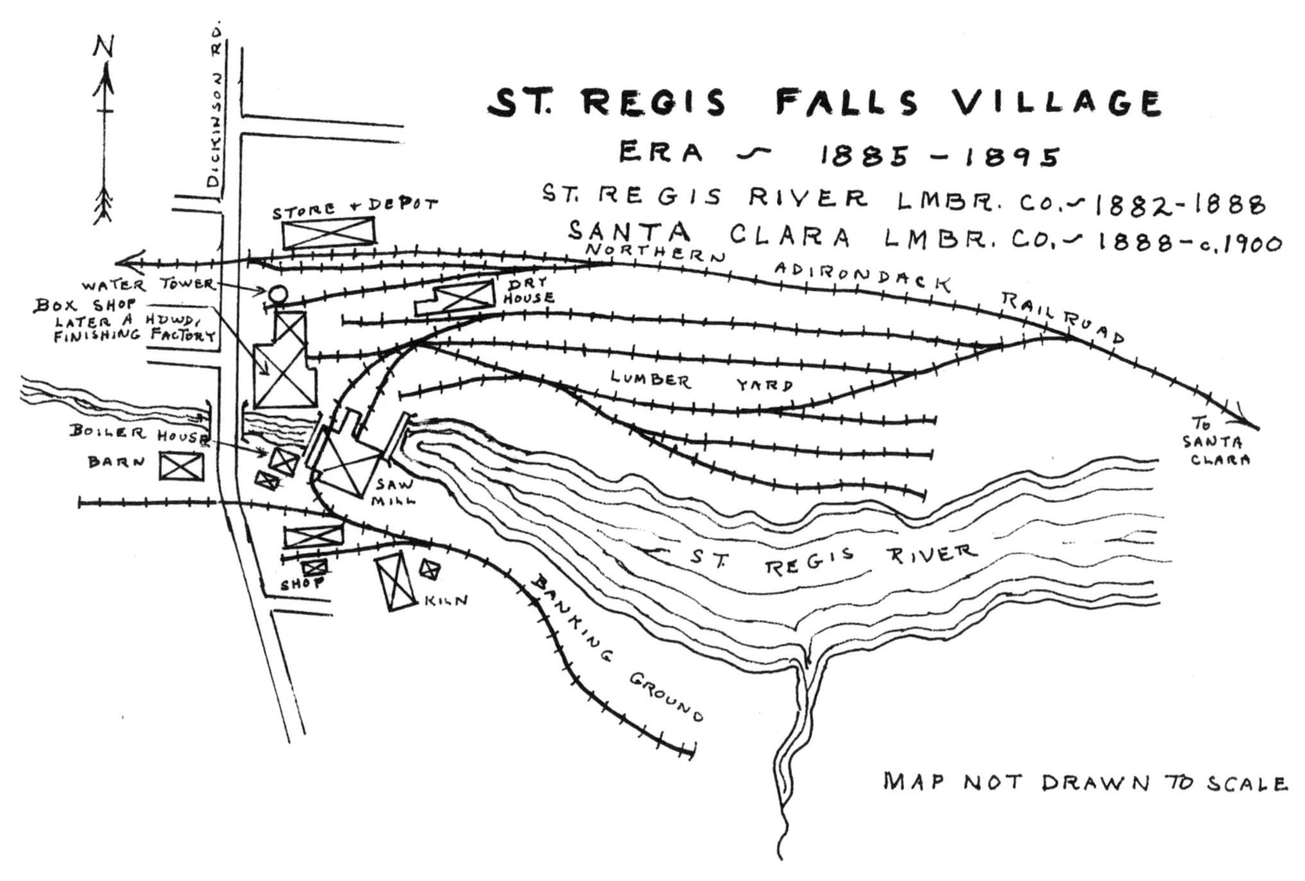

Map 19.

with as a business associate, and Peter MacFarlane eventually had enough of Hurd. He sold out his interest in 1884 in order to form a partnership that was to establish a lumbering enterprise in nearby Everton. Charles Hotchkiss would also eventually part company with John Hurd, leaving Hurd alone to enforce his will upon whatever lay before him. It would prove to be a burden, both financial and worrisome, that would eventually be the downfall of John Hurd. But his efforts were also to leave some permanent changes in the maps of the Adirondacks.

186. Santa Clara during its healthier years had a fine depot for the New York and Ottawa Railroad. The Windsor Hotel can be seen on the left. Circa 1900.

Santa Clara Village

The principal freight support for "Hurd's road," as it was popularly called, was lumber and logs, plus limited passenger service. Therefore, the next bold move for the firm of Hurd and Hotchkiss was the establishment of a sizable sawmill village in 1886 at Hurd's chosen location, Santa Clara. From a meager settlement of a few shacks, the village soon grew to a population of two thousand, with a railroad depot, boardinghouses, machine shop, store, general office, engine house, and over thirty dwellings, which Hurd rented to his employees for $3.50 to $4.50 per month. The new Santa Clara House opened in 1889 with a capacity of seventy-five guests paying $2.00 per day or $5 to $14 per week. Company employment vacillated between three hundred and five hundred.

The activity that put Santa Clara on the map was the two sawmills operated there by Hurd and Hotchkiss under the name of St. Regis River Lumber Company. In addition to the small mill sawing twelve thousand board feet daily, the firm built a much larger one capable of sawing one hundred thousand board feet in an eleven-hour day. Records show that the mill sawed out twenty-five million board feet of softwood lumber in 1887; the smaller sawmill concentrated on hardwood lumber. Production at that rate meant a huge quantity of logs to move in and final product to move out on "Hurd's road." The river was impounded for nine miles back from the mill for storage of softwood logs. The lumber sawn was about one-third white pine, a little birch and maple, and the remainder spruce.

185. The Northern Adirondack Railroad engine house at Santa Clara village. Circa 1890.

The large sawmill at Santa Clara was of a magnitude never seen before in that part of the country. As the logs entered the mill, they passed through a pair of circular saws that slabbed them on opposite sides. The log cant then flopped over onto a flattened side and made a one-way pass through one of the gang saws. Each gang saw had from twenty-six to thirty-two saws. Oversized logs were sent to the other side of the mill, where they were broken down by a standard log carriage and large circular saw. The carriage, which carried the log back and forth as it was conveyed through the saw, was a wide platform with two men riding on it. Their function was to set the dogs that firmly held the log and to advance the log for the next cut through the saw. They had a wild ride in the process. It was said that it took about fifteen minutes from the time the log entered the mill until the sawn lumber was

187. The tail end of the interior of a Santa Clara Lumber Company sawmill shows spruce lumber moving out on transfer chains. In the center rear is a sash gang saw, the edger is in the left foreground. Courtesy of The Adirondack Museum.

piled on the lumber stacks in the yard. "Mill run" spruce lumber had an FOB value of $10.25 per thousand board feet and the box grade of spruce $7. Log costs averaged $4.40 delivered to the mill pond or $3.26 yarded railside at Madawaska, a logging camp ten miles south of Santa Clara on the railroad.

Although John Hurd was an aggressive and blustering driver of his hired help on work days, he was just as aggressively religious on Sundays. For a period of time, he operated a "Sunday School and Church Train" over a part of his railroad as well as serving as a "lay reader" at the meeting house. His religious compassion did not extend throughout the working week, however. In June 1891 all twenty-five of the workers at the car shop in Santa Clara walked out one Thursday because of a lack of a regular payday. Hurd gave them their pay to date and discharged them all, although most eventually had little choice but to return.

Railroad construction of Hurd's road continued southerly, reaching Brandon in 1886. It was at Brandon that Pat Ducey had established another sawmill village in the wilderness, a mill reputedly capable eventually of sawing 125,000 board feet daily of pine and spruce cut on the sandy plains and low ridges bordering the St. Regis River. Historians tell us that Ducey and his contemporaries were credited with introducing into this region of New York the practice of felling the standing trees with cross-cut saws. Until this time, it had been the practice of loggers to fell the trees with ax only. The settlement of Brandon was to add more lumber freight for Hurd's Northern Adirondack Railroad. A telephone line was also put through from Moira to Brandon in 1886.

Other providers of lumber freight soon became established along Hurd's railroad. Former partner Peter MacFarlane had allied with new partners in 1886 to establish a sawmill enterprise in Everton, plus a six-mile railroad to connect with Hurd's road in St. Regis Falls. Lumber freight from this mill began to arrive in 1886, adding more freight for the Northern Adirondack Railroad. Other new sawmills were set up at locations on Hurd's railroad, such as Shanley's in 1884 and Derrick in 1896.

John Hurd's other partner, Charles Hotchkiss, was eventually to conclude, as did Peter MacFarlane, that Hurd had extended his limited financial resources too far and at too fast a pace. Hotchkiss was also concerned that Hurd operated with the thoughtless opinion that lumbermen could never cut all of the endless supply of timber and denude the land; Hotchkiss felt otherwise. He thus sold out as a partner and left John Hurd to pursue the rest of his dream alone. Hurd had not floated any bonds to finance all of this construction, and this venturesome financing was now beginning to overwhelm him. Then, in 1888, an opportunity was presented to him that would relieve him of some of his burden.

188. The former sawmill settlement of Derrick was established on the Northern Adirondack Railroad but was abandoned about the time of World War I. The village once supported six hundred to seven hundred inhabitants. Courtesy of the St. Regis Falls Historical Society.

SANTA CLARA LUMBER COMPANY

The firm of Dodge, Meigs, and Company had been in the wholesale lumber business since the 1860s and was apparently purchasing much of Hurd's lumber. The owners were seeking an opportunity to expand, to use their experience as an aid to enlarge into lumber manufacturing. Negotiations between Titus B. Meigs and John Hurd resulted in the formation of the Santa Clara Lumber Company in March 1888.

What Hurd had to offer was his control of about seventy-five thousand acres of timberland plus a couple of operating sawmills somewhat deeply in debt. But Hurd's assets were enough to provide him with 45 percent of the capital stock of the newly formed corporation. Also taken into the new corporation were the smaller hardwood sawmill at Santa Clara owned by the St. Regis River Lumber Company and the box factory at St. Regis Falls. John Hurd's relationship with the Meigs family was probably amicable at first, but it was not to remain so for long.

Railroad construction continued southerly, extending in 1889 to its terminus at a small clearing made by a pioneer settler on the east shore of Raquette Pond. This remote location would soon grow, thanks to John Hurd, to be the largest lumbering community in the entire Adirondacks—Tupper Lake. The Northern Adirondack Railroad had now reached what would prove to be the end of its growth, a total of fifty-four miles from Moira on the Ogdensburg and Lake Cham-

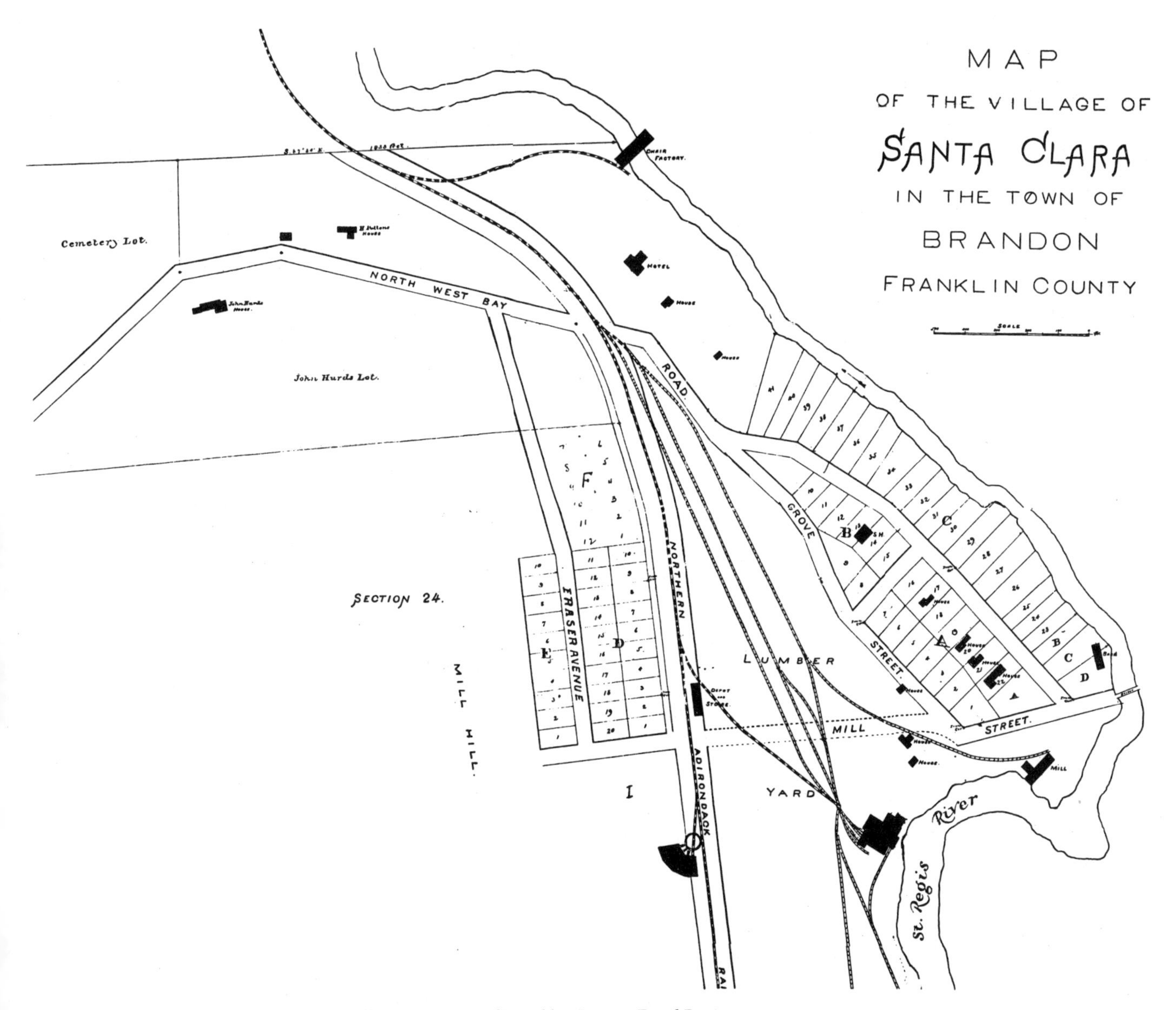

Map 20. *An early map of Santa Clara Village. Courtesy of Franklin County Deed Registry.*

plain Railroad to the Tupper Lake village. Other expansions were proposed but never realized.

No other railroad came near the area of Tupper Lake, and that was the way that John Hurd wanted it. He apparently wanted to avoid connection with any other line, assuring him a monopoly in hauling out the huge quantities of forest products from the region. As matters turned out, his hopes would be dashed within a couple of years.

The *Boonville Herald* reported in 1890 that John Hurd was planning to extend his Northern Adirondack Railroad southward from Tupper Lake to a junction with the New York Central at Utica. Although it probably excited Tupper Lake lumbermen, the construction of W. Seward Webb's Mohawk and Malone Railroad and the precarious financial position of Hurd soon squashed any hopes of that proposal.

The roadbed from Moira south as far as Santa Clara, eighteen miles, was said to be of good construction. But from Santa Clara southward, it had the reputation of being the crookedest railroad in existence. For the engineers hauling log trains, this presented no problem, of course, with their slow speeds. Rail sidings were established at a number of now unknown locations for the loading of logs on flat cars.

During the year 1886 the railroad hauled at least six million board feet of white pine, fifteen million board feet of spruce, and a half-million board feet of beech, birch, and maple. At the sawmill, the yard crew was loading eight to ten cars daily, each with eleven thousand board feet of sawn lumber. The reported cost to load a lumber car was about seventy-five cents.

Logs and lumber were the principal freight items, but a variety of other forest products were hauled on the Northern Adirondack. Charcoal kilns were located at a number of sites, such as Spring Cove, five miles south of Santa Clara; the coke was shipped out for use as fuel. Hemlock bark was taken to the St. Regis Leather Company at the Falls, for which the Santa Clara Lumber Company received $6.10 per ton. Other outgoing freight included fuel wood to Montreal, rossed pulpwood, clapboards, box shooks, softwood lath, and spruce fiddlebutt logs for manufacturing piano sounding boards. For the year 1891, the following carloads were shipped out: 1,313 of lumber, 105 of lath, 30 of shingle, 55 of leather, and 15 of fertilizer and other items to make a total of 1,518 cars. At the St. Regis Falls depot, 6,528 passenger tickets were sold.

Passenger service was provided through to Tupper Lake, subject of course to irregular scheduling and rough handling while being connected to a log or lumber train. It was even possible to ride a Wagner Palace sleeping car from Grand Central Station in New York to the North Country, leaving New York daily at 6 PM and arriving at Moira at 7:12 AM, where connection could perhaps be made with a coach on the Northern Adirondack. Cost was $12 one way, $20 round trip from either Boston or New York. Station stops on Hurd's road included a few scenic tourist dropoffs, such as the Paul Smith's Station, which provided access for some summer resorts.

Hurd's railroad offered a large convenience for the lumbering operators who had sawmills in Potsdam. When it came time to transport the logging crews and teams up to their remote camps for the winter logging season, the companies formerly had to walk the crews along forty miles of woods roads to reach the area. Now Hurd's road could quickly move them to Tupper Lake. Early lumbering in that area involved log drives down the Raquette River to the mills at Potsdam.

However, riding Hurd's Northern Adirondack Railroad could not be considered a joyride. On a cold day in February 1891, the passenger train from Tupper Lake rammed into a freight consist standing on the track at the Madawaska water tank. The engineer, fireman, and brakeman all leaped for their lives when they saw it coming, saving lives if not equipment.

Two days later, a train left Santa Clara at 6:30 AM bound for Tupper Lake over a track so blocked with ice and snow that nothing had run over it for two days. The train didn't reach Tupper Lake until 5:00 AM the following day. That must have been a bad week. On Thursday, an axle broke on the baggage car of the regular passenger run as the train was climbing LeBoeuf's Hill. It took two hours to obtain another car at Santa Clara and transfer the passengers.

No passenger was immune, either, to the dangers of traveling the Northern Adirondack Railroad. One day in August 1892, the regular mail train collided with the locomotive Pioneer near Childwold. The Pioneer was traveling at a high rate of speed, carrying John Hurd and his son to the family home at Santa Clara. The crew of the Pioneer, plus John Hurd and son, "took to the birds." John Hurd landed on a stump and was unconscious for a while. He was treated by a doctor at Brandon and sent on to Santa Clara the next morning in his private car.

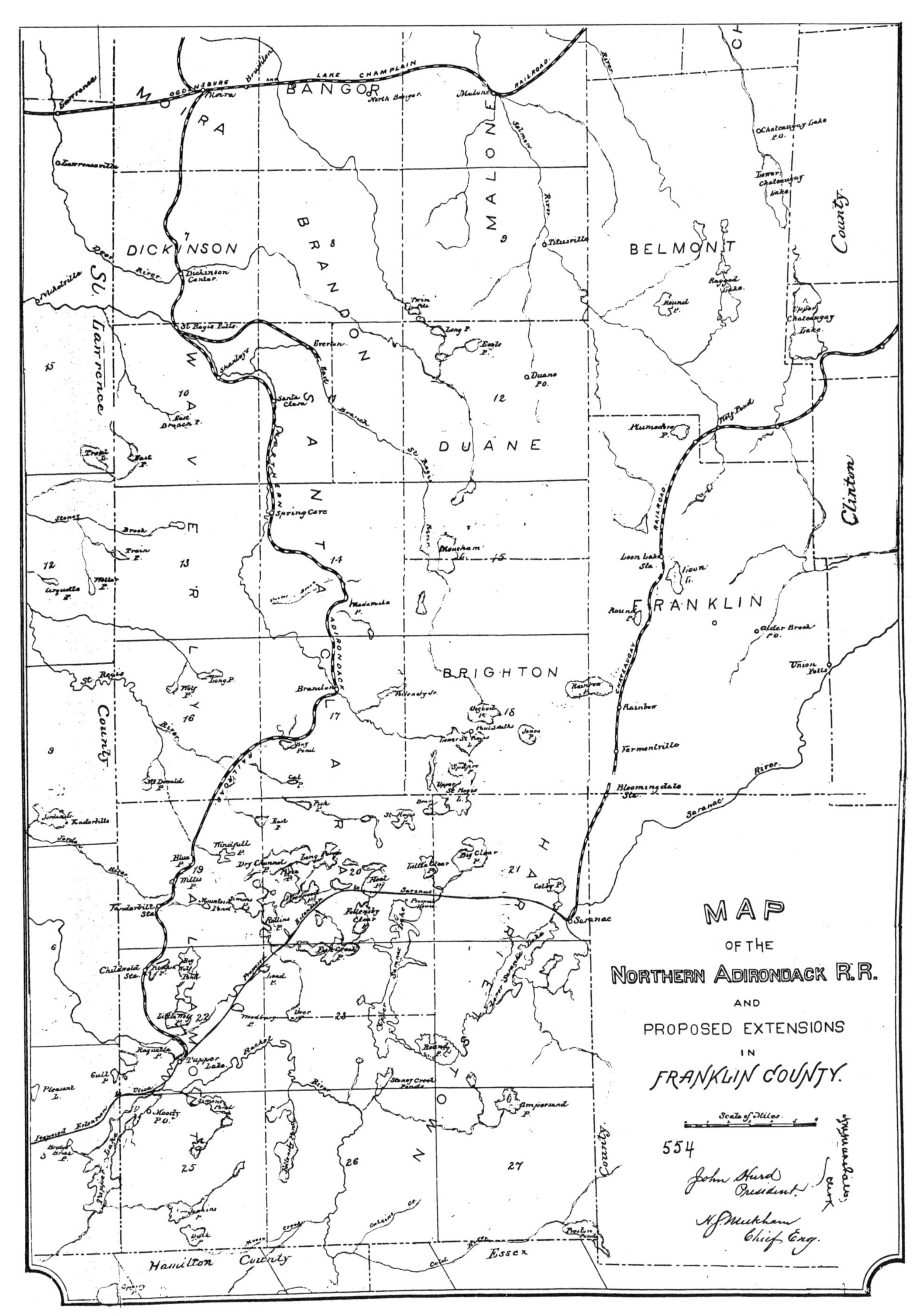

Map 21. An 1880s map of the Northern Adirondack Railroad. Courtesy of the Franklin County Deed Registry.

189. Tupper Lake depot in about 1920. The rail in the foreground is the track over which traffic was transferred from the New York Central to the old New York and Ottawa track. Courtesy of the Tupper Lake Public Library.

190. Tupper Lake Junction (Faust) was a busy railroad community in the early 1900s. The arriving southbound train is surrounded by about twenty sets of tracks in this westerly view. The engine house is on the left and the depot beyond.

Total cost of building the railroad was estimated in 1886 to reach $400,000, but many opinions had it that the figure was probably higher. The original locomotives obtained in the 1880 were thirty-five- and forty-five-ton Americans (4-4-0) built by the Schenectady and the Rhode Island companies, and a couple of "pony" engines said to be used in construction. One of the little steamers was called by the name of Little Logger. The locomotives were so light that when a tender jumped the track, the crew was said to be able to hoist it back on track with the use of a pry bar. A team of horses was used to rerail a locomotive that had wandered off.

By the end of 1889, the Northern Adirondack had acquired six locomotives and a total of 171 freight and passenger cars. About 100 of them were platform or log cars. To reduce initial investment, all of the locomotives and cars were leased equipment except for the first two locomotives, purchased used from the New York Central in 1881 and 1882, and 45 of the log cars.

John Hurd's legendary devastation of the timberland was ruthless and had no regard for extended operations. An 1891 traveler along the North Country rails wrote in an article for the *New York Herald* that Hurd had practically denuded the country of timber; by May of that year there were seventy-five million board feet of logs stacked railside ready for transport. The traveler was dismayed to note the blackened countryside around the charcoal kilns and the miles of nothing but slash on both sides of the railroad.

The thick blanket of slash left after the clear-cutting of the white pine on the sand plains created a fuel source that made forest fires a serious threat. In June 1891 a surface fire roared across the tracks at Bay Pond, burning ten freight cars

NORTHERN ADIRONDACK R. R.—MAY 14, 1893.

7	5	3	1	51	M	STATIONS.	2	56	4
	P. M. [illegible]		P. M. 1:45	A. M. 7:30		TUPPER LAKE.	P. M. 1:15	P. M. 6:20	P. M. 12:50
	3:45		1:49	7:34	2	TUPPER LAKE JC.	1:11	6:16	12:45
			1:58	7:43	6	CHILDWOLD.	1:02	6:07	
	4:40		2:20	8:05	16	BLACK RAPIDS JC.	12:41	5:46	11:50
P. M.	[illegible]:32	A. M.	2:35	8:20	22	BRANDON.	12:27	5:32	11:25
1:00	[illegible]25	5:30	3:07	8:52	36	SANTA CLARA.	11:55	5:00	10:10
			3:14	8:59	39	SHANLEY.	11:45	4:50	
3:00		6:10	3:25	9:10	42	ST. REGIS FALLS.	11:35	4:40	9:10
			3:34	9:19	46	DICKINSON CENTER	11:25	4:30	
8:00		6:50	3:55	9:40	54	MOIRA.	11:05	4:10	8:00 A. M.
P. M. 4:26		A. M. 7:45	P. M. 4:20	A. M. 10:10	68	MALONE.	A. M. 9:1[illegible]	P. M. 3:40	

191. *Northern Adirondack timetable, 1893.*

Table 18
Northern Adirondack Railroad Company
Operations, 1889

Train mileage	
Passenger	13,534
Freight	35,338
Total	48,872 miles
No. of passengers carried	29,867
Average rate per mile	4.9 cents
Freight carried	61,125 tons
Gross earnings	$99,779.03
Operating expenses	$52,700.10
Net	$47,078.93
Paid taxes and rentals	$41,170.90

192. *Charcoal production consumed large areas of timber and left the countryside denuded of tree growth. The above kilns were located at the Narrows on Chateaugay Lake. Courtesy, Special Collections, Feinberg Library, SUNY at Plattsburgh.*

on a siding and large quantities of bark piled along the track by Pat Ducey's crew. Two of the cars were loaded with bark.

On the logging operations, the company made use of a Lidgerwood high-lead cable system for a short while but was not satisfied with the results. It was reported that Hurd abandoned the system because he found it to be too destructive to young tree growth. Maybe so, but that reason is difficult to accept in view of the written eyewitness accounts of the massive forest devastation he left along the Northern Adirondack Railroad tracks. Whether this steam skidder was rail mounted or skid mounted is unknown.

It wasn't many months after the formation of the Santa Clara Lumber Company that the relationship between John Hurd and his partners, the Meigses and the Dodges, began to break down. In the original agreement, John Hurd, as vice president, was to devote his energies to railroad construction and operation, and Dodge, Meigs, and Company, as the majority stockholder, was to have control of lumber operations. But the new railroad had no sooner reached the Tupper Lake location than John Hurd began making plans for a new sawmill at that location. This action also violated another agreement, which had stated that the final decision on the location of the railroad terminus had been granted to the offices of the company and not to John Hurd alone.

Ferris Meigs once expressed this unflattering assessment of John Hurd: "He was a visionary; impractical, unreliable but hard working, a poor executive constantly overreaching in his business enterprises. He was always up to his ears in debt, without good standing among his bankers, and with his judgment warped by his ambitions, he was a hard man to get along with. His fellow officers found him a shrewd but not always reliable trader. These characteristics soon led to a break in friendly relations with his associated stockholders in the lumber company" (from "Mostly Spruce and Hemlock," by Louis Simmons, in the *Tupper Lake Free Press*, 1976).

Hurd's stock had been pledged as collateral in several banks against loans he was unable to pay, and he soon lost all of his Santa Clara Lumber Company stock. In that same year, 1890, the Santa Clara Lumber Company ownership was reorganized, and the company went on to become a major player for over four decades in the Adirondack lumber industry.

193. Tupper Lake's Big Mill, built by John Hurd in 1890 and pictured here in the late 1890s, when it was operated by Norwood Lumber Company. Courtesy of the Tupper Lake Public Library.

The "Big Mill" at Tupper Lake

Hurd was once again operating alone, and preparations were made for a new sawmill at Tupper Lake village that would be a crowning achievement among his many enterprises. But he still wasn't free of his confrontations with Titus Meigs. Hurd insisted that the lumber pile-bottoms in the two lumber yards at Santa Clara were not included in the sale of the mills to the Dodge and Meigs stockholders. Hurd demanded and managed to get $15,000 for the structures. Meigs, henceforth, moved quickly in getting retribution. It seems that Hurd had selected a site at the edge of Raquette Pond on which to build his huge sawmill and had spent many thousands of dollars preparing the site, including a large number of pilings sunk into the mud flats at water's edge. The Santa Clara Lumber Company was also purchasing land in the area for possible expansion of operations, and Meigs's attorney discovered that a portion of Hurd's mill site was actually on an unnumbered lot that was not owned by Hurd. Meigs was able to purchase the lot after an extensive search to find the title holder, and in a heated squeeze play in the office of the bank that held Hurd's mortgage, agreed to sell the property for $15,000.

The sawmill then built by John Hurd in 1890 on the shore of Raquette Pond was said to be the largest sawmill ever seen in the State of New York and possibly in the northeastern United States. At least in physical size it was. The two-story two-hundred- by four-hundred-foot main building contained an outlay of machinery designed to saw three hundred thousand board feet per day. But apparently the potential was never sustained. The highest known yearly production was thirty-three million board feet of pine and spruce, and the average annual output wasn't much over ten million board feet.

The population of Tupper Lake began to boom. The original crude and ugly shacks that had housed a tough and lawless population had all been destroyed in the fire of July 30, 1899, which swept the village—a blessing in disguise. Rows of new cottages appeared as the population rose to a thousand.

It was here in Tupper Lake that the reckless John Hurd was to meet the insurmountable. W. Seward Webb was building his Mohawk and Malone Railroad northerly through the treasured central Adirondack wilderness, much to the dismay of the conservationists and the delight of the fringe-area residents. His construction supervisor had just spent many grueling months pushing the roadbed northerly from Remsen, and the survey party had reached the Tupper Lake area in 1891. Coupled with his accelerated rate of construction and the refusal of permission to build across state-owned forestland, Webb saw an opportunity to make use of Hurd's financially troubled railroad.

It's not surprising, therefore, that Webb made a reasonable offer to John Hurd to buy his railroad in order to use it for the final link northward to the St. Lawrence River and

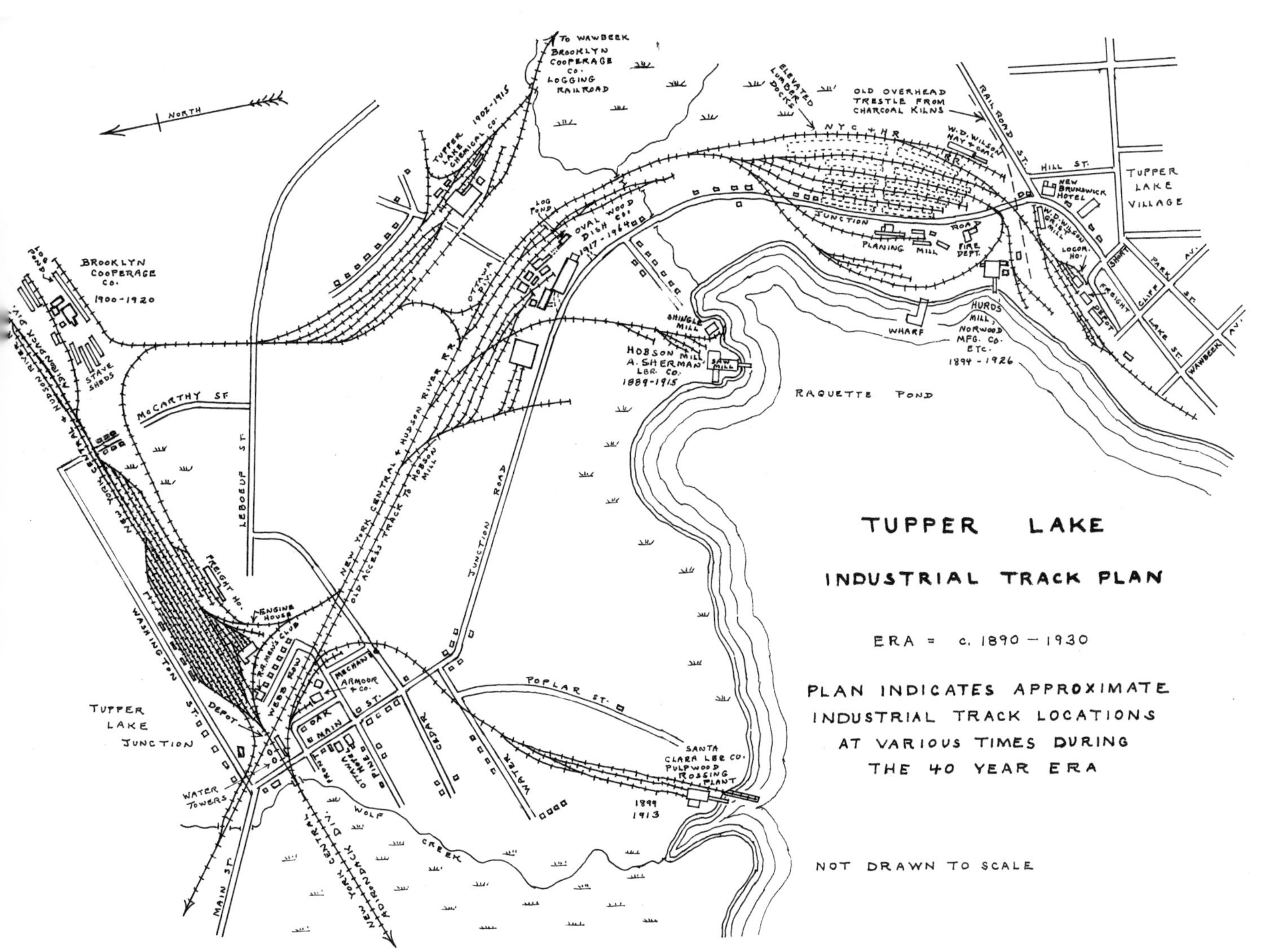

Map 22. After 1890, Tupper Lake went on to become a major industrial site for the wood industry of the Adirondacks. The above plan is a composite of the various industrial track locations over a forty-year period.

Canada. Hurd at first agreed, but greed then got the better of him. The evening of the same day that he had made a oral agreement with Webb to sell the Northern Adirondack to him, Hurd was visited by a Mr. Paine of the Pope-Paine Lumber Company, who upped Webb's price by $50,000, provided Hurd would give him an option agreement. Webb was afterward incensed when Hurd reneged on their verbal deal, and one can imagine his angry reaction when Paine later approached him and offered to sell the option at an inflated sum.

W. Seward Webb was not a man to be jerked around; dismissing any interest in Hurd's road, he continued construction of the Mohawk and Malone Railroad almost parallel to Hurd's, then deviated northeasterly to Malone instead of Moira. The railroad approached the Tupper Lake area on the west side of Raquette Lake and bypassed Tupper Lake village, crossing over Hurd's Northern Adirondack Railroad a mile and a half north of the village and creating what became known as Tupper Lake Junction or Faust. The two roads crossed on the same grade, with semaphore signals that gave precedence to neither road. However, the Junction was eventually to become the Adirondacks' leading railroad community.

To add to the many activities and accomplishments of John Hurd, some obviously successful, was one that came as a result of his association with former governor Alonzo B. Cornell in 1890. The two of them experimented with the lighting of railroad cars by means of electricity generated by the revolution of the car wheels, successfully doing so at the Santa Clara location.

John Hurd had been carrying his financial burden alone since his departure from the Santa Clara Lumber Company, and it became overwhelming. The sacrifice of properties that had been pledged as debt security did not reduce his interest charges to a manageable level, and a toll on his health was exacted. An attempt was made in 1894 to float a bond issue that would have discharged his obligations, but it failed. Mortgages and judgments against him reached a point where a receiver had to be appointed for the railroad and bankruptcy was established in 1895. The entire railroad was put up for bid, including the existing rolling stock of four locomotives and nine hundred cars, all of it mortgaged at the time to Shephard and Morse Lumber Company.

John Hurd's fifty-two-mile railroad was sold to a syndicate, which promptly changed the name to the Northern New York Railroad and then changed it again in 1897 to the New York and Ottawa Railroad. In 1896, the management made an effort to extend the railroad southerly to North Creek to junction with the Delaware and Hudson Railroad system, but the attempt was thwarted. State-owned lands had to be crossed, and the courts enjoined or prevented their doing so. The case was taken to a court of appeals, where the railroad company again lost.

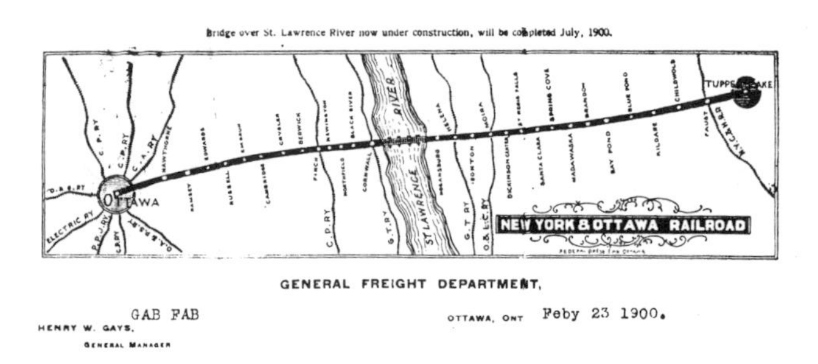

194. Letterhead.

NEW YORK AND OTTAWA RAILROAD
[IN EFFECT AUG. 28, 1899]

No. 61 leave P. M.	No 63 leave A. M.	STATIONS.	No. 60 arriv P. M.
6.05	1.30	Tupper Lake .	12.00
6.12	1.56	Central Junction.	11.54
†6.28	†1.50	Childwold	†11.45
†6.36	†2.08	Kildare	†11.35
6.50	2.17	Blue Pond	11.25
†7.08	†2.55	Bay Pond	†11.11
7.20	2.47	Brandon	11.02
†7.34	†3.02	Madawaska	†10.50
†7.54	†3.22	Spring Cove	†10.34
8.08	3.36	Santa Clara	10.22
8.30	3.57	St. Regis Falls,	10.05
8.40	4.11	Dickinson Ctr.	9.58

195. New York and Ottawa Railroad, detail of timetable, 1899.

An effort was also started in 1896 to extend the railroad northerly from Moira to Ottawa, and this project was accomplished. The St. Lawrence River was bridged at Cornwall, Ontario, but the construction was marred with a disaster. On September 6, 1898, the first bridge built collapsed, killing fourteen workers.

Passenger service was established through to Ottawa in the fall of 1900. But the new owners could not hold onto the railroad for any length of time. On December 22, 1904, the New York Central and Hudson River Railroad took control of the road at a bondholder's foreclosure sale. "Hurd's road" became the Ottawa Division of the New York Central and Hudson River Railroad. When it became known that the D&H was also contending for control of NY&O at the time of the foreclosure sale, hopes arose for a better future for the line, and there was a rise in land prices.

Snowstorms in the Adirondacks created some difficult operations for the New York Central. In March 1908 a passenger train became stalled in drifts between Dickinson Center and St. Regis Falls. The locomotive was detached and, after repeated attempts at breaking through, found that it could not get back to the cars. The passengers were stranded with no heat. It was past midnight before a rescue crew arrived from the Bishop Hotel and transported the passengers to St. Regis Falls. The train eventually arrived at Tupper Lake, twelve hours late.

Logs and forest products continued as a principal support for the Ottawa Division, along with supplies and passengers to the small settlements and lumber camps, enough to provide another thirty years of sometimes profitable life. When the Oval Wood Dish Company set up their extensive logging railroad operation at Kildare in 1916, the company's log trains operated seven miles over the Ottawa Division tracks into Tupper Lake. Other logging railroad that had junctioned with the Ottawa Division over the years had all passed into oblivion by that time.

Freight and passenger traffic steadily declined on the

TUPPER LAKE TO OTTAWA

NORTHWARD — FIRST CLASS

Miles from Tupper Lake Jct.	STATIONS	61 Passenger Daily except Sunday	63 Passenger Daily except Sunday
	LEAVE	A. M.	P. M.
	Tupper Lake........	5 25	
0.00	Tupper Lake Jct....	s 5 50	
7.11	Kildare	s 6 06	
11.06	Derrick	s 6 16	
17.00	Bay Pond...........	s 6 29	
18.82	McDonald	s 6 34	
20.04	Brandon	f 6 37	
24.13	Madawaska	f 6 47	
25.13	Meno	f 6 50	
30.00	Spring Cove........	f 7 03	
34.20	Santa Clara.........	s 7 13	
40.48	St. Regis Falls.....	s 7 29	
43.90	Dickinson Center...	s 7 38	P. M.
52.29	Moira	s 8 00	L 4 15
60.07	Ironton	f 8 16	f 4 33
62.61	Helena	s 8 20	s 4 40
66.08	Nyando	s 8 27	s 4 47
67.28	Int. Boundary.......		
68.13	Uscan	f 8 32	f 4 51
69.29	Cornwall	s 8 45	s 5 05
70.13	Cornwall Jct........	s 8 49	s 5 08
77.29	Black River........	f 9 01	f 5 22
78.89	Harrison	s 9 05	f 5 26
81.72	Northfield	s 9 11	s 5 33
84.78	Newington	s 9 17	s 5 39
88.91	Finch ....A	s 9 25 (60)	A s 5 48 (62)
	Finch L	9 30	L 5 55
91.28	Berwick	s 9 36	s 6 03
94.59	Crysler	s 9 45	s 6 09
97.16	St. Albert...........	s 9 50	s 6 14
98.57	Cambridge	f 9 54	f 6 18
102.32	Embrun	s10 01	s 6 25
105.93	Russell	s10 13	s 6 32
109.25	Pana	f10 22	f 6 38
112.67	Edwards	s10 30	s 6 46
115.49	Piperville	f10 36	f 6 52
118.89	Ramsayville	f10 44	f 6 59
120.57	Hawthorne	f10 50	f 7 03
124.26	Hurdman	11 00	7 10
126.00	*Ottawa*	*11 05*	*7 15*
	ARRIVE	A. M.	P. M.

On single track, southward trains are superior to northward trains of the same class, unless otherwise specified.
Time shown at *Ottawa* is for information only.
Northward trains stop at Cornwall for Customs inspection.

196. Timetable for the New York Central Railroad Ottawa Division, 1929.

Ottawa Division, and on January 9, 1936, the New York Central Railroad petitioned the Interstate Commerce Commission for permission to abandon the entire line from Tupper Lake Junction to Helena on the Canadian National Railway, a distance of 62.6 miles. Freight trains were going through only three times a week, and passenger service operated once a day each way between Tupper Lake and Ottawa, all at an estimated loss of $72,900 per year. Despite the opposition by the village of Tupper Lake and the timberland owners such as Oval Wood Dish Company and International Paper Company, the ICC examiners recommended the abandonment, and the last passenger train to traverse Hurd's old Northern Adirondack Railroad pulled out of Tupper Lake Junction on May 6, 1937. The rail was taken up the same year, starting on the north end and working toward Tupper Lake. The last seven miles between Kildare and Tupper Lake were left for a later date in order to accommodate a logging operation then active.

197. A former lumbering settlement on the Northern Adirondack Railroad, long since abandoned and identified only as "Charley's." Courtesy of the Tupper Lake Public Library.

Hurd's Big Mill at Tupper Lake had continued to operate under a succession of owners, even the Santa Clara Lumber Company, which bought it in 1913 for $20,000 to saw timber harvested in the Cold River area and driven down the Raquette River. The mill was closed down permanently in 1927 and torn down two years later. The site is now a village recreation park with a monument commemorating the former Big Mill. The Santa Clara Lumber Company continued as a major manufacturer for many years; their final liquidation came in 1941.

Thus, there were many John Hurd enterprises that continued and thrived far beyond the lifetime of the courageous and reckless John Hurd. He had undertaken his Adirondack ventures at a point in his career well beyond his prime of life. And with the loss of his property in 1895, he returned to his home state of Connecticut, where he died in poverty in August 1913 at the age of eighty-three. He made his marks in the Adirondacks, the most enduring of them being the village of Tupper Lake.

CHAPTER TWENTY-ONE

St. Regis Falls and Everton Railroad

LISTED AMONG THE FORMER Adirondack sawmill villages that have disappeared into oblivion is the once-thriving settlement of Everton. Although there is no evidence of any habitation today, the village roots go far back in history. Back in 1831, a man named Sanford had built a water-powered mill on the East Branch of the St. Regis River, and then, just before the Civil War, the firm of Sanford, Skinner, and Holmes constructed a larger sawmill on the site. Enlargement of the settlement occurred in 1883 when a Robert Douglass built a larger water-powered circular sawmill a mile below the Sanford mill. Douglass's mill included clapboard and shingle manufacture plus a store. This, then, constituted the little industrial village of Everton in the Town of Santa Clara in the mid-1880s. Everton was located about seven miles east of St. Regis Falls.

In 1883, the firm of Hurd, Hotchkiss, and MacFarlane had become established in the St. Regis Falls area, buying large tracts of timberland south of town, enlarging a sawmill at the falls, and building a new railroad south from Moira, ultimately to reach fifty-four miles to Tupper Lake. John Hurd had eventually shown himself to be a difficult man to have as a business associate, and as a result, Peter MacFarlane, who had been in partnership in Michigan with John Hurd before coming to New York, sold out in 1885 to start his own enterprise. In a new partnership with W. J. Ross and H. W. Stearns, MacFarlane purchased the new sawmill of Robert Douglass at Everton after it had run for only one year.

Table 19
St. Regis Falls and Everton Railroad

Dates	1886–c. 1898
Product hauled	softwood logs, lumber, passengers
Length	6.5 miles
Gauge	standard
Motive power	1 Porter saddletank locomotive, 16-ton

In addition to the Douglass mill, the firm built, at a cost of $35,000, a new combined water-and-steam-powered sawmill upriver at the site of the former Sanford mill, now abandoned. Three Lane water wheels helped to power the twin fifty-six-inch circular saws that slabbed the logs and sent them on to a double gang saw, which was cutting at the rate of two hundred strokes per minute. The new mill required thirteen employees. The mill dam backed up stillwater for fourteen miles on the East Branch of the St. Regis River for the storage of softwood logs.

The combined maximum output of the two sawmills was seventy thousand board feet in the customary eleven-hour work day. To supply these mills, MacFarlane's firm had purchased sixteen thousand acres of land in the towns of Santa Clara, Waverly, and Duane at a cost of $50,000. The normal employment totaled about thirty but during the winter logging season increased to one hundred. Population of the village had grown to 175.

In conjunction with the rejuvenated sawmill village, MacFarlane built a six-and-a-half-mile railroad in 1886 to connect Everton with the outside world at St. Regis Falls. Thirty-pound rail was used in construction; total cost was $40,000. The single locomotive was a sixteen-ton Porter saddletank, but nothing more is known about the equipment used on the little railroad. The line was officially known as the St. Regis Falls and Everton Railroad.

The lumber product was hauled down to St. Regis Falls, where the Everton Railroad junctioned with John Hurd's Northern Adirondack Railroad. Passengers were hauled too, and logs were included in the freight, according to an 1887 edition of *Northwest Lumberman*. The average load of the

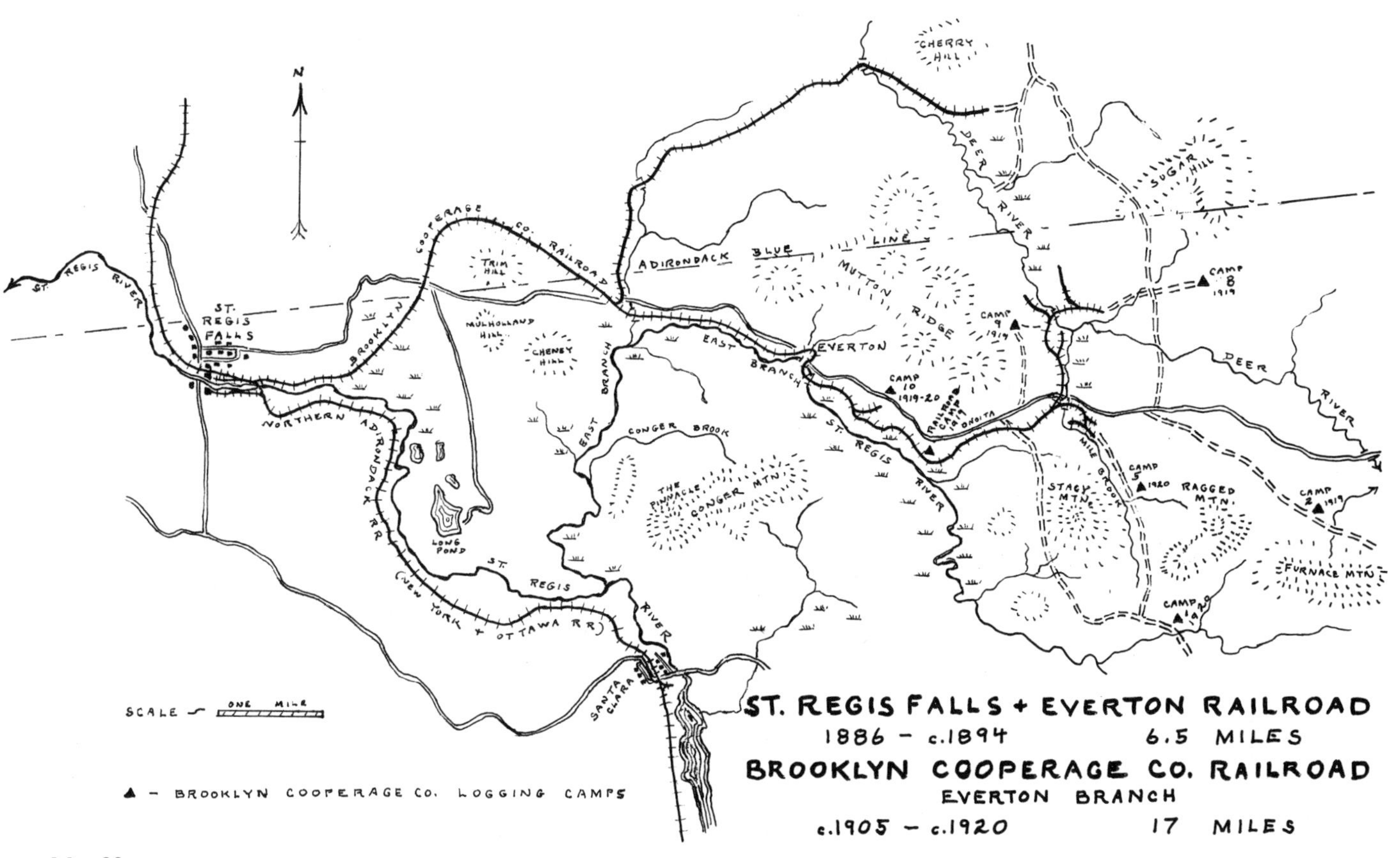

Map 23.

type of log car used was twelve hundred board feet, with a total annual haul of twelve million board feet of logs.

The firm of MacFarlane, Ross, and Stearns ran the operation for three years before dissolving the company. The three years of profit had totaled almost $80,000, and the company was reported to be in flourishing condition. But the business was sold in April 1888, to Henry and David Patton of Albany, who incorporated the Everton Lumber Company. The sale price of $130,000 included the next winter's supply of sawlogs (ten million board feet), sixteen thousand acres of land, two sawmills, twenty horses, and the railroad.

198. The St. Regis Falls and Everton Railroad saddletank stands out well amidst a fresh snowfall. The locomotive was a sixteen-ton Porter. Circa 1890. Courtesy of the St. Regis Falls Historical Society.

The Pattons continued to operate the railroad as a common carrier. But the little railroad had its problems with spreading rails and insufficient maintenance, especially during the spring thaws. In March 1892 engineer Woodbury took advantage of the spring shutdown to overhaul the little Porter saddletank locomotive named Hiawatha, install a Detroit lubricator, and give the engine a repainting, with some artistic lettering applied. Another of the engineers was Phillip Winters.

Area residents had been given hope that the short railroad would some day become a connecting link between a Canadian Pacific line and a rail line planned to be built east from Carthage. Visions of a great railway line through Everton came to naught, however, as so often happened in the days of railroad expansion.

The Pattons kept the operation going for about seven years. But the timber supply was beginning to run out and mortgage payments began to be missed. Legal action was taken in February 1895, and businessman William T. O'Neil of St. Regis Falls was appointed receiver—an all-too-familiar story in the lumber and railroad business. O'Neil assumed management and continued to saw out the winter supply of logs. The timber supply wasn't exhausted, however, because John Fraser of Albany took a contract in 1894 to log eight million board feet yearly for the next four years and set up four logging camps to accomplish it.

The Everton sawmills, by then operating under the name Forest Land and Mill Company, continued to operate for two more years, but in February 1897 another familiar event shocked the declining community. Fire destroyed one of the sawmills on a Sunday morning. The mill had been sawing a large hardwood order for the Santa Clara Lumber Company.

That proved to be the last winter season of sawmill operation. The railroad was still alive, however, the following summer. In August, Mr. Woodbury of the Everton Railroad organized a picnic excursion to Everton by rail, leaving St. Regis Falls at 9 AM. The locals were reminded that this would be their last opportunity to enjoy a visit to Everton; the town would be dead and buried in the course of another week. The rails weren't taken up until two years later in September 1899, which is why the New York and Ottawa Railroad still listed a connection with the Everton Railroad in 1898.

St. Regis Paper Company had acquired ownership of the property by then, including the buildings, mills, machinery, and railroad right-of-way. In 1900, St. Regis Paper sold twenty-two thousand acres of Everton land to a man who was expanding his vast ownership in that part of the Adirondacks, William Rockefeller. The price of $88,350 included what was left of the buildings, sawmill machinery, and the railroad right-of-way.

Not many years later, in 1904, the Brooklyn Cooperage Company came to St. Regis Falls and set up a barrel stave and heading manufactory. Land with hardwood timber was needed, and William Rockefeller had almost twenty thousand acres for sale; in 1907 Brooklyn Cooperage purchased the old Everton Lumber Company tract.

To transport the logs to St. Regis Falls, Brooklyn Cooperage laid rail again on the old Everton right-of-way, extended it eight miles further into the Town of Brandon, and made use of it for another thirteen years.

The former sawmill village of Everton was never revived, and the site is now barely recognizable among the forest growth. Even back in the early 1920s the site was difficult to find, according to John Arfmann of the Brooklyn Cooperage Company, who occasionally hunted in the area.

CHAPTER TWENTY-TWO

Watson Page Lumber Company Railroad

The local folks called it the "Toonerville Trolley," and what a strange sight it must have been, a little boxlike, single-truck electric locomotive hauling large hardwood logs out from the Adirondack forestland. But the operator proclaimed the little railroad a success, even though the original plans were not for an electric railroad and even though the life span of the railroad was quite short.

The railroad's creator, the Watson Page Lumber Company, had its origin in 1896 when Watson Page and R. W. Babcock leased the old Hammond mill in St. Regis Falls village. The mill had formerly been the site of the Santa Clara Lumber Company box shop and then the sawmill to which Santa Clara Lumber Company had transferred their lumber manufacturing in 1892 from Santa Clara village. In 1895 a railroad track had been built along the south side of the river from John Hurd's railroad to accommodate the log trains into the sawmill. The Northern Adirondack Railroad thenceforth delivered all of the Santa Clara Lumber Company logs to the St. Regis Falls sawmill.

Shortly after the lease by Page and Babcock, the business was incorporated as the Watson Page Lumber Company by Page, William T. O'Neil, and H. E. O'Neil. The principal product was hardwood lumber. However, sawmilling by Watson Page Lumber Company didn't last for long.

Table 20
Watson Page Lumber Company Railroad

Dates	1906–1909
Gauge	narrow, 3-foot
Product hauled	hardwood logs
Length	4.8 miles
Motive power	electric (electric locomotives, single-truck)
Log cars	16

In June 1900 the Weidmann Stave and Heading Company was formed at St. Regis Falls. Louis Weidmann had just suffered a large fire loss at his factory in Brooklyn and had come north to take over the Watson Page factory with a twenty-year lease. The Watson Page Lumber Company signed a contract with Weidmann in June 1901 to deliver Weidmann four million feet of hardwood logs annually for the next six years.

Weidmann's barrel manufacturing didn't last for long either. In September 1903 his long-term leases of mills and lands were picked up by the Brooklyn Cooperage Company. The Watson Page Lumber Company continued to harvest the hardwood timber on their land and started selling to a new firm in St. Regis Falls, the Cascade Chair Company. Cascade Chair had been formed in 1901 by the same William T. and H. E. O'Neil and set up in an old tannery factory below the falls. After a couple of years of sledding logs to the Cascade Chair plant, Watson, Page, and associates de-

LOGS WANTED.

1899

We will pay the following in cash for hard wood logs delivered at the mills in St. Regis Falls:—

Birch, per thousand,	-	**$7.00**
Hard Maple,	" - -	**6.00**
Ash,	" - -	**7.00**
Elm,	" - -	**6.00**

The Watson Page Lumber Co.

199. Advertised lumber prices.

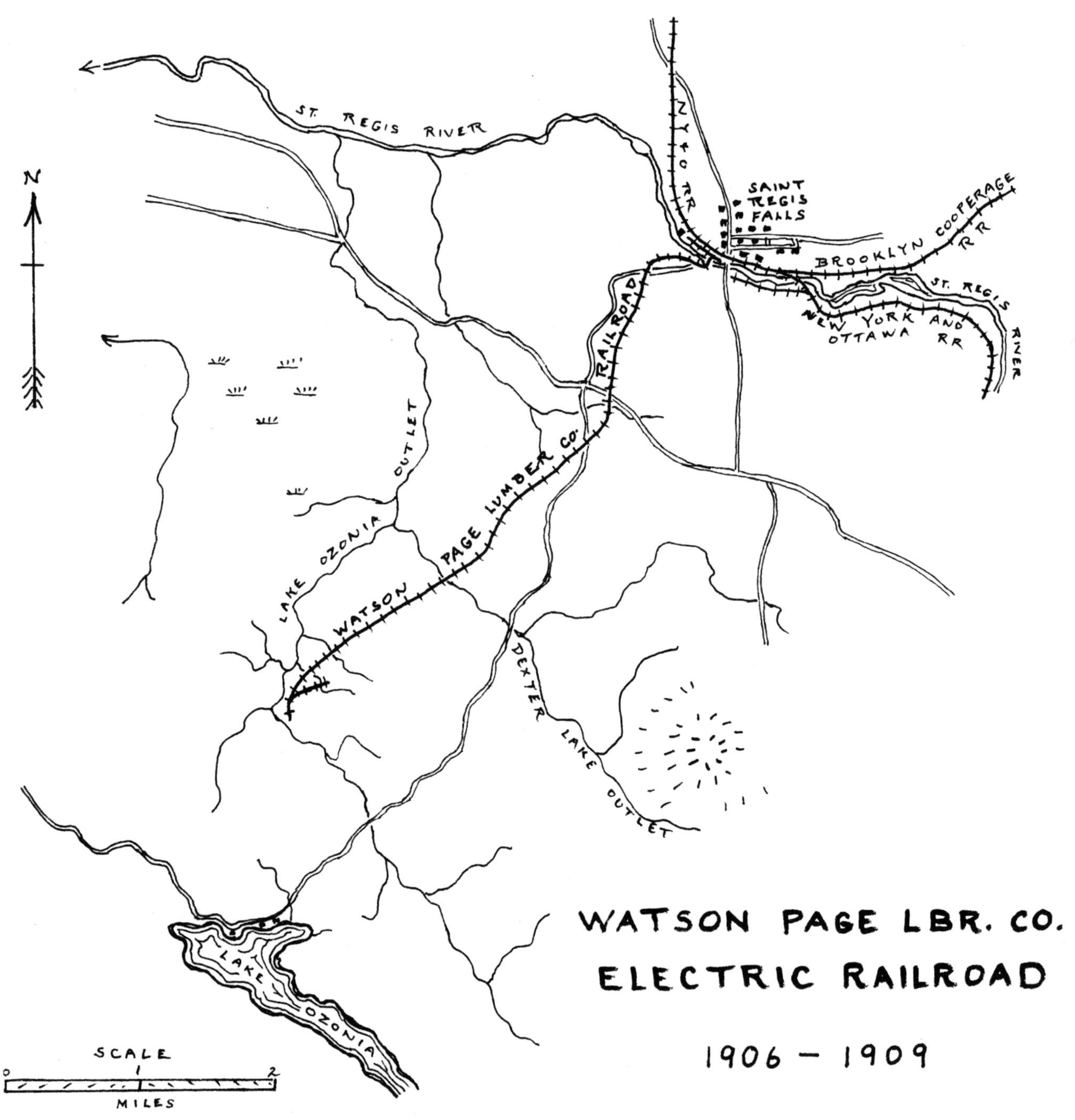

Map 24.

cided to build a narrow-gauge railroad southwesterly from St. Regis Falls into their timber along the outlet of Lake Ozonia.

In April 1906 the Watson Page Lumber Company had a construction gang start the grading of a roadbed, beginning at the River Street bridge in the village. By the third week in April, they had begun to lay down light sixteen-pound rail in anticipation of acquiring a small locomotive. The company's original plan was to use a small gasoline-powered engine, and one was obtained to assist in the track laying.

By September of that same year, they came to the realization that the gasoline locomotive would be unable to pull the heavier loads of log cars, and the little locomotive was returned. Also realized was the fact that a steam locomotive would be too heavy for the light rail that had been laid, and the company turned to another type of motive power then coming into popularity—the electric railroad. Installation of poles and wire began in September, and after a couple of months a successful electric railroad was in operation.

200. The Watson Page Lumber Company electric locomotive was parked on the north side of the St. Regis River. The railroad river crossing is on the left, beyond which the track ascended the bank and continued toward Lake Ozonia. In the background is the old River Street highway bridge, and the switchback track leading to the chair factory is just in front of the engine. Circa 1908. Courtesy of the Tupper Lake Public Library.

201. Sawn lumber for the Cascade Chair Company plant en route by rail and horse power. The lumber scaler on the right still had his lumber scaling stick and tally sheet in hand. Courtesy of the St. Regis Falls Historical Society.

The equipment cost apparently put a critical strain on the company's limited resources; by late winter the following year Watson Page Lumber Company approached its sister company, the Cascade Chair Company, for relief. On March 1, 1907, the Watson Page Lumber Company ceased to exist as a corporation, having been merged into Cascade Chair.

The Cascade Chair Company operation was capable of producing six hundred chairs daily with about thirty employees. The plant obviously made a good market for hardwood logs. With the merger, H. E. O'Neil became the president.

Having acquired an unfinished railroad in the merger, the Cascade Chair Company completed the construction, a total of 4.8 miles to the Lake Ozonia outlet. In April 1907 a new dynamo was installed to power the railroad, and the electric road began to haul logs to the Cascade Chair plant.

The electric locomotive was a small, one-truck, four-wheel trolley type of engine about eleven feet in length. There was said to be another electric locomotive used, although it's unknown what it was. Locally, the little railroad was known as the Toonerville Trolley, named after a comic strip of that era. Some also called it Ed O'Neil's electric railroad; the O'Neil family had a majority interest in the company, and Ed was general manager. The motorman was Frank Ploof, and employed as brakemen were Art Fadden, Ernie Brabon, and Pete DeLaire.

A story repeated locally for many years concerned one of the small boys in a gang of youngsters who were fascinated by the little railroad. With the encouragement of some grown-ups, he was induced to grease the tracks on Tannery Hill to see what would happen. When the fully loaded log train appeared, the spectators no doubt stayed well hidden as they watched the train slide wildly down the heavily greased rails, unable to brake the descent. It finally drifted over onto

202. Although the little electric railroad was taken over by the Cascade Chair Company, the name Watson Page Lumber Company remained on the locomotive. Circa 1908. Courtesy of the Tupper Lake Public Library.

River Street and stopped without any injuries or serious damage. The only item hurting was the little culprit, who was punished, according to the tale.

The Cascade Chair Company plant was lost to fire in 1909 with a $25,000 loss, and that also was the end of the railroad. A small pulp mill was built on the site in 1911 and a new firm formed, known as the Cascade Wood Products Company. Cascade Chair Company sold the timberlands and the abandoned railroad right-of-way to the Brooklyn Cooperage Company.

CHAPTER TWENTY-THREE

Brooklyn Cooperage Company Railroad

The Brooklyn Cooperage Company was once the world's largest producer of wooden barrels, manufacturing containers principally for the shipment of sugar and other dry goods. The company had its origin in the 1850s, although it was not until 1881 that the company became incorporated in New York State with offices in Brooklyn.

The peak of company operations came in the early 1900s with barrel manufacturing plants in the states of New York, Pennsylvania, Maryland, Louisiana, Massachusetts, and Missouri, as well as additional remote lumber manufacturing facilities. Company employment reached ten thousand in the year 1917.

The two major components of a slack barrel (container for dry contents) were the hardwood staves and hardwood headings. The staves and headings were often manufactured at a location close to the hardwood timber supply, a location separate from the barrel assembly plant. Maple was the preferred log species for the barrel heads, while beech was considered satisfactory for the staves. The logs were cut into short bolts for sawing the staves and heading, thirty-

Table 21
Brooklyn Cooperage Company Railroads

Dates	1900–1920
Product hauled	hardwood sawlogs, chemical wood, occasionally softwood pulpwood and hemlock bark
Length	total of about 68 miles for all branches combined
Gauge	standard

Motive Power

No.	Builder	Type	Builder No.	Year Built	Weight in Tons	Cylinders Drivers	Source	Disposition
1	Lima	Shay 2-truck	633	1901	20	8" x 8" 27.5"	new	Brooklyn Coop., S.C.
2	Lima	Shay 2-truck	634	1901	25	8" x 12" 28"	new	Brooklyn Coop., S.C.
	Climax	2-truck	455	1903	35		new	scrapped 1919
	Climax	2-truck		1909			new	scrapped 1919

Equipment

Loaders	Barnhart
Log cars	flat cars with iron trip stakes

Note: These are the only locomotives known to have been used in the Adirondack operations. Other geared locomotives were purchased for operations in southern United States.

Lowell M. Palmer, President. C. R. Heike, Secy. A. Ward Brigham, Treasurer.

#20/1/561

Office of The Brooklyn Cooperage Co.

CABLE ADDRESS ENGELDON, NEW YORK.

Manufacturers of Flour, Sugar, Fruit, Oil & Syrup Barrels, also Molasses, Rum, Wine & Alcohol Shooks, FOR EXPORT.

AND DEALERS IN STAVES, HEADING, HOOPS SECOND-HAND MOLASSES HHDS. AND COOPERAGE MATERIAL GENERALLY.

184 Front St. cor. Burling Slip. New York, Nov. 19th. 1900.

BROOKLYN COOPERAGE COMPANY

CABLE ADDRESS PARBROOK, NEW YORK

117 WALL STREET
NEW YORK

Brooklyn Cooperage Company

St. Regis Falls

New York

GENERAL OFFICES
117 WALL STREET
NEW YORK

Dec. 28, 1921
File 540

ALL CONTRACTS SUBJECT TO FIRES, STRIKES AND CAUSES BEYOND OUR CONTROL

203. Letterheads.

two inches being the most common length. As many as sixty thousand staves would be cut daily, after which they were stacked to dry. The barrel components would then be bundled and shipped to the assembly plant. Brooklyn Cooperage operated a number of these stave and heading plants.

In addition to operating several mills in Pennsylvania and Missouri, Brooklyn Cooperage became established in the Adirondacks just prior to 1900. Stave and heading mills were set up at Santa Clara, St. Regis Falls, Tupper Lake, and Salisbury Center. For over twenty years the company operated on 123,000 acres of both company owned timberland and purchased timber (stumpage) in the North Country. Brooklyn Cooperage was a company experienced in railroad logging; thus, all four of the New York plants were fed logs by logging railroad operations.

Santa Clara and St. Regis Falls Mills

Santa Clara was the location of the first mill set up by Brooklyn Cooperage in New York State. In 1895, or shortly thereafter, the company leased one of the sawmill buildings belonging to the Santa Clara Lumber Company. The site was short-lived, however. On November 3, 1903, a raging fire in the small village destroyed a large company storage

204. A house of freshly cut staves at Brooklyn Cooperage, waiting the next step of steaming and shaving. Courtesy of the St. Regis Falls Historical Society.

205. Finishing and bundling the staves in 1906. Courtesy of the St. Regis Falls Historical Society.

shed containing a vast quantity of staves and also burned down a number of homes. Loss was estimated at $50,000.

The village of Santa Clara apparently held no further appeal as an industrial site after the fire; both Brooklyn Cooperage and the Santa Clara Lumber Company chose to dismantle the remaining machinery and move elsewhere. Brooklyn Cooperage transferred operations to a mill building in St. Regis Falls village, which they leased from the recently organized St. Regis Paper Company. The mill had been occupied for the previous three years by the Weidmann Stave and Heading Company, set up in June 1900 by Louis Weidmann after he had suffered the huge fire loss of a barrel factory in Brooklyn.

Santa Clara village didn't fare well after that 1903 fire. Another fire in 1915 destroyed the railroad machine shops, and then another fire the same year leveled both of the hotels. That brought an end to any industrial activity in a village that had once supported some large sawmill enterprises. Population shrunk from 2,000 to 675 in 1915. Remaining today is a small cluster of a few populated older houses that give no hint of the extent of former activity.

St. Regis Falls had already become an active village for the forest products industry after John Hurd's railroad arrived in 1883. With the expansion of facilities by the Brooklyn Cooperage Company, the employment force took another leap. Brooklyn Cooperage provided work for three hundred men in the mill and woods operation. Company operations were to last until about 1920 or shortly thereafter, when the factory closed down.

206. A loaded log car had been placed at the log pond, located in front of the sawmill. Circa 1905.

Railroad Operations at St. Regis Falls

Brooklyn Cooperage obtained their hardwood log supply not only by the purchase of timberlands but also by the purchase of standing trees as well as logs delivered railside. Standard-gauge railroad lines were built by the company to

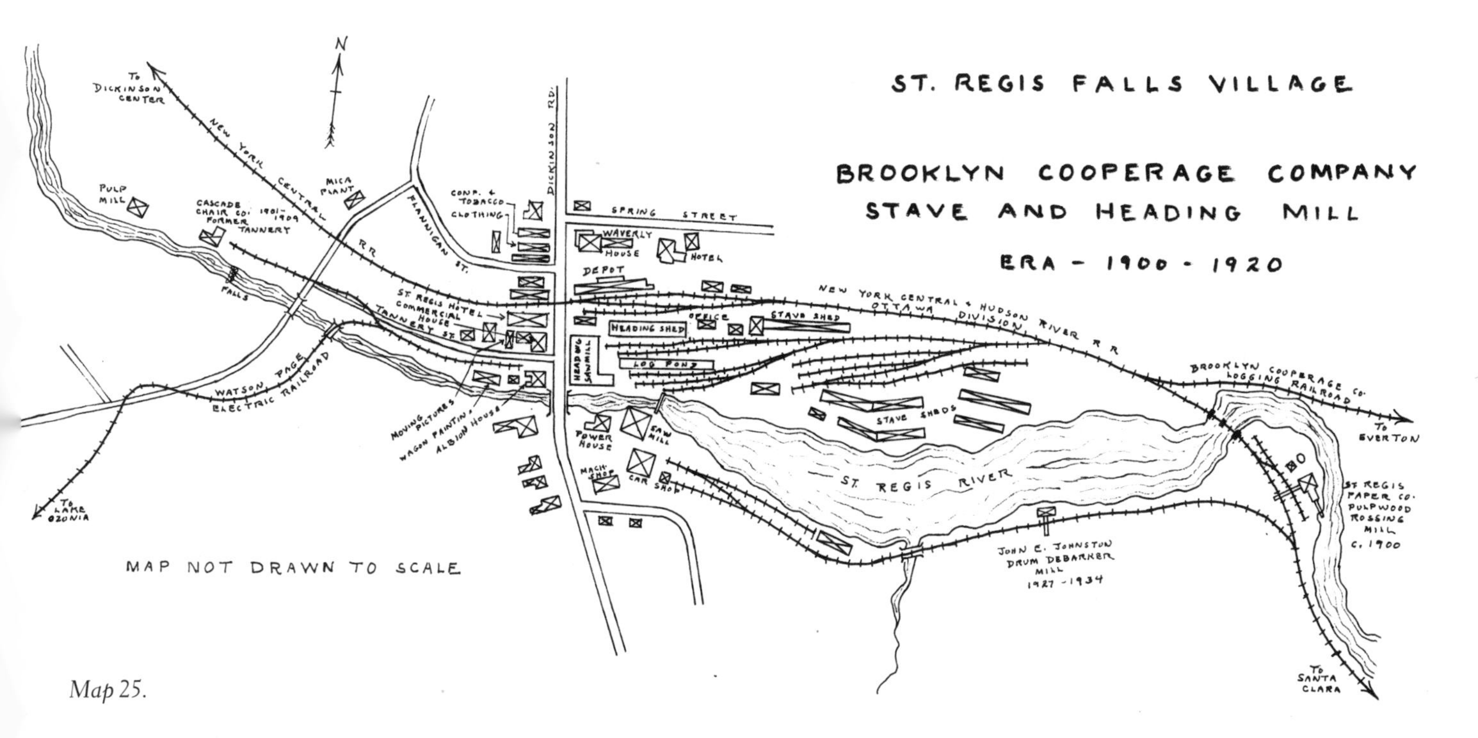

Map 25.

207. Some interesting poses can be noted in this gathering of Brooklyn Cooperage Company workers. Note the young workers lined up in the front row. Courtesy of the St. Regis Falls Historical Society.

208. The log cars appear to be all loaded now that the Barnhart loader is on the flat car next to the Climax. Identified in the photo are engineer Fred Bailey, fireman Floyd Fullerton, log train foreman Cap Lyman, loader engineer Frank Ploof, and tong hookers Joe Ploof and Orren Mallette. Courtesy of the St. Regis Falls Historical Society.

access most of the timber supply they had acquired. All of the company railroad lines junctioned with the New York and Ottawa Railroad (the Ottawa Division of the New York Central after 1906). Brooklyn Cooperage made arrangements to operate its own company locomotive and log trains over the New York and Ottawa tracks, to haul the logs directly into the mill at St. Regis Falls.

The following were the four different railroad branch lines built by the company to supply the St. Regis Falls mill.

Everton Branch Line

In September 1907 the Brooklyn Cooperage Company purchased the old Everton Lumber Company property from William Rockefeller, including twenty thousand acres of land and the old railroad right-of-way. Rail was laid again on the old Everton railway line, and hardwood logs began to roll out. Before long it was decided to extend this branch line, and in 1910 the surveyors laid out a new line, diverging at Cady Brook and extending northeast to Deer River. Construction of the line began in May 1910. This track was put in to reach a large tract of timberland in the Town of Brandon owned by the Reynolds Brothers. They had a contract with the company to deliver three to five million board feet of logs railside each year.

The Everton branch connected with the NY&O at the east end of the St. Regis Falls railroad yard. Total mileage in the Everton branch line was about seventeen miles, and logging activity on this branch extended over the period 1908 to 1920.

Meno West Branch

Meno was a small station stop located about nine miles south of Santa Clara, listed by the New York Central timetables as a stop in the early 1900s. The company built the logging line westerly from Meno into St. Lawrence County, the lower spur reaching over to the west branch of the St. Regis River, where the company had purchased property known as the Vilas Tract. Retired chief accountant John Arfmann stated many years ago that he didn't believe that the company ever built the lower spur into the Vilas Tract. Other reports and old maps suggest, however, that possibly this line was built through to the Cooperage Club on the west branch of the St. Regis River.

Construction on the Meno west branch began about 1908–10, the upper spur reaching Center Camp in 1911. Center Camp was a large logging camp on the Frank Cutting property that served as Cutting's logging headquarters. There were a number of buildings there and a variety of livestock kept for the seasonal loggers and the year-round resi-

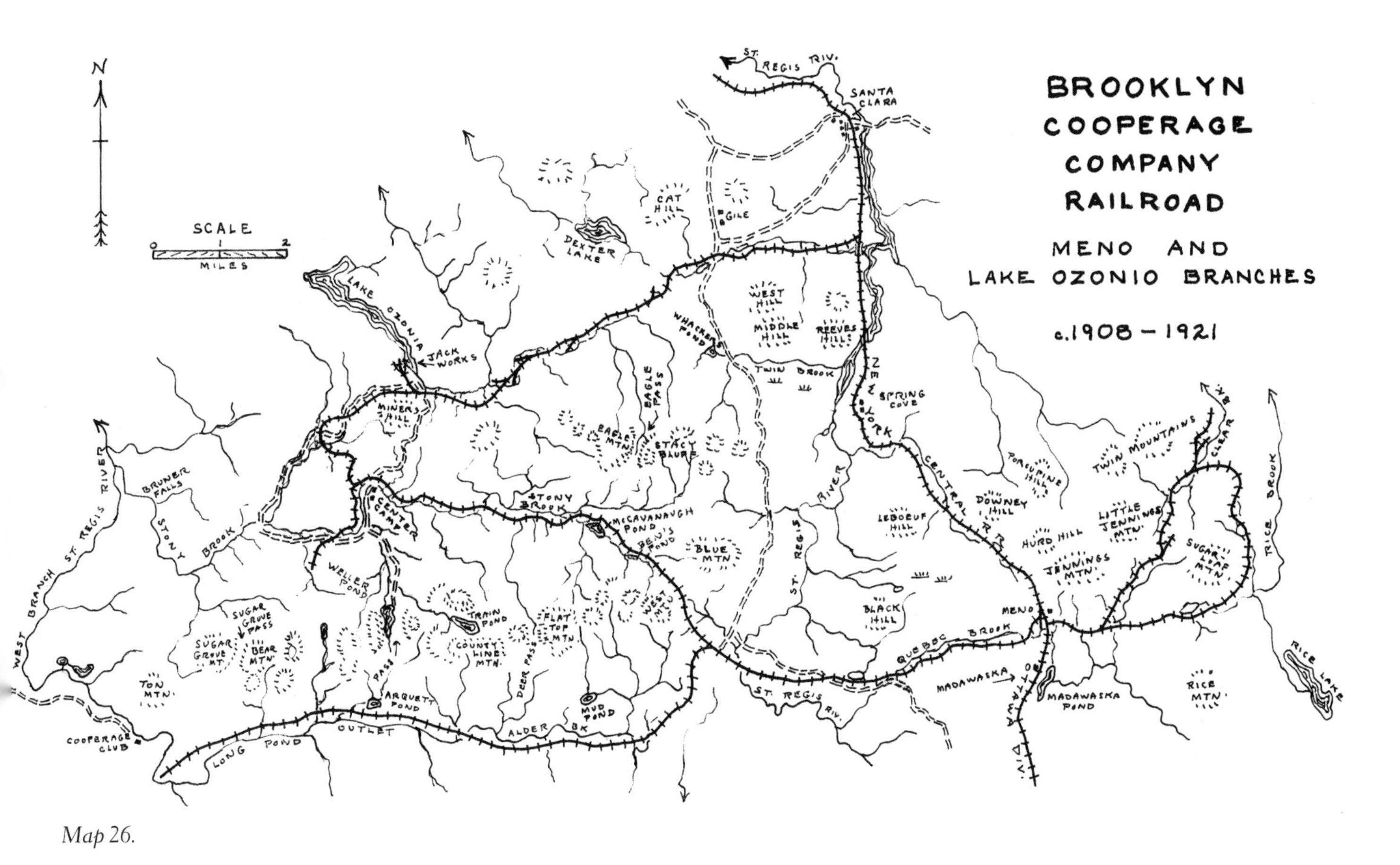

Map 26.

dents at the settlement. About a mile and a half further along the company railroad there was a fifteen-acre cleared field called the potato patch where the Center Camp residents grew vegetables. A small Lane number 2 sawmill was set up by Frank Cutting near Center Camp to saw railroad ties for Brooklyn Cooperage, but its short life came to an end on October 31, 1919, when it was destroyed by fire. Tracks were extended further west into the Frank Cutting property after the logging contract was made with Cutting in 1912.

Total mileage on the Meno west branch was about twenty-four miles. Connection with the New York Central was severed in 1921 and the tracks taken up in 1924.

Meno East Branch

The logging line east from Meno was put in at the same time as the west branch, about 1910. It extended for a total of ten miles, reaching to the Brighton town line and looping around Sugarloaf and Daniel Mountains.

Lake Ozonia Branch

Possibly the last of the four branch lines to be built by Brooklyn Cooperage, this ten-mile rail line went to the south end of Lake Ozonia and looped around to connect with the Meno west branch. It junctioned with the New York and Ottawa 1.8 miles south of Santa Clara village. This line was established in 1917 at the time that the Downy-Snell Logging Company began a large operation on Lake Ozonia.

The total railroad trackage built for the St. Regis Falls operation was about sixty-eight miles, including the nine-mile spur to the Vilas Tract, whose existence is questionable. The abandonment of the Brooklyn Cooperage railroad operations occurred by 1921, if not sooner on some branches.

Railroad Logging for St. Regis Falls Mill

Almost all of the logs delivered to the St. Regis Falls mill during the seventeen years of operation were brought in by

209. A Brooklyn Cooperage Company Shay was arriving at a landing of hardwood logs with a string of empty flats, near Lake Ozonia. Circa 1908. Courtesy of Virginia McLoughlin.

the company railroad. Brooklyn Cooperage operated four different locomotives at this location, two new Shays and two new Climaxes, all gear-driven locomotives. The logs were hauled on company-owned flat cars, loaded with car-mounted loaders.

In April 1912 the Brooklyn Cooperage Company executed a log purchase contract with Frank Cutting, one of the largest landowners in the region. Cutting was a dealer in hemlock bark, which he shipped to a number of leather tanneries located in the Boston area. In the 1880s Cutting had purchased nearly nine thousand acres in the Town of Hopkinton because of the prevalence of old-growth hemlock, and he made plans in 1887 to build a sawmill on the Trout River and construct a railroad to junction with the Northern Adirondack at Spring Cove. Preliminary surveys were made, but the line was never built. For many years afterward, Cutting's loggers hauled the two-foot by three-foot sheets of hemlock bark on horse-drawn sledges in the winter to Meno and Santa Clara for loading on the railroad.

Cutting had even designed and had built one hundred of his own railroad bark cars, open-top boxcars with slatted sides. The railroad would pay Cutting for the use of these privately owned cars, and of course, there was no demurrage charge when they sat at the loading locations in the woods for extended periods of time.

Discouraged by the fires of 1908, which burned over about two thousand acres of his land, Cutting decided in 1910 to log extensively both the hardwood and softwood timber. The Brooklyn Cooperage railroad operation gave him the opportunity. Brooklyn Cooperage had built the branch westerly from Meno and had reached Cutting's Center Camp in 1911. A 1912 log purchase agreement made the provision for Brooklyn Cooperage to extend the railroad for another four miles into Cutting's land.

Under the timber contract, Cutting was to sell all of the hardwood logs on his tract, thirteen inches and up in size, and would arrange to have it logged and delivered to within seventy feet of the railroad line. Logs were to be cut between September 1 and April 15, and Brooklyn Cooperage would pay $11 per thousand board feet delivered railside. Frank Cutting was to supply three to four million board feet of timber yearly.

Brooklyn Cooperage also agreed to haul out the softwood pulpwood and bark to Meno for Cutting at $5 per car and to haul in camp supplies at $7.50 per loaded car. When the contract was signed, it was expected that the operation would last for twenty years, but Cutting had no way of knowing that in two years the world would be thrown into World War I, upsetting the economy with a risk that fell entirely on Frank Cutting. Cost of operations rose drastically; wages of loggers in the northern Adirondacks increased from three to five or six dollars a day in 1918. His contract with Brooklyn Cooperage became a losing effort.

The banks gradually lost patience with Frank Cutting as his logging activities slowed down. Then in 1921, Brooklyn Cooperage gave up on Cutting and broke the rail connection between the logging line and the New York Central at Meno. Among the matters that then angered Frank Cutting was the fact that Brooklyn Cooperage knowingly stranded twelve of Cutting's bark cars back in the woods with no way out. Cutting's attempted legal action claimed a six-thousand-dollar loss. Before long, his creditors forced sales of much of his property, and in 1927 Frank Cutting died.

When Horace W. Downey and Kimball J. Snell of Pots-

210. Climax No. 2 waits the loading of the flats, decked with white flags above the boiler. Circa 1910.

211. The Frank A. Cutting letterhead illustrates the type of hemlock bark car he had built.

dam purchased the large tract surrounding the upper part of Lake Ozonia in 1917, they acquired thirty-five hundred acres of virgin timber growth. As experienced lumbermen, they expected to harvest three to five million board feet yearly for a number of years. With such a large quantity of old-growth timber, the Downey-Snell Logging Company persuaded the Brooklyn Cooperage Company to construct a new railroad branch to receive the logs. In 1917 Brooklyn Cooperage laid down a new rail line to reach the lower end of Lake Ozonia.

Log slides were built by Downey and Snell on the slopes facing Lake Ozonia to dump the softwood logs down into the lake. Logs were boomed and towed down the lake to a jack-works set up near the southern end of the lake. Hemlock bark was towed down on barges for loading onto Frank Cutting's bark cars. The hardwood logs were shipped to Brooklyn Cooperage and the softwood logs pulled from the lake, cut into four-foot lengths at a slashing mill, and sold to St. Regis Paper Company. At the peak of operations, Downey and Snell were operating four camps with over a hundred men in the woods. The sudden end of World War I brought a market decline and the Downey-Snell Logging Company likewise went into bankruptcy in 1919.

Tupper Lake Plant

At the same time that the Brooklyn Cooperage Company had established the plant in St. Regis Falls village, the company was induced to establish a presence in the village of Tupper Lake. In one of those strange chapters of Adirondack forest history, an ill-advised forestry project created a storm of controversy in the north woods and involved Brooklyn Cooperage in a losing effort.

A Runaway Train

In his book *Lake Ozonia: An Informed History*, William G. McLaughlin relates the story of the near tragedy when John and Ambrose Stark took out one of the last loads of bark from Lake Ozonia. Whenever the story was told to McLaughlin by either of the Stark Brothers, they got to laughing so hard that the facts seemed to come out a little different each time. But the incident apparently occurred about like this.

"Frank Cutting called John and Ambrose to help him load the last stacks of hemlock bark onto his cars on the Brooklyn Cooperage Railroad in about 1921 or 1922. There was an engine and four cars. Ambrose was a young fellow then and wanted the privilege and honor of taking charge of that last trainload of bark. So they told him he could handle the brakes. There was a steep hill from Sunshine Pond down toward St. Regis Falls, and the brakeman had to keep a tight hold on the brake or the train would run away. There was also a culvert or tunnel with a very low overhead, and the brakeman, after applying the brake from the top of the cars, had to get down between the cars or he would get knocked off the train.

"They got the four cars loaded. It was very cold, and the wheels were frozen to the tracks. The men had to take iron bars and crack each wheel loose before the engine could start pulling. The train started down the steep grade, and Ambrose began to pull on the brakes. But the brake was either rusted or frozen, and it would not budge. The train began to move faster and faster. Frank Cutting and John shouted, 'Put on the brake, the brake!' and Ambrose answered 'I can't, it's stuck,' 'Then jump, jump,' they yelled, but Ambrose had made too much of a fuss about how competent he was to manage the train to jump off. He felt responsible, and he would not jump. He just kept tugging on the brake. Frank Cutting grew alarmed and shouted to his son Bo, 'There will be a crash; I lose the load and Ambrose will be hurt.' Bob sprang into action, 'I'll try to head them off at the switch.' He jumped into his model T Ford and swung around through the woods on the road; he beat the train to the switch, pulled it, and sent the train racing down another track which went up a hill. The train came to a halt by itself.

"After the ride, Frank Cutting said to Ambrose, 'I knew you wouldn't be gone long if you came back as fast as you went!' "

German-born forester Bernhard Eduard Fernow had first come to the attention of the public in 1886 when he was appointed to a new forestry position in the federal government, the first academically prepared forester to go to Washington. Then in 1898 Fernow became the first director of the New York State College of Forestry, founded at Cornell University. The new college was the first professional forestry school in the United States, starting out in 1899 with sixteen students. In that same year, Fernow began a controversial forestry experimental project near Tupper Lake, new and untried in its concept.

The goal of the project was to demonstrate that cut-over northern hardwood timber land could be rejuvenated with spruce timber when properly treated. The newly established College of Forestry had received the authorization and the funds from the state to purchase thirty thousand acres of land in Harrietstown for $165,000, land formerly owned and heavily cut for softwood timber by the Santa Clara Lumber Company. College ownership was to extend for thirty years before deeding the property to the State of New York.

The arrogant and dogmatic Bernhard Fernow was employed to supervise the project of removing much of the hardwood cover, replanting softwood seedlings where needed, and establishing a market for the large quantity of hardwood material that was to be cut. To make a success of the experiment, it was essential to establish a commercial market that would be able to consume this large quantity of hardwood logs, even the small and low-grade material that Fernow decided must be removed.

The science of forestry was, at this time, still in its formative years, with the idea prevailing that extensive tree planting could be an effective management tool in the treatment of cut-over Adirondack timberlands. The foresters also thought, at the turn of the twentieth century, that the Adirondack "forever wild" provision affecting state-owned land would eventually be relaxed. Thus, they were preparing for

212. The pulpwood debarking mill set up by John E. Johnston in 1927 on the south bank of the St. Regis River. Log-length spruce was driven down the river from as far up as Mecham Lake, conveyed from the river, cut into four-foot lengths, and debarked by a rotating drum debarker. The debarked sticks were then conveyed over the adjacent railroad track by the high conveyor and loaded on New York Central cars. Six to eight cars were loaded per nine-hour shift, each holding sixteen cords. Courtesy of the St. Regis Falls Historical Society.

213. Bernhard E. Fernow, New York's first forester.

the time when state-owned Adirondack timberlands could be managed for the growth of spruce timber, still the preferred species of the lumbermen. This thinking motivated Fernow's experiment. Field headquarters were established in 1898 at Axton, the former location of the Santa Clara Lumber Company's logging headquarters.

To secure a means of transporting the hardwood material from the huge tract, Fernow was persistent in attempting to induce the establishment of a railroad branch. He had formerly suggested that the Santa Clara Lumber Company retain a right-of-way through the tract before selling the land to the college, and so he encouraged the New York and Ottawa Railroad to use the available right-of-way for their proposed extension of a railroad through to North Creek, where he hoped they could connect with the Delaware and Hudson Railroad and bypass the New York Central. Or, at least build south as far as Axton, he prompted.

Henry Gays, general manager of the New York and Ottawa, let it be known that it was their intention to build through to North Creek, but as events unfolded later in 1899, the state took action to prevent them from building through state-owned land. Further efforts by the New York and Ottawa to reverse the decision didn't prevail.

Undaunted, Fernow next tried to negotiate with the New York Central in an effort to establish a branch line into the school property. A five-mile line would be needed, switching from the NYC main line somewhere near Floodwood or Rollins Pond. Fernow had been negotiating with a wood alcohol manufacturer to build a forty-six-retort plant, probably on the college property; transportation was essential for the products: wood alcohol, acetate of lime, charcoal, and firewood for the urban markets. Negotiations with this manufacturer were not successful.

Again, Fernow received a disappointing reply. Edgar Van Etten, NYC general superintendent, could not foresee enough freight traffic to justify the investment and suggested the alternative of the state building and operating their own railroad on the branch. Fernow had no such funds to cover that proposition. By the end of 1899, both the NY&O and NYC made it definite that neither could build a short branch line into the state property. Fernow and his staff were dismayed but not thrown down.

Apparently Fernow did give some consideration to building his own railroad branch. A request regarding the purchase of some log cars brought a quote of $130 each for some 23.5-foot Russell cars, used but in top condition. Freight quotes were sought from the NY&O for shipping logs to Brooklyn Cooperage at St. Regis Falls and cord wood to lime kilns at Fonda and other locations.

Early in 1900, however, Fernow was able to convince the Brooklyn Cooperage Company that its presence in Tupper Lake village would be of benefit to both parties. In that year a stave and heading plant was built at a site northeast of McCarthy Street and adjacent to the NYC track. Two years later the company built another plant nearby, a wood chemical

214. The entire plant crew appears to have turned out for a photo of the Brooklyn Cooperage Company factory at Tupper Lake. Circa 1912. Courtesy of the Tupper Lake Public Library.

factory to utilize hardwood waste and cordwood for the manufacture of charcoal, wood alcohol, and acetate of lime.

By August Brooklyn Cooperage was in the first stages of building a railroad from the Tupper Lake plant into the college property. Eventually the track would extend for about seven miles, and Fernow's plan was to have the college finance the construction of short spurs connecting with the Brooklyn Cooperage railroad. Inquiries were made for available used lightweight rail and quotes received for forty-pound rail at about $29 per ton; then the NYC informed Fernow that it was disposing of some sixty-five-pound rail at a price of $25 per ton. Santa Clara Lumber Company at Tupper Lake chose not to sell what rail it had on hand because it was contemplating some future logging by rail, though it never happened.

The Brooklyn Cooperage railroad construction crew, made up of immigrant Italian laborers, under the company woods boss LeBoeuf, extended the new line as far as Meigs's boathouse on the Raquette River in the fall of 1900, but December brought unexpected troubles. The cold temperatures and eighteen inches of snow sent the Italian workers scurrying back to the city. As George Leighton, Brooklyn Cooperage superintendent, put it in a letter to Fernow in 1900, the "Italiens [*sic*] are leaving for New York to den up for winter. We are laying track with Canada French who for ten generations have been seasoning to this climate." Leighton had been complaining about the increased cost of building the railroad for Fernow. Common laborers were being paid $1.35 per day.

Under the agreement with the Brooklyn Cooperage Company, the college was to supply the railroad ties without cost for all of the new railroad. As for the short college spurs, Brooklyn Cooperage graded and formed the switch connection and then billed the college. By the summer of 1901, Brooklyn Cooperage had completed the railroad to the terminus at Cross Clearing. This location was almost out to Wawbeck Corners, where Routes 3 and 30 now junction. Route 3, in fact, is built over a portion of the former track bed.

Under a fifteen-year contract with the college, Brooklyn Cooperage paid $6 per thousand board feet for the hardwood logs and $2.04 per cord for the chemical wood delivered railside, a very attractive arrangement for the company. Cordwood was also shipped out to a lime kiln operation before Brooklyn Cooperage built their own wood chemical plant in 1902.

The spur tracks built by the college were not well built when it came to winter use. In November 1901 company official Robert Langlotz informed Fernow that they were having considerable trouble keeping the cars on the track; five cars had gone off on that particular day. Notice was given that no more rolling stock would be put on that track until repairs were made.

Fernow's efforts did establish some fine softwood plantations, which can be viewed today in the vicinity of Wawbeck Corners. His experimental goal, however, was doomed almost from the start. The college and forester Fernow were not only bucking nature in attempting to convert fast-growing hardwood sites into a softwood growth, they were also raising the wrath of wealthy camp owners on Upper Saranac Lake. The forestry crew was burning large piles of hardwood brush and slash, sending clouds of smoke over the camps. To add to the woes, logging costs eventually proved to be greater than financial returns.

Then a tragedy—after about six hundred acres had been heavily cut, a strong windstorm blew down or snapped off many of the residual trees. No sooner had Fernow's crew begun to clear this land than it was ravaged by the 1903 forest fires. The year was a dry one, but part of the blame was put on careless berry pickers. Brooklyn Cooperage had a large quantity of chemical wood stacked in the woods that was also lost in the fire.

Criticism came from many directions. Bernhard Fernow had applied some fundamental forestry practices that have

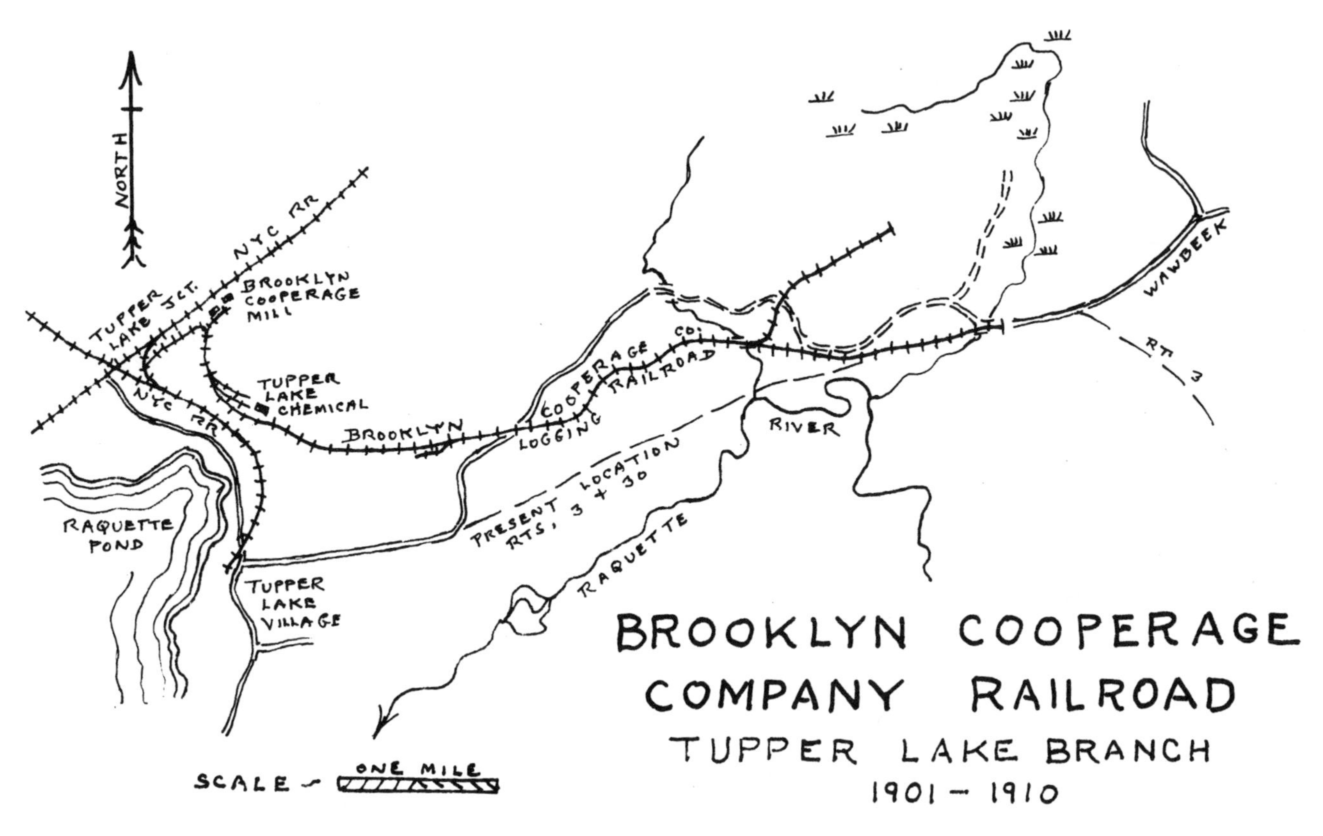

Map 27.

Railroad Construction Cost of Brooklyn Cooperage Company in 1900

The expense charged by the Brooklyn Cooperage Company for the two-day job of forming one of the connections of the College railroad spur with the company railroad:

Nov. 7:

38 men @ 7 hr. = 266 hours @ 13.5 cents = $35.91
2 foremen @ 7 hr. = 14 hours @ 25 cents = $3.50

Nov. 8:

34 men @ 7 hr. = 238 hours @ 13.5 cents = $32.13
2 foremen @ 7 hr. = 14 hours @ 25 cents = $3.50

Total cost: $75.04
These figures include the cost of distributing ties.

Source: Brooklyn Coop. Co. letter to Bernhard Fernow.

215. The Brooklyn Cooperage loader was placing maple logs onto company flats at Cross Clearing. The rack car in the background was used to haul small material, probably wood for the chemical plant. Circa 1910.

216. Bernhard Fernow's plantations in the area of Wawbeek Corners developed into a showpiece of white pine and Norway spruce stands. Photograph by author.

since proven to be quite valid. But he was accused of forest devastation in an era when forest conservation unfortunately meant forest preservation. Added to that, Fernow managed to alienate many legislators, states officials, and all the landowners because of his arrogant disposition. The results were that the legislature withdrew funds for the College of Forestry in 1903, and lawsuits were initiated that transferred title of the land from the college to the state. The college facility at Axton was closed down, and the faculty was sent back to Cornell University.

The college's woods crew continued to cut and deliver some timber to Brooklyn Cooperage until 1905, when the state stepped in and sued both the college and Brooklyn Cooperage. With the investment of building the railroad still to be recovered, the Brooklyn Cooperage Company delayed compliance of the order to cease and countered with a $1,200,000 damage claim against New York State. The courts eventually decided in favor of the state in 1910, and continued appeals after that by Brooklyn Cooperage did nothing to win back their source of inexpensive wood. The company abandoned the railroad and pulled up the rails.

To supply the Tupper Lake mill with logs, Brooklyn Cooperage had to look elsewhere for a source. Log trains were run into Tupper Lake over the Ottawa Division of the New York Central from the branch lines built to supply the St. Regis Falls mill. In 1903 Brooklyn Cooperage purchased five thousand acres of timber from William Rockefeller on his Tekene tract, and log trains rolled into Tupper Lake over the Adirondack Division of the New York Central Railroad. (See chapter 19 on the Kinsley Lumber Company railroad at Tekene.) This timber had to be cut immediately because the area had recently burned. The woods superintendent at this time was A. E. Underhill.

At one point in its history, the Brooklyn Cooperage Company tried an experiment of its own in reforesting some cut-over lands on the Vilas Tract. They planted sixty thousand white, Norway, and Scotch pine seedlings and two thousand locusts. They had planned to continue the effort if successful but met with similar failure. Not only were some of the tree species inappropriate, but survival must have been scant among the new natural hardwood growth.

The Tupper Lake operation of Brooklyn Cooperage at its peak had over 150 men working in the mill and the logging camps. This mill continued until about 1920, when the

DESTRUCTIVE LOGGING

The Brooklyn Cooperage Company controlled 123,000 acres of timberland by ownership or lease in the Adirondacks, and the company's logging practices showed little concern for future timber growth. Forester Gifford Pinchot once commented about the company.

"Logging done by this company is more destructive than any other which I am acquainted with in the eastern states, and the damage by fires, for which its carelessness is said to be responsible, will cost the people of New York large sums of money and long years of time to repair."

company ceased their Adirondack operations. The wood chemical plant was operated until 1915, when it was leased to a new company, the Tupper Lake Chemical Company. This new company kept the mill operating until the abandonment of the site in 1919. The location was later occupied by the Draper Corporation for many years.

The fourth stave and heading plant built by the Brooklyn Cooperage Company in the Adirondacks was constructed in Salisbury Center in 1914. However, the railroad that fed this plant was built and operated by the timberland owner, not Brooklyn Cooperage. (See chapter 32 on the Jerseyfield railroad.)

The Brooklyn Cooperage Company closed down its many New York operations in 1920–21. Operations were then shifted to a company operation in Georgetown, South Carolina, and another one built in 1927 in Sumter, South Carolina. The locomotives and other railroad equipment were sent south to log the southern hardwoods.

The Brooklyn Cooperage name continued in use for many years afterward. The company began to manufacture folding cartons in 1916 and established a carton plant in Versailles, Connecticut, in 1928. The use of wooden barrels, however, began to fade out; changes in food packaging and merchandising pushed them into oblivion by the mid-1940s. The parent corporation, Amstar, continued to call its carton manufacturing company the Brooklyn Cooperage Company for many years, until Amstar was merged with another company.

CHAPTER TWENTY-FOUR

Bay Pond Railroad

The sandy plains of the northern Adirondacks once supported a magnificent growth of white pine, lofty conifers that were craved by the early lumbermen. Though much of New York State's white pine had been picked over by the mid-1800s, it wasn't until the late 1880s that John Hurd's railroad laid iron among the pine stands scattered along the headwaters of the St. Regis River. Intense railroad logging over the ensuing thirty years, plus severe forest fires, were to leave a land vastly changed in character.

When logger John N. McDonald of Utica brought his Mac-A-Mac Corporation railroad from Brandreth up to Franklin County in 1924, there was only a sparse remnant remaining of those once-towering pine forests of the previous century. But pine was not the primary attraction for logger McDonald. Along with his partners, McDonald had acquired a huge area of spruce and hardwood forest, and it was the spruce pulpwood that they sought.

The Mac-A-Mac Corporation purchased a fifty-two-thousand-acre tract of land that had been put together back in the 1890s. The creation of that tract involves one of the strangest tales of Adirondack land ownership.

The tale began back in 1881 when a Michigan lumberman named Patrick Ducey built one of the first large sawmills in the Adirondacks. He carved out a village among the huge white pine along the upper portion of the St. Regis River, a village that soon grew to a population of twelve hun-

Table 22
Bay Pond Railroad

Owner	Bay Pond Inc. (John N. McDonald)
Dates	1924–1932
Product hauled	spruce and fir pulpwood, white pine sawlogs
Length	13 miles
Gauge	standard

Motive Power

No.	Builder	Type	Builder No.	Year Built	Weight (tons)	Cylinders Drivers	Source	Disposition
2	Alco-Schenectady	2-6-0	53845	1913			Mac-A-Mac, Brandreth	Whitney Ind.
33	Baldwin	0-6-2T	6264	1882	38	16" x 22" 38"	Emporium Forestry, Co., 1924	Whitney Ind.

Rolling Stock
Flat cars

dred and became known as Brandon. His sawmill on the bank of the St. Regis was originally water powered, later converted to steam, and was said to be one of the largest then in the Adirondacks. Eight-saw gang rigs were used to put out 125,000 board feet of pine daily, fed by his purchase of thirty thousand acres of virgin timber.

Ducey's enterprise had been able to attain such a proportion with the assistance of John Hurd's Northern Adirondack Railroad. Built initially to haul forest products, "Hurd's road" could be said to be the first logging railroad built into the Adirondacks. Lumber soon became the road's lifeblood, coursing north to a junction with the Ogdensburg and Lake Champlain Railroad and on to market. Hurd's newly built railroad created new lumber manufacturing centers—including the greatest of them all, Tupper Lake—as it pushed through virgin territory. The rails had reached the little settlement of Brandon in 1886.

A number of branch logging lines and spurs were built off of Hurd's railroad by other individuals. Even Pat Ducey built a short railroad spur, which junctioned with the Northern Adirondack midway between Bay Pond and Derrick and went in two and a half miles northwesterly to a jackworks on the west branch of the St. Regis River.

Patrick Ducey's village of Brandon grew rapidly, and employment mushroomed there after the appearance of Hurd's Northern Adirondack Railroad. There were four hotels, two churches, four general stores, a hardware store, a drugstore, a town hall, and a school along with a number of inexpensive homes. But lumber was the sole reason for its existence, and the appetite of Ducey's large sawmill and the ravages of forest fire soon wiped out this precarious resource base. By 1890 the sawmill boilers were cold, and Ducey was looking for a buyer. Paul Smith, the famous resort owner, was reportedly offered the land for $1.50 per acre but for lack of finances turned down the deal, which he was later to regret.

This was the era when those with wealth were able to amass large Adirondack acreage for a fashionable and secluded summer retreat. Thus, Patrick Ducey didn't have to wait long before an agent of William Rockefeller, brother to John D., was knocking at the door. Most of Ducey's ownership, twenty-seven thousand acres, passed to William Rockefeller's ownership for $49,000 in July 1898, to make a planned quiet mountain retreat, which he called Bay Pond. Succeeding events did not develop nearly as quietly as Rockefeller had planned.

It seems that a number of the Brandon residents had previously purchased the land title to their house lots from Pat Ducey, and when Ducey sold his acreage to William Rockefeller, they were suddenly isolated in the middle of a private preserve with no access short of trespass. Most accepted Rockefeller's fair offer to sell out, but a few resisted and thus began what was to be a battle of bitter feelings that stretched out for many years, even generations. Ensuing efforts to settle the differences, and the court battles over hunting and fishing violations on Rockefeller's posted land, intensified local resentment against him and even inspired national press coverage to the point where the original facts got lost in the scuffle. The dispute was popularly ballyhooed as the rich man persecuting the poor man.

Rockefeller attempted many strategies to clear out any remaining residents. By 1910 there were only four families left, a total of twenty people; at its prime the village had supported a prosperous population. Brandon had been one of the leading North Country communities. But now with the Rockefeller ownership entirely surrounding the sparse population, the only access was by railroad. The lone road to Brandon was posted "No Admittance" by Rockefeller. One of the residents, Oliver Lamora, was forbidden by court injunction to step foot on the road because of his defiance of William Rockefeller.

It was under these circumstances that the New York Central Railroad vice president Albert H. Harris requested permission from the Public Service Commission in April 1910 to close down the Brandon railroad station. Sparse passenger business and low freight receipts were cited as the reasons for the discontinuance. But not to be overlooked is the fact that Rockefeller was a power in the New York Central.

Only ten days were required for the commission to come up with a ruling in which they refused to consent to the request. As chairman Stevens stated, "to drive out these people from their homes it would be difficult to conceive a more efficacious plan to that end than to discontinue the station." The station and a very few residents remained.

There were two sides to the issue, as there usually is, but it soon became evident that local folks saw only the plight of poor Oliver Lamora, chief holdout and leader of the opposition. Threats were made on Rockefeller's life, and Bay Pond, the quiet summer retreat, became at one point an armed

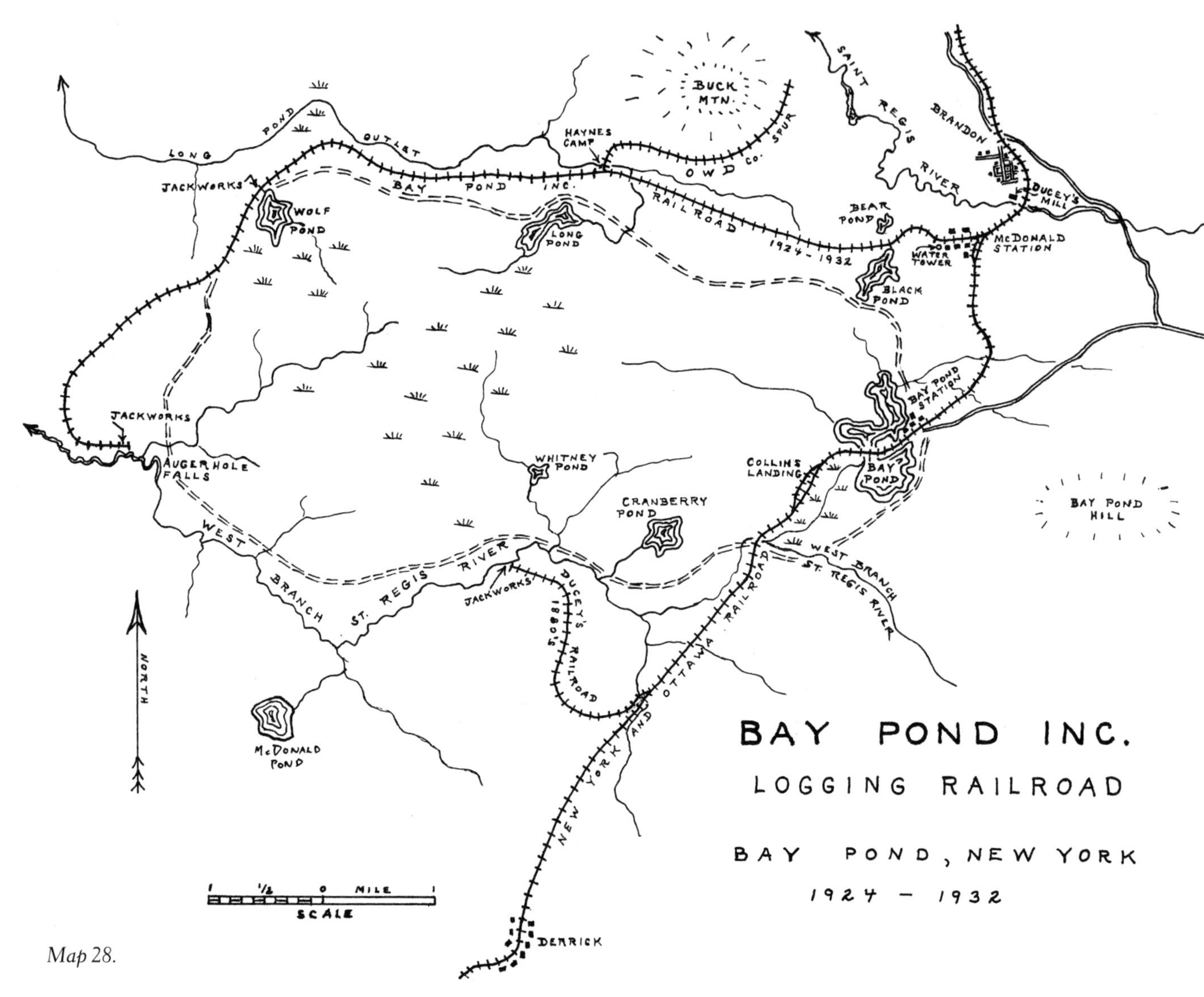

Map 28.

camp. It took many years before feelings abated, but Lamora's death in 1915 brought an end to the active struggle by everyone except one lone resident—Old Black Joe.

Joseph Peryea, Old Black Joe, had been the very first settler at the Brandon site, preceding Pat Ducey by a number of years and, as matters turned out, was the last settler to leave. Old Black Joe vowed that the Rockefellers would never get his land, and he never gave in, living out his life in that lonely and secluded location, catering to a few sportsmen and occasional logging camps with his small store and eatery. Joe died in 1941, and today the only visible evidence of the entire large sawmill village of Brandon are a few gravestones hardly noticeable in the brush.

Despite the lack of peace with the locals, William Rockefeller added to his holdings to amass a fifty-two-thousand-acre property at Bay Pond for the family's palatial summer residence. Before his death in June 1922, he had turned the property over to his son, William G., but William G. Rockefeller also died in year, in November, and to settle the estate, it became necessary to sell the entire property. William G. Rockefeller's son, William A., could not afford, it has been said, to take over the entire property. Therefore, the estate was sold in December 1923 to one of the Adirondack's outstanding logging operators and railroaders, the Mac-A-Mac Corporation.

Bay Pond Railroad

The five-man combine that purchased the huge property for the price of $700,000 included the same three investors who had previously formed the Mac-A-Mac Corporation at Brandreth: John N. McDonald, Vasco P. Abbott, and Benjamin B. McAlpin. The company named Bay Pond Inc. was formed with John McDonald, again the working member of the investment group.

The large purchase included the Rockefeller forty-room mansion, a railroad station, stables, and various and sundry homes and buildings, but it was logging that these speculators had in their plans. The physical assets of the Mac-A-Mac Corporation, including the railroad equipment, were moved in to do the job.

A headquarters settlement was established on the mainline railroad, Hurd's old railroad, which by this time had become the Ottawa Division of the New York Central. The site was about a mile south of the location that had once been Brandon village. In May 1924 a dirt road was cut across the estate from the state highway to the New York Central Railroad, and the village of McDonald came into existence. Included in the new settlement was a boardinghouse, railroad station, store, office, about seven residences, and a post office known officially as McDonald.

The company's logging railroad was surveyed by Parker Brownell and constructed under the supervision of Bill Pool, extending westerly from the settlement of McDonald and covering in total almost thirteen miles of track. Much of the construction work was done by immigrant Italian laborers shipped up from Utica.

The locomotive roster underwent a change from that in Brandreth. The Heisler geared locomotive (No. 1) was not brought up from Brandreth but was replaced by a small rod engine. This switch gave the company two rod locomotives, which allowed for faster speeds on the relatively level grades at Bay Pond than would be possible with a geared engine. John McDonald used to relate with pride that the president of the New York Central once rode over the logging railroad to view McDonald's lumbering activity and proclaimed it to be the smoothest roadbed in the state.

The other locomotive, the one designated as No. 2, was a mogul (2-6-0) built by Alco-Schenectady in 1913. It was sold to the Emporium Forestry Company in 1941, where it continued in service as a logging locomotive out of Conifer,

217. John McDonald purchased No. 33, a thirty-eight-ton Baldwin saddletank, from Emporium Forestry Company in 1924. The above photo was taken at the Emporium mill in Keating Summit, Pennsylvania. The young man on the left is Jesse Blesh, later to become Emporium's master mechanic. From the collection of Thomas T. Taber III.

218. The Alco-Schenectady mogul No. 2 bought new by John McDonald in 1913 for use at Brandreth. After use at Bay Pond and Whitney Park, he sold it in 1941 to the Emporium, who eventually scrapped it in 1951.

219. The locomotive had arrived at the landing to pick up a string of loaded cars. The railroad was shipping about a hundred cords of log-length pulpwood each day. Circa 1926.

being scrapped in 1951. Among the engineers used at Bay Pond was a familiar name from the Brandreth operation, George LaFountain, with his son George Jr., firing. Other engineers on the woods line were Fred Bailey and John Cuddy, who had previously operated a logging loco for Oval Wood Dish Company at Kildare.

Logging operations for the first couple of years were managed directly by Bay Pond Inc., using jobbers Eugene A. McDonald, Kenneth Hunter, Fred Gautremant, Duncan McDonald, and Bill McCarthy under the watchful eye of George McDonald, woods superintendent and scaler. The jobbers contracted to cut, peel, skid, and sled-haul the softwood to a number of different railside landings or to the jackworks. Contractor Eugene McDonald received $7 per cord for pulpwood delivered railside in log length, about $1 of which he realized in profit. His crew put twenty thousand cords of wood to the landings in 1925, not a bad profit in those days.

The log sleds were loaded from tall skidway decks, often built up to a height of eight feet above the sled bunks. To get a large load, the loggers would build a double-header skidway with the rear portion higher for topping off the sled load. Probably about six cords at most could be rolled on from the lower level, but after completing the load from the upper level, it wasn't uncommon to put ten cords on a sled. A load of this size required a good team of horses and well-maintained sled roads.

There were two jackworks on the Bay Pond operation, the one at Wolf Pond operating through much of the summer season. The jackworks at Augerhole Falls was located below the man-made sluiceway.

The first season of operation (1924), which extended from the end of May until the end of the following March, put about forty thousand cords of peeled spruce pulpwood into the St. Regis Paper Company yard at Deferiet. The next year saw almost one hundred thousand cords of pulpwood logged, and then a change was made in the operation. St. Regis purchased the remaining softwood stumpage and put their own crews on the land. The Oval Wood Dish Company from Tupper Lake did likewise when they made arrangements to purchase the hardwood timber, putting in their own jobbers after the softwood had been cut.

There was not a great amount of white pine left on the tract by this late point in history, the previous logging by Ducey and the serious fires of 1903 across the sandy pine plains having removed much of it. There were occasional scattered patches of pine, which were railroaded to Santa Clara Lumber Company's "Big Mill" at Tupper Lake before it closed down in 1926. Much of the sawn pine lumber was marketed to the Eastman Kodak Company. This remnant of the old-growth pine was large, however, sometimes a single log making a carload. Logger Jim Camp recalled one log that contained twenty-two hundred board feet.

Contractors used by St. Regis and Oval Wood Dish included Gasper LaPort, Billy Smith, Frank Duplintz, and A. C. Bricky. There were a number of Indians employed in the camps, mostly from the reservation near Hogansburg.

220. The smoke plume of the pusher engine is quite visible on the other end of the trainload of pulpwood headed for McDonald Station. As was customary, the company mogul is on the front end. Circa 1928.

Although horses were, of course, still the primary woods power, an occasional tractor began to appear on the log jobs. In the winter of 1926 jobber Eugene McDonald was drawing logs out of a cut northwest of Buck Mountain and was faced with an uphill pull too difficult for horses. He solved the problem by purchasing a ten-ton Linn tractor, one of the first used in the Adirondacks. The price tag new—$5,000.

At the Haynes camp, McDonald bunched the logs with a small Ford tractor and made the long hauls with a Monarch tractor, a large cumbersome machine that took plenty of room to operate. Howard Haynes was a teamster who worked for Eugene McDonald during the winter of 1924–25 and who was made a subjobber in the spring of 1925. His camp was at the base of Buck Mountain, where there was a railroad spur to the camp from the main line of the logging railroad.

The Bay Pond camps were tar-paper covered, often with holes big enough for the snow to blow inside. Blankets were said to be a thick wool flannel made from damaged wood-chip belts discarded at the St. Regis Paper Company mill. If a logger wanted clean blankets and a pillowcase he had to bring them in himself.

It was customary in the logging camps for the men to help themselves to available venison, regardless of season or limits. Thus, the railroad engineer had a scheme to provide advance notification to the camps whenever he had a game warden aboard who was hoping to nab a lawbreaker or two. When close enough for the camp residents to hear, the engineer would pretend to be whistling a critter off the track, using a whistle combination known as a warning by men at the camp.

Mechanized loaders were making an appearance, although much of the wood was still loaded by hand. On the railroad spur that Oval Wood Dish built around the southerly and easterly flanks of Buck Mountain, there was a track-mounted gasoline loader used for loading hardwood logs.

The pulpwood was hauled in sixteen-foot lengths on the Bay Pond railroad and loaded in the woods or at the jack-works on flat cars with high stakes. It is said that there was a "slasher" set up at the junction with the New York Central to cut the wood into four-foot lengths and drop it into rack cars.

Although the companies that had purchased the stumpage had their own jobbers on the tract, John McDonald still maintained supervision from his little office at the small settlement of McDonald. Office and quarters were in one of the residences, where he would spend four days of the week and then commute by train to his Utica home. Office manager at the settlement was Fay Soultz, and Henry McDonald was storekeeper.

Operations at Bay Pond continued until 1932. Total cut during the nine years of operation was reported to be 350,000 cords of spruce and fir pulpwood, 40 million board feet of white pine, and 45 million board feet of hardwood. As often occurred after logging, fire ravaged much of the tract in 1934, leaving even to this day large barren sandy plains, especially noticeable along the abandoned NYC railroad right-of-way. The New York Central Railroad later abandoned John Hurd's crooked railroad in 1937.

Bay Pond Inc. had no further use for the fifty-two-thousand-acre tract after the logging was completed, and a purchase offer was made to the State of New York, which was turned down, although they did purchase about six square miles around St. Regis Mountain. Then, in December 1935, the southern portion of the tract was sold back to the Rockefeller family, involving this time about twenty-three thousand acres, and soon after, the northern portion of the twenty-eight thousand acres was sold to Donald Ross of Delaware, one of the Dupont heirs.

The Rockefeller purchase was made by William A. Rockefeller, grandson of the William who had made the original purchase in 1892 and who had to incur so much wrath from the national press and the local sportsmen for wanting privacy on his summer estate. Although the 1937 purchase back into the family encompassed only about half the acreage originally owned by the family, the property transfer did include the spacious buildings and grounds around Bay Pond.

But it also included something else—Old Black Joe Peryea, Rockefeller's nemesis, who was still living out his days at the Brandon site, still vowing that the Rockefellers would never chase him out. Black Joe is gone now, as are the next three generations of William Rockefellers after the original William, and the estate is still maintained by the Rockefeller descendants. There has been some forest management applied professionally in recent years, though much of the property has been badly burned over since the days of the Bay Pond railroad. The land is slowly recuperating with a new growth of spruce and pine.

CHAPTER TWENTY-FIVE

Oval Wood Dish Corporation Railroad

EXCITEMENT SWEPT THE VILLAGE of Tupper Lake in 1916—a new large corporation was said to be moving in that would bolster the sagging sawmill industry in the village. Tupper Lake village had, years before, reigned supreme in the state as a center for spruce lumber manufacture, leading the state in 1907 with a cut exceeding forty million board feet of spruce.

However, times had changed in the years just prior to World War I. The diminishing supply of the old-growth spruce resource and the establishment of the spruce pulp and paper industry elsewhere had forced the closure or slowdown of much of Tupper Lake's mainstay industry. The A. Sherman Lumber Company mill on Demars Blvd. had finally ceased operation in 1915. Thus, the word spread quickly when it was revealed that a major hardwood timber industry, the Oval Wood Dish Corporation (OWD), would be moving in. It would prove to be the largest mill ever to be set up in the lumbering community of Tupper Lake.

Oval Wood Dish Corporation had its origin in Mancelona and Traverse City, Michigan, in 1883 when Henry S. Hull began to manufacture hardwood bowls for food containment. As the available timber supply began to diminish in Michigan, the company looked elsewhere and noted the large stands of virgin hardwood in New York's Adirondack Mountains. Before there was any consideration of exactly where a New York mill should be located, the Hull

Table 23
Oval Wood Dish Corporation

Dates	1918–1926
Product hauled	hardwood sawlogs
Total length	47 miles
Gauge	standard

Motive Power

No.	*Builder*	*Type*	*Weight (tons)*	*Builder No.*	*Year Built*	*Source*	*Disposition*
1	Heisler	2-truck	63	1352	1916	new	Dist. of Columbia, Dept. of Corrections
2	Heisler	2-truck	63	1368	1917	new	Sisson White Co.

Equipment

Log cars	65 flat cars, wooden-silled, truss rod underframe
Loaders	2 Raymond gasoline-powered, car-mounted
Derrick	Raymond self-propelled, car-mounted
Skidder	Lidgerwood steam skidder

Note: Weight of Heislers was 63 tons with 5 tons coal and 2,500 gallons water. Maintained 200 pounds steam pressure.
[a] It's questionable that No. 3 Heisler was used much, if at all, by OWD.

221. *Letterhead.*

222. *Oval Wood Dish Company factory, Tupper Lake.*

family began to purchase Adirondack timberlands as an investment. That was in late 1914. The Follensby Tract and a portion of the Mount Morris Tract, 11,805 acres, were purchased from the Santa Clara Lumber Company for $225,000. The following year, 1915, an option was acquired on lands of the A. Sherman Lumber Company. Eventually a total of 75,000 acres of timberland was accumulated, supporting a reported 275 million board feet of hardwood timber. The purchased land had been logged over for softwood timber by the previous owners.

The State of New York obviously was targeted for a new plant. The company's initial choice for mill location was Utica, with its good rail connections. The New York Central Railroad, however, preferred otherwise; their struggling Ottawa Division in the Adirondacks needed new business. Pressure was put on Oval Wood Dish to locate in Tupper Lake, closer to the timber supply, where, incidentally, the New York Central had no railroad competition for hauling outbound products.

To increase the pressure, the New York Central quoted freight rates from Tupper Lake that were only about one-third the rate of hauling logs all the way to Utica, provided

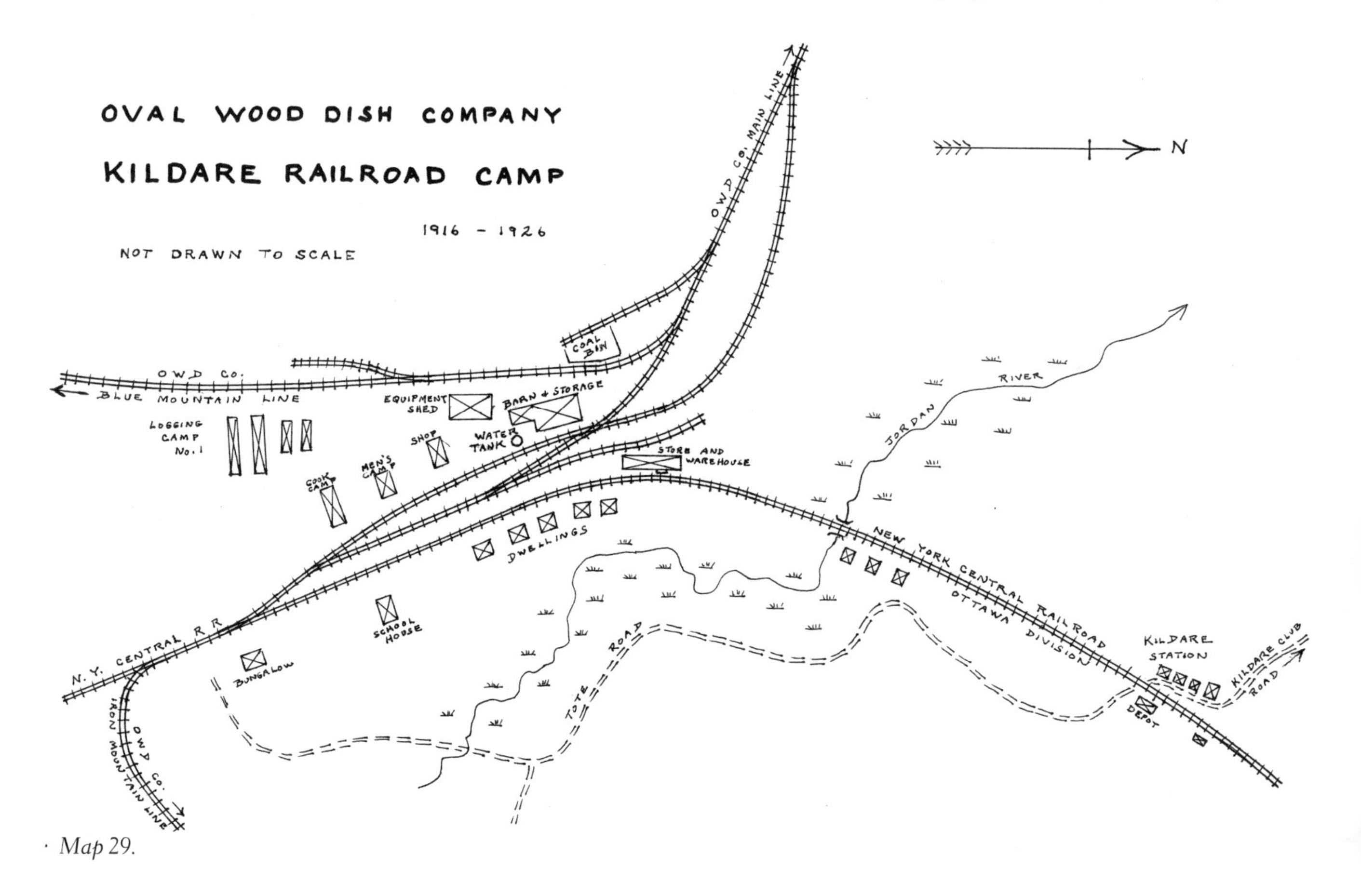

Map 29.

there was a minimum of four thousand cars outgoing per year. Thus, freight rates became an important factor in Oval Wood Dish's announcement to locate in Tupper Lake.

The site of the idle A. Sherman Lumber Company sawmill on the shore of Raquette Pond between Tupper Lake village and Tupper Lake Junction was purchased in 1915. Formerly a large softwood lumber mill, the mill was rebuilt to produce the variety of hardwood products made by Oval Wood Dish: throw-away wood veneer dishes, wooden spoons and forks, ice cream sticks, and hardwood flooring and lumber. The 350,000-square-foot factory and a second plant built in Potsdam employed 500 in addition to 150 workers on the woods operations. Oval Wood Dish quickly became the largest single industry in Franklin County. The Tupper Lake plant included a double-band sawmill equipped with a Wick's sash gang saw, capable of sawing twenty-one million board feet of hardwood yearly at peak production.

Plant construction was accomplished during World War I, which meant dealing with tough times and a scarcity of labor. Thus, the company brought in a labor recruiting train from Green Bay, Wisconsin, hauling in any company labor willing to relocate and even picking up potential new employees en route.

Workers have never hesitated to use the shortage of skilled labor as a lever to boost wages. In September 1918 the Emporium Forestry Company foreman of lumber handlers attempted to lure the Oval Wood Dish foreman away with the tempting wage offer of $1.00 for each thousand feet of lumber piled. The Oval Wood Dish foreman didn't accept but used the offer to force up his own wages, much to the consternation of general manager William Cary Hull. He promptly wrote to Emporium president William Sykes, informing him of the offer and adding that he was sure the Emporium would not approve of such underhanded action. Hull also reminded Sykes that he had been approached by a number of Emporium employees for positions or contracts and had always turned them down.

Railroad Operations

Construction of an extensive standard-gauge railroad logging operation began in 1918 on the company's large timber holdings north of Tupper Lake. The base of logging operations was established at Kildare, a stop since 1896 on the old New York and Ottawa Railroad, by this date a division of the New York Central. Kildare was about seven miles north of Tupper Lake Junction. The establishment of the new mill at Tupper Lake and the logging headquarters at Kildare definitely sparked new life in the NYC Ottawa Division, at least for a few years.

223. The camp boss had his dog along for a pose at Oval Wood Dish's Kildare railroad camp. The New York Central track is on the left, behind which are some crude dwellings. Beyond the Oval Wood Dish tracks on the right are the barn, water tank and, at the rear, the crew quarters. Circa 1920. Courtesy of the Tupper Lake Public Library.

The logging and railroad headquarters at Kildare, also known as Camp No. 1, had a large central supply store and warehouse, barn and storage, equipment shed, shop, men's camp, cook camp, several small dwellings, and a schoolhouse. The settlement was about 350 feet south of the old Kildare Station building, which is now on permanent exhibit at the Adirondack Museum in Blue Mountain Lake.

The Kildare school district operated for about ten years. But because the only access to the camp was by rail, the school teachers either had to return to Tupper Lake on the late evening train or stay at the settlement during the week.

The logging railroad was built by contractor James Angello of Ridgway, Pennsylvania, using crews of Italian immigrants. Little care was taken to give the roadbeds much of a permanent nature. Hollows were filled in with old rotten logs and brush when the short-life spurs were built, creating a somewhat springboard condition for the rail. Ties were often young trees cut along the way, hewed flat only at the

224. The Kildare school in May 1920 with the complete student body. Courtesy of J. T. Giblin.

225. Cutting ice at Kildare for food storage in the log camps. From left: woods superintendent Roy LaVoy; Tom Olds, a camp cook for many years; and Lawrence Cheverette, a Tupper Lake postal clerk who helped cut ice on Sundays. The ice cakes were pulled off the lake on the sled. Courtesy of J. T. Giblin.

226. Ferdinand Jensen, on the left, was a Swedish sailor who came to the Adirondacks and worked for Oval Wood Dish for many years. After the Kildare operation was completed, he lived as a hermit deep in the Oval Wood Dish woods. Courtesy of J. T. Giblin.

spot where the rails were fastened down with a couple of spikes.

Although many logging railroaders used lighter rails to decrease costs, Oval Wood Dish chose to use seventy-pound rail. The heavier rail gave better resistance to deflections over voids and soft spots in the roadbed, reducing track maintenance. The rail was obtained from the New York Central under a lease-rental agreement that included the provision that upon completion of the operation, Oval Wood Dish would pick up the iron and return it to the New York Central.

When laying out the track location, the Oval Wood Dish surveyor would sometimes avoid the flat, more obvious locations for the roadbed. Grading on a sloping terrain would increase the initial costs over the level, wet ground, but subsequent upkeep costs were lower without the continual resupporting and propping needed in the swampy areas that are frequent in the Adirondacks. Even with a thick corduroy bed formed of poles supporting the track in wet areas, it was noticed that the Heisler locomotives sometimes sank the track six to ten inches under the engine weight. A locomotive toppled over into a swamp was an expensive item to salvage.

227. The Oval Wood Dish Heisler is spewing black smoke as it picks up motion with a full load. Circa 1922. Courtesy of the Tupper Lake Public Library.

228. The Oval Wood Dish Heisler arrives for work, sporting a large spark arrester stack for use in the woods on this summer day. Circa 1922.

Logging operations out of Kildare began in 1918 under the supervision of logging superintendent Thomas Creighton, a very large Irishman who had come down from Canada. Most of the initial logging was on both sides of the New York Central tracks near Kildare, in Township 19, Franklin County. Then new rail was laid easterly from Kildare as well as up on the slopes of Mount Matumbla west of the NYC. The climb up the easterly slope of Mount Matumbla was accomplished by a series of three switchbacks that gained an elevation of five hundred feet above Kildare. It was the only switchback built to gain steep grades on any Adirondack logging railroad.

Also at this time new rail was being laid northerly from Kildare to the northwest corner of Township 19 and then northwesterly across Township 8 and 9 of St. Lawrence County. The main logging trunk line extended about sixteen miles to the further end. Numerous spurs were established, some of them short in length as well as short in duration. In all, about forty-seven miles of railroad track were put down by the Oval Wood Dish Woods Department in the Kildare operation.

For motive power on their Adirondack operation, Oval Wood Dish ordered two new sixty-three-ton Heisler geared locomotives, one in 1916 and the other the following year, at a cost of $20,000 each. In addition, a rod locomotive was leased from the Emporium Forestry Company but reportedly could not be kept on the tracks and was thus returned after a few weeks.

Oval Wood Dish was very satisfied with the performance of their Heisler engines. The company had operated other types of geared locomotives and, was, therefore, speaking from experience. Oval Wood Dish president W. C. Hull wrote the Heisler Locomotive Works that "considering the adverse track conditions, we have no reason to regret our purchase." Operating speed of the Heisler topped off at about twelve miles per hour, running with a steam pressure of two hundred pounds. Fuel consumption was two tons per day of bituminous coal.

During the night the hostler kept the fires banked in the boilers of both locomotives, periodically injecting steam into the water tank to prevent freezing. Two of the more active locomotive engineers were John Cuddy and Wild Bill Cudhea.

229. The loader crew used dogs (hooks) to grip each log end rather than tongs. This method required two men on the ground, each with a cable to a log end to help guide the log in load placement. Circa 1920.

Log cars were the older wooden-sill flat cars with a truss rod underframe. Sixty-five of the flats were purchased for about $900 each during the World War I years, when newer more suitable log cars were almost impossible to obtain.

Loading of the log cars was done with two Raymond gasoline-powered loaders, which cost $4,000 each and which differed from the steam Barnhart loader in that only the boom would swing, not the entire operator's cabin. The Raymond had three friction clutch drums to accomplish separate functions: hoisting, swinging the thirty-foot boom, and transferring itself from car to car along rails on the car deck. Powered by a Raymond twenty-four horsepower, four-cylinder, make-and-break gasoline engine, this loader was capable of loading ten cars per day with about fifty thousand

230. Oval Wood Dish's Raymond self-propelled derrick was used to unload the log cars at the Tupper Lake sawmill log pond. Circa 1920. From the collections at College Archives, SUNY ESF.

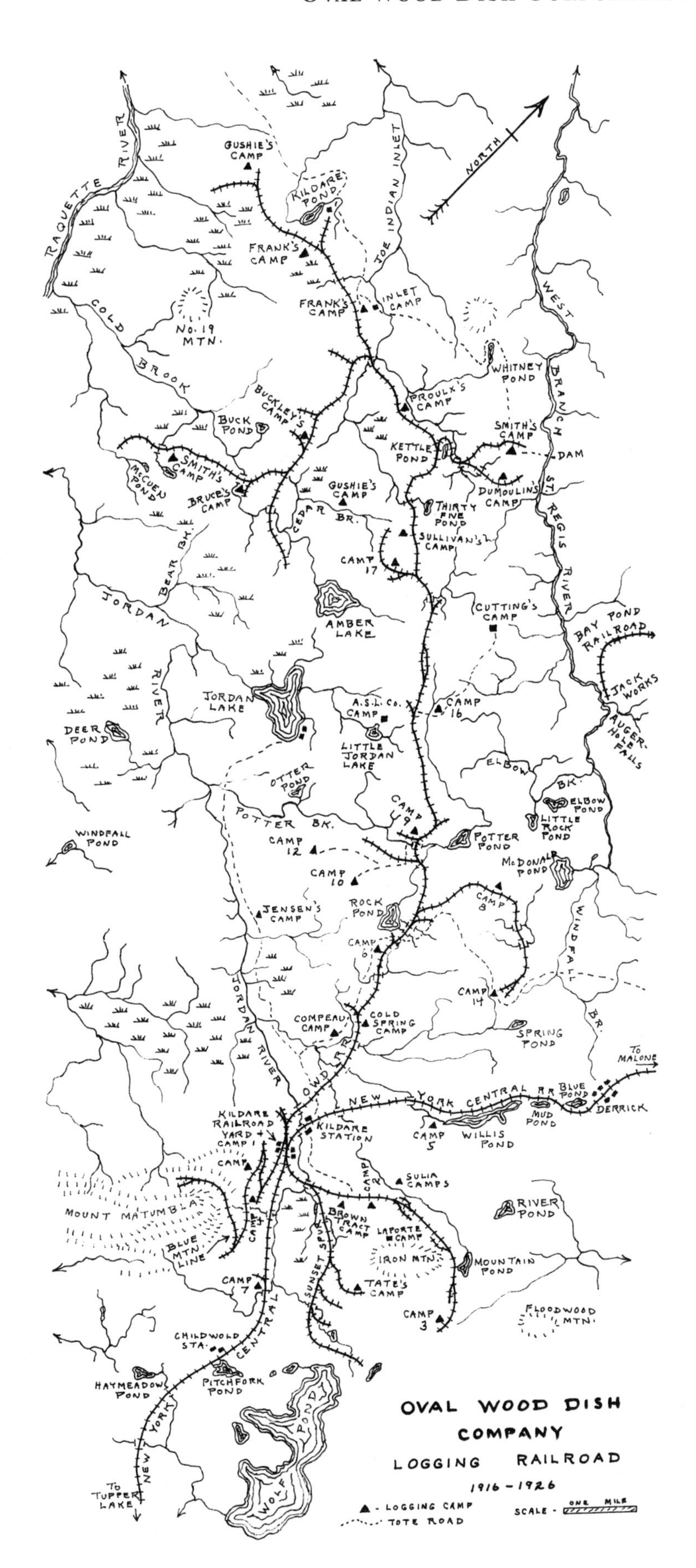
OVAL WOOD DISH
COMPANY
LOGGING RAILROAD
1916 - 1926
▲ - LOGGING CAMP
TOTE ROAD
SCALE - ONE MILE
NORTH
RAQUETTE RIVER
GUSHIE'S CAMP
KILDARE POND
TOE INDIAN INLET
FRANK'S CAMP
INLET CAMP
No. 19 MTN.
COLD BROOK
WHITNEY POND
WEST BRANCH ST. REGIS RIVER
PROULX'S CAMP
BUCKLEY'S CAMP
BUCK POND
SMITH'S CAMP
KETTLE POND
DAM
McCUEN POND
BRUCE'S CAMP
DUMOULIN'S CAMP
THIRTY FIVE POND
CEDAR BR.
SULLIVAN'S CAMP
CAMP 17
BEAR BK.
JORDAN
AMBER LAKE
CUTTING'S CAMP
BAY POND RAILROAD
JACK WORKS
JORDAN LAKE
A.S.L. Co. CAMP
CAMP 16
AUGER HOLE FALLS
DEER POND
LITTLE JORDAN LAKE
ELBOW BK.
OTTER POND
ELBOW POND
LITTLE ROCK POND
POTTER BK.
CAMP 9
POTTER POND
WINDFALL POND
CAMP 12
McDONALD POND
CAMP 10
CAMP 8
JENSEN'S CAMP
ROCK POND
WINDFALL BR.
CAMP 6
CAMP 14
COMPEAU CAMP
COLD SPRING CAMP
SPRING POND
JORDAN RIVER
OWD RR
TO MALONE
NEW YORK CENTRAL RR
BLUE POND
DERRICK
KILDARE RAILROAD YARD + CAMP 1
KILDARE STATION
CAMP 5
WILLIS POND
MUD POND
CAMP 2
SULIA CAMPS
MOUNT MATUMBLA
BROWN TRACT CAMP
LAPORTE CAMP
RIVER POND
BLUE MTN. LINE
IRON MTN.
MOUNTAIN POND
CAMP 7
SUNSET SPUR
TATE'S CAMP
CAMP 3
FLOODWOOD MTN.
CHILDWOLD STA.
NEW YORK CENTRAL
HAYMEADOW POND
PITCHFORK POND
WOLF POND
TO TUPPER LAKE

Map 30.

Table 24
Cost of Operations for a Log Train, 1919
Oval Wood Dish Company

Locomotive crew	
Engineer	$6.00–7.00 per day
Fireman	$3.25–4.25 per day
Conductor	$3.50–4.25 per day
Job foreman	$150 per month
Log loader crew	
4 tong men	$3.25–4.25 per day each
4 top loaders	$3.50–4.50 per day each
Loader engineer	$4.25–7.00 per day
Wages for crew per day	$44.00
Meals for 11 men	$11.00
Total crew cost per day	$55.00
Fuel cost	
10 gallons of gasoline for loader	$2.50
1.5 tons of bituminous coal for locomotive	$11.00
Total fuel cost	$13.50
Total daily operating cost for one train	$68.50

Source: Data from Frederick H. Alfke and Charles Ten Eick, *Wood and Mill Operations in the Adirondacks in the Vicinity of Tupper Lake, New York.* New York State College of Forestry Pamphlets, no. 20-1-401. Division of Rare Manuscript Collections, Cornell Univ. Library, Ithaca, N.Y., 1919.

231. There appears to be a fishing trip in store for some of the Oval Wood Dish employees at Kildare. A small speeder such as this one once left the track en route to Tupper Lake from Kildare, throwing Tom Creighton and Roy LaVoy down the bank. Creighton suffered leg injuries from which he never fully recovered. Courtesy of J. T. Giblin.

board feet of logs; gasoline consumption was ten gallons. The Raymond loading crew used cables dogged to each log end instead of tongs to lift the log. The disadvantage was that two men were required on the ground to position the log with guide lines for alignment and placement on the load. But loading went more smoothly and possibly faster than tong loading, which required only one man on the ground. Time to load a car varied from twenty-five to forty minutes. The log carload averaged fifty-five hundred board feet, about fifty to sixty-five hardwood logs.

There were frequent complaints from the train crew about the placement of the skidway decks. The spacing between the decks did not line up with the cars and frequent shifting was required. Other complaints were voiced that some skidways had an excessive number of logs while others had very few, again requiring extra log car movement.

The company also had a $6,500 Raymond car derrick, mounted on a self-propelled flat car. The log cars had air brakes, maintained with 110 pounds of air pressure, but accidents and derailments occurred often enough to make the derrick an essential piece of equipment. Damaged cars were hauled to the shop and tagged with a card reading "cripple," along with a description of repair work needed. Also on the roster was a boxcar, shop-built on a flat car, for hauling supplies to camp.

The Oval Wood Dish Company received rights from the New York Central to operate their log train over the Ottawa Division main line, and one of the Heisler locos would usually stay busy taking the train over the seven-mile haul from Kildare to Tupper Lake Junction. The tariff was $1.00 per thousand board feet of logs. Oval Wood Dish obviously had to employ an engineer and a conductor who both met the qual-

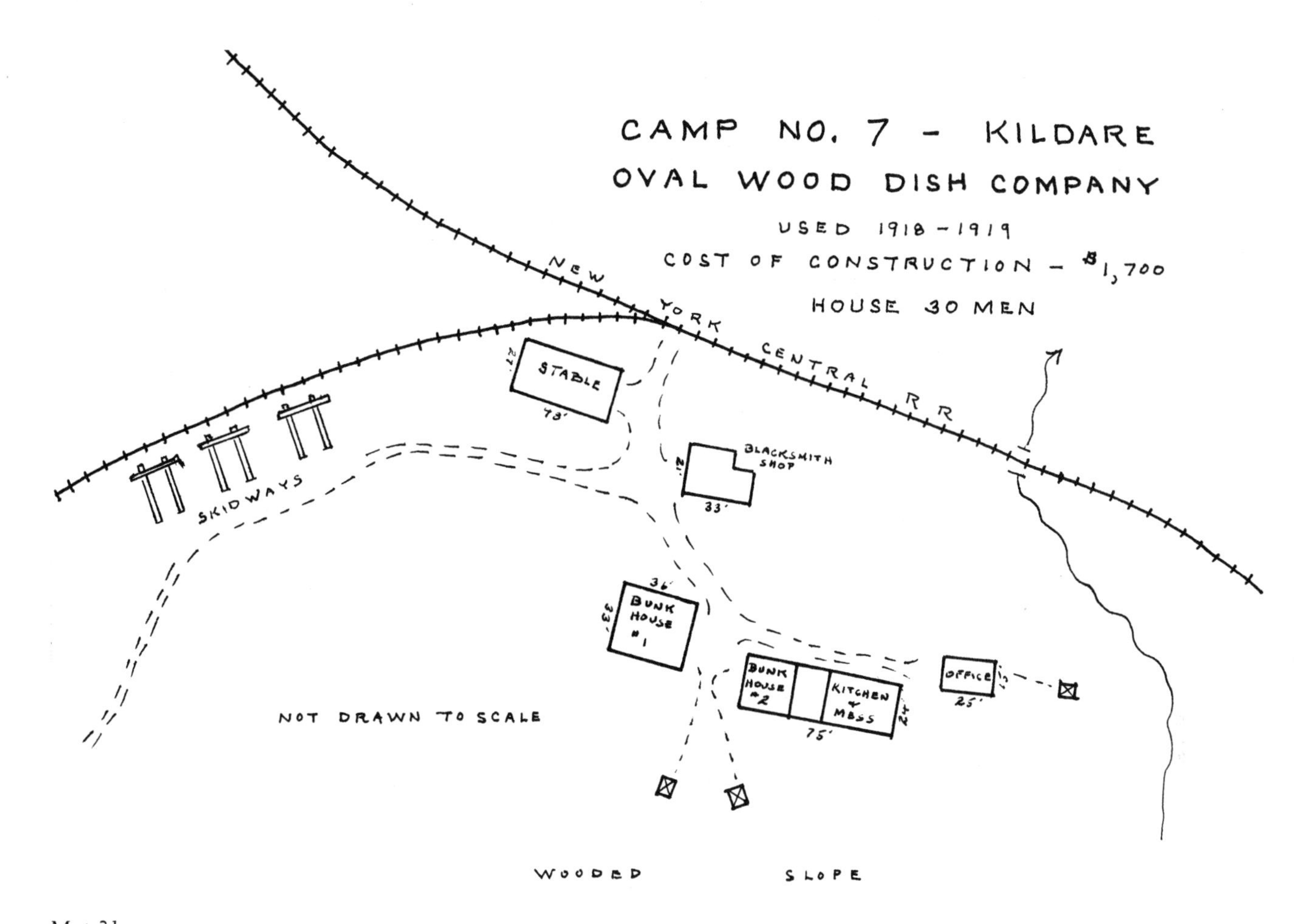

Map 31.

ifications for mainline operation. There was one occasion during the late afternoon when the Oval Wood Dish conductor mistakenly left a mainline switch open into the logging line. Later that evening when the night train from Ottawa came through, the erring turn was made onto the Oval Wood Dish tracks, and the train started across the country on a very shaky, unballasted track definitely not built for such a heavy locomotive. The incident's conclusion was never recorded, but one can be sure there was a severe reprimand at the least.

Oval Wood Dish had a few labor problems, both in the mill and in the woods. At one point in 1919, the working conditions in the mill were causing a 70-percent labor turnover in just two months, despite paying an average wage. The average for a ten-hour day was $3.00–$3.50 for men, $2.00–$2.50 for boys, and $2.00 for girls.

In the woods one camp foreman reported that only one man in fifty was the responsible sort of employee. At one point in the summer of 1919 both the engineer and the conductor of logging train number 2 were off on a seven-day drunk.

Logging operations covered a large area across three townships. Timberland maps prepared by civil engineer F. A. Hutchins indicate at least thirty-four different logging camps established, although some appear not to be Oval Wood Dish contractors. Most of the camps were obviously on the company railroad lines. Camps 1, 5, and 7 were on the main line of the New York Central Ottawa Division. About seven camps were operated each season, each holding forty to fifty men.

Only hardwoods were cut and hauled on the Oval Wood Dish railroad. Some of the areas had already been logged for softwoods by the Santa Clara Lumber Company; other tracts were relogged afterward for the spruce and fir pulpwood by companies such as the St. Regis Paper Company.

Each crew of four men, two fellers and two buckers, would cut fifty to sixty logs per day. One observer noted that there was too much waste in the woods. Logs were cut only out of the main bole of the tree; there were no lower-grade logs cut out of the crown, where branches had to be lopped off.

Woods boss Thomas Creighton had a brother, Henry,

232. *Camp 7 on what appears to be a Sunday judging by the number of idle men in view. Circa 1920. From the collections at College Archives, SUNY ESF.*

who was killed in a logging accident on the Kildare operation. Because he actually was working for an independent contractor and not Oval Wood Dish, Thomas had to do a little fudging on the paperwork but managed to get his widow a weekly pension of $13.88 from Oval Wood Dish. Since the widow lived next to the railroad track in Tupper Lake Junction, the company train would occasionally stop and roll off a couple of reject logs for her firewood supply.

At one point in history, the company operated a rail-mounted, hundred-ton Lidgerwood steam-powered skidder to log some of the areas, especially up on the slopes of Mount Matumbla in 1918. Set up by Floyd Hutchins, it was a true high-lead operation using a slack-line system. The Lidgerwood was not used to load the log cars, however. That was done with a two-cylinder, gasoline-powered jammer mounted on a flat car.

There was one occasion when the heavy Lidgerwood skidder left the track and piled up in a heap while being moved by rail between setup locations. The track conditions were so poor that the car with the heavy skidder on board had sunk the track enough to cause the couplers to part.

The Lidgerwood skidder turned out to be an absolute failure. The reasons stated were the scattered condition of the timber and the need for a large crew to operate it. The company couldn't keep any experienced help on the machine because of low pay, and efficiency suffered.

Camp 5 used a log slide, a well-greased trough with a slight uphill grade, to yard logs. Camp 5 was at Derrick on the main line of the New York Central, about four miles

FORESTRY STUDENTS IN CAMP AT OVAL WOOD DISH

The first forestry school in New York State was set up at the New York College of Agriculture at Cornell. Summer camp for the senior students was held in various locations to provide the students an opportunity for training in the field. The first camp held was near Saratoga in 1916, but in 1919, after the war, the college made arrangements to have the twenty students on the woodland property of the Oval Wood Dish Company near Kildare.

The camp was an unoccupied logging camp on the Turner Tract from which the students had access to virgin timber areas as well as land currently being logged. Studies included an analysis of logging railroad costs and the advantages of railroad logging.

233. *Cornell forestry students at Camp 7.*

Table 25
Crew and Wages for a Twenty-Man Camp

Position	*Number*	*Wages ($)*
Cutters (3 in a crew)	9	1.75–2.25 per day
Landing men	2	1.75 per day
Grab drivers	2	1.50–2.00 per day
Teamsters	4[a]	45–50 per month
Blacksmith	1	2.25 per day
Saw filer	1	2.25 per day
Camp boss	1	90 per month plus board
Cook and assistant	1[b]	60–90 per month plus board for both

Source: Data from Frederick H. Alfke and Charles Ten Eick, *Wood and Mill Operations in the Adirondacks in the Vicinity of Tupper Lake, New York.* New York State College of Forestry Pamphlets, no. 20-1-401. Division of Rare Manuscript Collections, Cornell Univ. Library, Ithaca, N.Y., 1919.
Note: The cutting crew would harvest about 7,000 BF per day. They were given a daily goal of 100 logs to cut.
[a] Teamsters boarded themselves and their horses.
[b] The cook hired and paid his own assistant.

north of Kildare and just below Willis Pond. Derrick was a former stage stop established in the late 1890s and was for a brief time a sawmill boom town supporting almost a hundred families.

The company made frequent use of log slides, using two types of slides, a pole slide and the Sykes-patented slide. The pole slide was made of two or three poles fastened into a trough and cost $300 to $400 per mile to build. The portable Sykes slide cost almost $1,600 per mile but would last much longer.

The Kildare-based operation lasted until 1926, when Oval Wood Dish closed the job, although not all of the timber had been cut. By that time logging superintendent Roy LaVoy had replaced Thomas Creighton, who had died from "dropsy." In 1926 twenty-five hundred acres of uncut land known as the Cutting Block were sold to the Sisson-White Company, who then operated their own logging railroad on much of the same rail that Oval Wood Dish had laid.

Bay Pond Spur

In 1923 the large William G. Rockefeller estate at Bay Pond was sold to a lumbering group headed by John N. McDonald (see chapter 24). Three years later, McDonald sold the hardwood timber to Oval Wood Dish, and logging camps were set up to harvest the hardwood logs after the softwood timber had been removed.

Table 26
Railroad Equipment and Construction Costs, 1919
Oval Wood Dish Company

Equipment	*Number*	*Price per Unit ($)*
Heisler locomotives	2	20,000
Flat cars	50	900
Raymond loaders	2	4,000
Derrick car	1	6,500
Construction of Railroad Track		
Preparation of rights-of-way, corduroy, grading, etc.		$2,000.00 per mile
Rails and ties		$2,500.00 per mile
Total expense to operate two trains, including interest on investment and depreciation, for one year (200 working days)		$67,260.00

Source: Data from Frederick H. Alfke and Charles Ten Eick, *Wood and Mill Operations in the Adirondacks in the Vicinity of Tupper Lake, New York.* New York State College of Forestry Pamphlets, no. 20-1-401. Division of Rare Manuscript Collections, Cornell Univ. Library, Ithaca, N.Y., 1919.

234. The Oval Wood Dish Heisler locomotive was pushing a steep grade on Mount Matumbla, delivering a load of machinery that appears to be the steam-powered log skidder. Photograph from Heisler catalog.

Table 27

Total Cost for Logs Delivered to Tupper Lake Mill, 1919

Oval Wood Dish Company

Logging costs, stump to railside	$11.00
Operation of log trains	3.36
Operating rights over New York Central RR	1.00
Miscellaneous costs	1.64
Total cost, stump to millpond	$17.00

Source: Data from Frederick H. Alfke and Charles Ten Eick, *Wood and Mill Operations in the Adirondacks in the Vicinity of Tupper Lake, New York.* New York State College of Forestry Pamphlets, no. 20-1-401. Division of Rare Manuscript Collections, Cornell Univ. Library, Ithaca, N.Y., 1919.
Note: All costs are per thousand board feet (MBF) and are based on a yearly haul of 20,000,000 board feet, or 100 MBF/day for a 200-day period.

235. A two-pole slide on an Oval Wood Dish operation. The team was dragging a train of hardwood logs down the steep grade. Circa 1922. From the collections at College Archives, SUNY ESF.

A $2\frac{1}{2}$-mile railroad spur was built by Oval Wood Dish around the south and easterly side of Buck Mountain. The hauling was done by McDonald's Bay Pond railroad, but Oval Wood Dish brought over one of their gasoline-powered log loaders. See the map of the Bay Pond railroad to locate the Oval Wood Dish spur.

In December 1929 Oval Wood Dish purchased three hundred-horsepower Linn tractors and put them to work on the long log haul from the north end of the Rockefeller property. The Linns were each hauling sixteen sleds loaded with twenty-five thousand board feet of hardwood logs for a six-mile haul into Santa Clara village. At the village, the logs were loaded on the New York Central Railroad. The last of the Oval Wood Dish logging rails at Kildare were pulled in 1937.

The Oval Wood Dish Corporation continued for many years as a mainstay of the Tupper Lake economy. The son of company founder Henry S. Hull, William Cary Hull, had become president upon the relocation of the company to Tupper Lake in 1916, and he was succeeded by his son Gerald P. Hull in 1941. In 1964 the plant was sold to Adirondack Plywood Corporation, and Oval Wood Dish ceased to be a producer of hardwood products.

236. The legendary logging contractor John E. Johnson, on the right, paid a visit to the Kildare operation. Camp foreman Edmund Tope, third from left, was conferring with woods superintendent Roy LaVoy, second from left. Courtesy of J. T. Giblin.

237. Logging superintendent Roy LaVoy on left conversing with a jobber (contract logger) for Oval Wood Dish. Courtesy of J. T. Giblin.

238. An Oval Wood Dish Linn tractor with a train of sleds loaded with hardwood logs. Circa 1924. Courtesy of the Tupper Lake Public Library.

CHAPTER TWENTY-SIX

Sisson-White Company Railroad

The Sisson-White Lumber Company was a logging partnership composed of Stanley Sisson and Donald White, both of Potsdam. Stanley was the son of George Sisson, president of the Raquette River Paper Company, and Donald White was George Sisson's son-in-law. Pulpwood logging was their principal livelihood.

Table 28
Sisson-White Company Railroad

Dates	1926–1928
Product hauled	hardwood sawlogs and softwood pulpwood
Length	7 miles
Gauge	standard
Motive power	Heisler; 2-truck; 63-ton; builder no. 1368; built 1917; source, Oval Wood Dish Co.; disposition—Southern Iron & Equip. Co., Atlanta, Ga.

When the Oval Wood Dish Company closed down their railroad at Kildare in 1926, not all of the timber had been cut. The twenty-five-hundred-acre tract known as the Cutting Block was yet to be logged; Oval Wood Dish sold this uncut land to Sisson-White in 1926.

Oval Wood Dish pulled all of the iron except seven miles of the main line from Kildare to the Cutting Block. Sisson and White bought the No. 2 Heisler from Oval Wood Dish and operated the railroad themselves to haul both sawlogs and pulpwood. The operation lasted about two years at the most. As a consideration of the sale of the locomotive by Oval Wood Dish to Sisson-White and the use of the railroad track, Sisson-White agreed to pull up the track and return it to the owner, the New York Central Railroad. As with so many of the former railroad beds, the right-of-way was later turned into a first-class woods truck road.

Logging Railroads That Connected with the New York Central and Hudson River Railroad, Carthage and Adirondack Branch

THE CARTHAGE AND ADIRONDACK RAILROAD was built easterly from the city of Carthage in the 1880s, entering the Adirondacks in 1886. Carthage was located on the Rome, Watertown and Ogdensburg Railroad. The Carthage and Adirondack reached its terminus at Newton Falls in the late 1890s and soon after became absorbed by the New York Central, as was also the Rome, Watertown and Ogdensburg Railroad. The rail line into the Adirondacks became known as the Carthage and Adirondack Branch of the New York Central Railroad.

There were five Adirondack logging railroads built off of the Carthage and Adirondack.

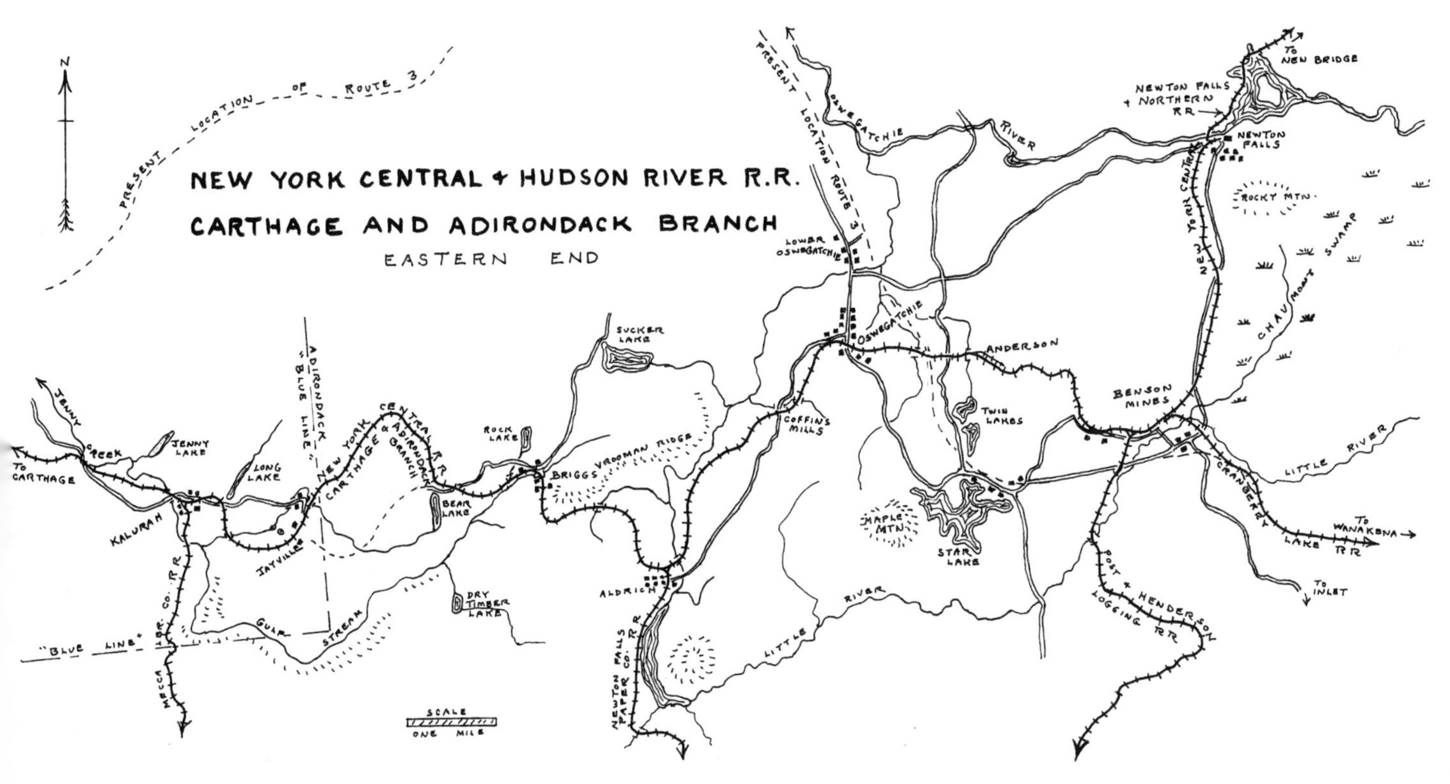

Map 32.

CHAPTER TWENTY-SEVEN

Mecca Lumber Company Railroad

The Mecca Lumber Company was formed in 1903 by three investors from Utica, Harold Nellis, Carl Amos, and Charles Swift. Frederick S. Kellogg was treasurer. In that same year a sawmill was built at a small settlement known as Little Mill on the Carthage and Adirondack Railroad. The company renamed the settlement Kalurah, the name of a Masonic Lodge in Binghamton. The settlement was never large at all, described by one person as six shacks and a sawmill. The sawmill was relatively small, going through about thirty thousand board feet of logs daily.

The company had obtained over seven thousand acres of land in the Town of Fine, and an agreement was made in July 1903 with the Raquette River Pulp Company for the marketing of the timber. Mecca Lumber Company agreed to cut and deliver spruce, pine, fir, and hemlock pulpwood, loaded in four-foot lengths onto cars in the Carthage and Adirondack Railroad. Mecca would receive $3 per cord. Hemlock bark was to be piled railside, for which Mecca received $3.25 per cord. Raquette River Pulp Company paid $7 per thousand board feet for any softwood sawn into lumber at the Mecca sawmill and loaded on cars.

The Mecca Lumber Company retained control of the hardwood timber, which was apparently sawn at the Kalurah sawmill. They produced large quantities of railroad ties.

A seven-mile railroad, built by Italian laborers with fifty-six–pound rail, was constructed southerly to a location known as Scuttle Hole on the St. Lawrence/Herkimer County line. The Raquette River Pulp Company granted permission for Mecca Lumber to build the railroad across this land.

Two used Baldwin elevated railway locomotives were

Table 29
Mecca Lumber Company Railroad

Dates	1903–1910
Product hauled	softwood sawlogs and pulpwood, hardwood sawlogs, hemlock bark
Length	7 miles
Gauge	standard

Motive Power

Builder	*Type*	*Builder No.*	*Year Built*	*Cylinders Drivers*	*Source*	*Disposition*
Baldwin	0-4-4T	4488	1878	10" x 14" 38"	Manhattan Ry #78, bought 1903	Penn Water & Power Co., 1910
Baldwin	0-4-4T	6443	1882	11" x 16" 42"	Manhattan Ry #150, bought 1906	O'Brien Construction Co., 1910

Rolling Stock
12 log cars

239. *The Mecca Lumber Company log train was unloaded by hand at the sawmill in Kalura. Circa 1906.*

bought from Manhattan Railway in New York City for use on the relatively level grade.

The first wood cut was in an area that had recently been burned over, but that cut was completed within two years. All of the timber was to be cut by January 1, 1911, and it was. The Mecca Lumber Company ceased operation in 1910, and reportedly packed up and moved on to New Hampshire.

It appears that the sawmill continued to be operated by Alfred Kilbourne, primarily cutting railroad ties. The timber property was sold by the Mecca Lumber Company to the state in 1920 for $45,000.

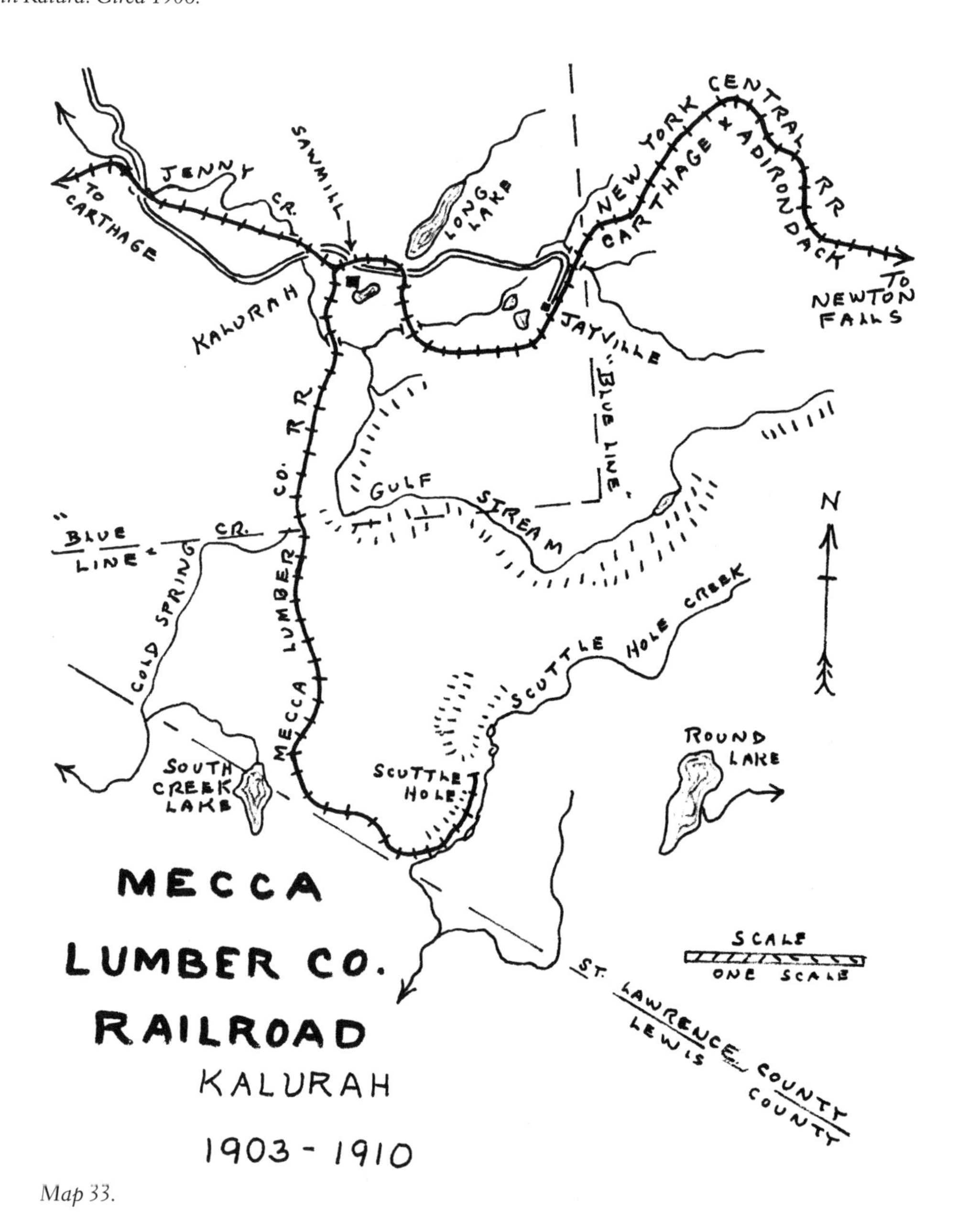

Map 33.

CHAPTER TWENTY-EIGHT

Newton Falls Paper Company Railroad

The Newton Falls Paper Company was founded in 1894 by James L. Newton, a Watertown industrialist, in alliance with a group of prominent investors. In the early years the pulp and paper mill consisted of a sulfite plant, a groundwood plant, and two paper machines.

NEWTON FALLS PAPER COMPANY

BLEACHED SULPHITE
SULPHITE BOND
LEDGER
WRITING
OFFSET
COVERS
COATING PAPER

MILLS AT
NEWTON FALLS, N.Y.

NEWTON FALLS PAPER CO.
Strength with Tear

ARTHUR J. BALDWIN
PRESIDENT
W. D. GREGOR
VICE PRES. & GEN. MGR.
C. H. THOMPSON
TREASURER
H. J. REDFIELD
SECRETARY

SALES OFFICE
10TH AVE. & 36TH ST., NEW YORK, N.Y.
TELEPHONE, CHICKERING 0700
HARRY H. CONKLIN
SALES MANAGER

240. *Letterhead.*

The attorney and secretary of the corporation was a man named V. K. Kellogg. On the basis of his recommendation, the company hired his brother, William N. Kellogg, as the stock clerk in March 1903. Woolsey Glasby was the woods superintendent for the company at that time, having been given that position after the previous woods boss, Nelson Jarvis, fled to Canada to avoid retribution for some wrongdoing.

In June 1904 Glasby drowned while night hunting from a boat out on Moosehead Pond. Cause of the accident was never ascertained; foul play was ruled out. Bill Kellogg was no doubt disappointed in not being his replacement, though he later stated that he probably wasn't even considered, being only twenty-seven years old. James O'Neil of Harrisville was named the new woods boss. It wasn't many years, however, before Bill Kellogg did become the woods superin-

Table 30
Newton Falls Paper Company Railroad

Dates	1915–1922
Product hauled	sawlogs (hardwood and softwood) and pulpwood
Length	7.6 miles
Gauge	standard

Motive Power

No.	Builder	*Type*	*Builder No.*	*Year Built*	*Cylinders Drivers*	*Source*	*Disposition*
1		0-4-0				manufactured by Penn. RR, class A3	
2	Lima	Shay 2-truck	2881	1916	12" x 12" 36"	new	sold to Basic Refractories Co., Natural Bridge, N.Y., then to Emporium Forestry Co., 1934, scrapped 1942

Rolling stock
Flat cars

241. William N. Kellogg with his family at a Cranberry Lake cottage in 1912. Courtesy of Bill Kellogg.

242. The crew at the Newton Falls Paper Company sawmill.

tendent for Newton Falls Paper. And in this position, Kellogg was told in 1915 to build a railroad.

Newton Falls Paper Company had been receiving much of their pulpwood from the Cranberry Lake region, floating the wood down the Oswegatchie River to Newton Falls. Then, in 1914, arrangements were made to purchase a seven-thousand-acre tract in the southern part of the Town of Fine, south of Aldrich. There was no way that the wood from this tract could reach the Newton Falls mill by a water route, prompting Bill Kellogg to spend considerable time looking over the tract with a railroad in mind.

The company railroad would junction with the New York Central's Carthage and Adirondack Branch at Aldrich. The village of Aldrich originated in about 1901 when a lumberman by the same name invested in a sawmill built at the site. At one time as many as twenty-five families lived in the village. The Carthage and Adirondack Railroad had been built through Aldrich in 1887 and before many years became a division of the New York Central and Hudson River Railroad. Freight items on the Carthage and Adirondack were principally wood products, products of the local iron mines, and pulp and paper after the Newton Falls paper mill opened up.

Construction of a logging railroad into the company property began the following year, 1915, building south from the village of Aldrich and flanking the west edge of Aldrich Pond. Construction crews consisted initially of Italian immigrants but were later supplanted by German and French Canadian workers.

The existing sawmill at Aldrich Pond was owned by Ball Brothers and was purchased at that time by the Newton Falls Paper Company. Also purchased was a block mill, so called, a facility set up to cut two-foot lengths of pulpwood for shipment to a paper mill. Bill Kellogg set up an office in Aldrich and established residence there in 1916.

At the start of the Aldrich operation the company had hired a young forestry graduate named Theron Dalrymple, who was just out of the New York State Ranger School. But he enlisted in the army at the outbreak of the war and was, unfortunately, killed in one of his first battles. As Bill Kellogg noted in his diary, "Dalrymple was a fine man and good with his fists."

Railroad construction was interrupted in June 1915 by a serious forest fire that threatened to consume the village of Aldrich. The fire had started on the back side of Maple Mountain, west of Star Lake, and within three days had blackened over two thousand acres as it swept rapidly toward Aldrich. Only a change in wind direction saved the village. This fire occurred just four years after the worst forest fire ever to hit the Town of Fine had devastated the hamlet of Upper Oswegatchie, including four businesses and the railroad station, before racing westward toward Aldrich. The NYC train came through with its whistle blowing to alert the residents, but the fire stopped just short of the village.

In November 1915, Bill Kellogg traveled to New York City to make arrangements for the purchase of a small locomotive, a used 0-4-0 built by the Pennsylvania Railroad in the early 1900s. The new trackwork was completed that year, preparatory for winter logging, although in December the

243. Boys never ceased to be fascinated with steam engines and wanting to be engineers when they grew up. Engine No. 1, an 0-4-0 manufactured by the Pennsylvania Railroad was purchased second-hand by Newton Falls Paper Company in 1915. Picture taken at Aldrich. Courtesy of Bill Kellogg.

construction crews were still working on the new engine shed and other structures.

The total length of the railroad was about 7.6 miles, with no spurs other than to sidings for log loading or to the camps. The line terminated at Scanlon's Camp on the Middle Branch of the Oswegatchie River. There was a jackworks there, the only one on the line.

The little 0-4-0 certainly could not handle all of the logging chores once the operation was underway; it probably was not intended to be the sole source of power. As 1916 began, the company made use of a leased locomotive, a Climax belonging to the Emporium Forestry Company, but in April the drive shaft on the Climax broke, and Emporium had to ship another one over from its Conifer shop. Newton Falls Paper Company ordered a new two-truck Shay locomotive that year, and it was delivered in August.

The logs were carried on flat cars rigged with side-pocket wooden stakes. Logs were loaded with a car-mounted log loader. Luther Shaw, Jr., worked as a conductor and at other times ran a logging camp for the company.

When the locomotive was traveling into the woods with the empties, it would encounter a fairly steep pull after passing the inlet to Aldrich Pond, lasting until reaching the height of land shortly before Streeter Lake. Thus, it was customary to take only half of the string of empties from Aldrich to the height of land, leave them on a siding, and return for the remainder. From the siding at the height it was a fairly level run to the jackworks at the end. The siding at the height thus became known as the "double ender."

244. Loading spruce logs. Circa 1918.

One day a week was known as "tote day." This was the day for supplying the camps with needed food, clothing, and tools as requested by each camp boss. A railroad car with the supplies would be spotted at a siding for each camp, and if

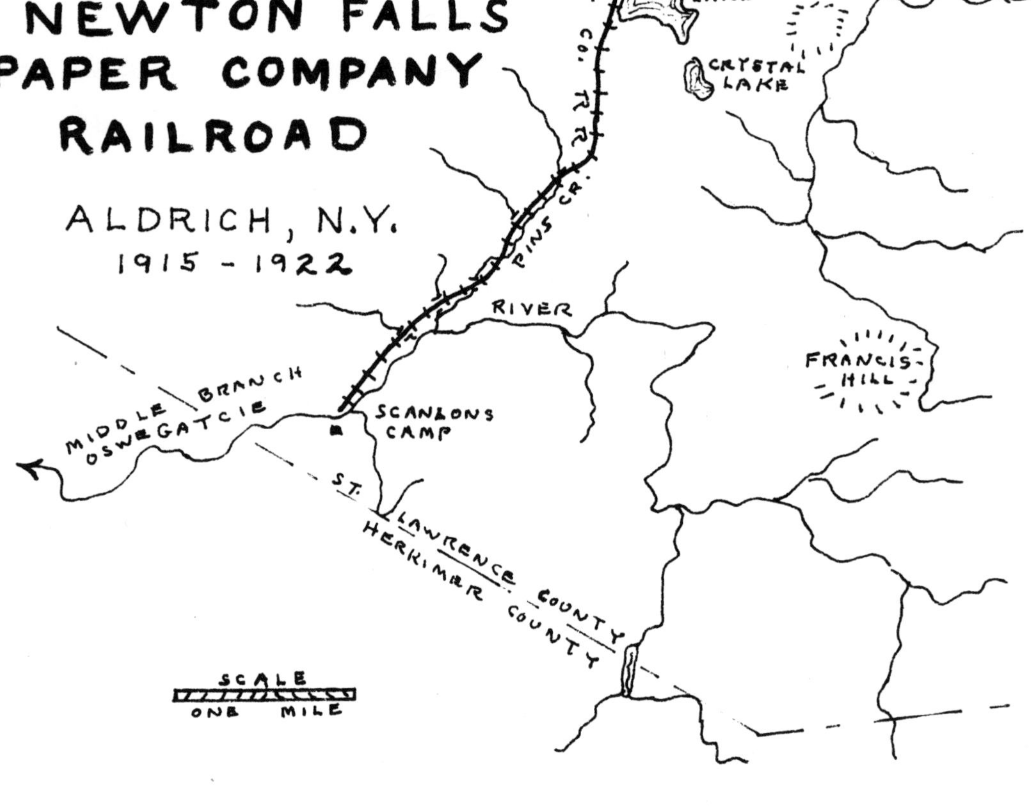

TO NEWTON FALLS
COLLINS MILLS
NORTH
LITTLE RIV.
NEW YORK CENTRAL
VROOMAN RIDGE
BRIGGS
TO CARTHAGE
ALDRICH
BLOCK MILL
SAWMILL
ALDRICH POND
LITTLE
RIVER
CREEK
TAMARACK
MUD CREEK
NEWTON FALLS PAPER CO. R.R.
STREETER LAKE
CRYSTAL LAKE
NEWTON FALLS
PAPER COMPANY
RAILROAD
ALDRICH, N.Y.
1915 - 1922
PINS CR.
RIVER
FRANCIS HILL
MIDDLE BRANCH OSWEGATCIE
SCANLONS CAMP
ST. LAWRENCE COUNTY
HERKIMER COUNTY
SCALE
ONE MILE

Map 34.

245. A large maple log made an ideal prop for the loading crew when showing off. Circa 1918. Courtesy of the Shaw family.

the camp was not on the railroad there would be a team of horses with a tote sled there to move the goods to camp.

Spruce sawlogs were hauled to the sawmill owned by the Newton Falls Paper Company at Aldrich. The sawmill was operated by Peter Matthis and employed from thirty to thirty-five workers sawing thirty thousand board feet of rough softwood lumber daily. In 1920 the company was also railroading sawlogs to a sawmill at Briggs. Briggs was a small sawmill village on the New York Central, about two and a half miles west of Aldrich. Pulpwood was shipped to Newton Falls or to the block mill in Aldrich.

Some of the pulpwood being cut during this time was not taken out by railroad. One of the company's jobbers was cutting pulpwood east of Streeter Lake and used a circular cut-up saw on the bank of Tamarack Creek. The drop blade on the saw was powered by a one-cylinder stationary gas engine. After being cut, the four-foot was dumped into Tamarack Creek and driven down to Little River and then to Aldrich, where it was loaded on railroad cars.

In 1920 Bill Kellogg went with company president F. L. Moore to inspect some timberland near Saratoga Springs. Against Kellogg's advice, Moore arranged for the purchase of the property, and it was Moore's undoing. In 1921 the owners of Newton Falls Paper Company came up from their New York City offices, did an investigation, and fired Moore on the spot for his poor judgment in making an unsatisfactory land purchase.

Railroad operations apparently ceased in 1922; the rails were pulled up the following year. Pulpwood cutting shifted to other areas, especially to locations where the wood could be floated down the Oswegatchie River to Newton Falls. The company operated an old steel tug on Cranberry Lake to boom pulpwood down to the log chute at the outlet into the Oswegatchie River.

Bill Kellogg parted company with Newton Falls Paper Company in 1923, following which he worked as woods superintendent for the Emporium Forestry Company for a couple of years before leaving the Adirondack area.

CHAPTER TWENTY-NINE

Post and Henderson Logging Railroad

One of the oldest firms to log by rail in the Adirondacks was the Post and Henderson outfit, originally from Oswego. To say they originally logged by rail is stretching the facts a bit, however; their initial railroading was actually a pole road, using logs for rails.

The partnership of Robert G. Post and Washington T. Henderson made its first venture into the Adirondacks in the early 1890s with a sawmill at Jayville. Production in the mid-1890s was four million board feet of lumber yearly. Then in 1898 the firm built a larger sawmill in the village of Benson Mines, St. Lawrence County, between the main line of the Carthage and Adirondack Railroad and Little River. A crude pole railroad was built southward from Benson Mines in a location unknown today. Motive power was a steam traction engine with flanged wheels to ride on the log "rails." In 1900 the two partners formed the Post and Henderson Company.

246. The original depot and office of Carthage and Adirondack Railroad at Benson Mines.

The sawmill doubtless needed a better logging system because in about 1905 a genuine railroad was laid out in the Town of Fine southerly from Benson Mines to the headwaters of the Oswegatchie River, where logging camps were established. Possibly the grade of the earlier pole road was used for part of the roadbed. Along Alice Brook the railroad fol-

Table 31

Post and Henderson Logging Railroad

Dates	c. 1905–1908
Product hauled	softwood sawlogs
Length	8.5 miles
Gauge	standard

Motive Power

Builder	*Type*	*Builder No.*	*Year Built*	*Weight (tons)*	*Cylinders Drivers*	*Source*	*Disposition*
Baldwin	0-4-4T Forney	12599	1892	20	14" x 16" 42"	Chicago and South Side Elevated Railway	1. Emporium Lumber Co. in 1908, Galeton, Pa., and Conifer, N.Y.; 2. Tupper Lake Chemical Co. in 1923

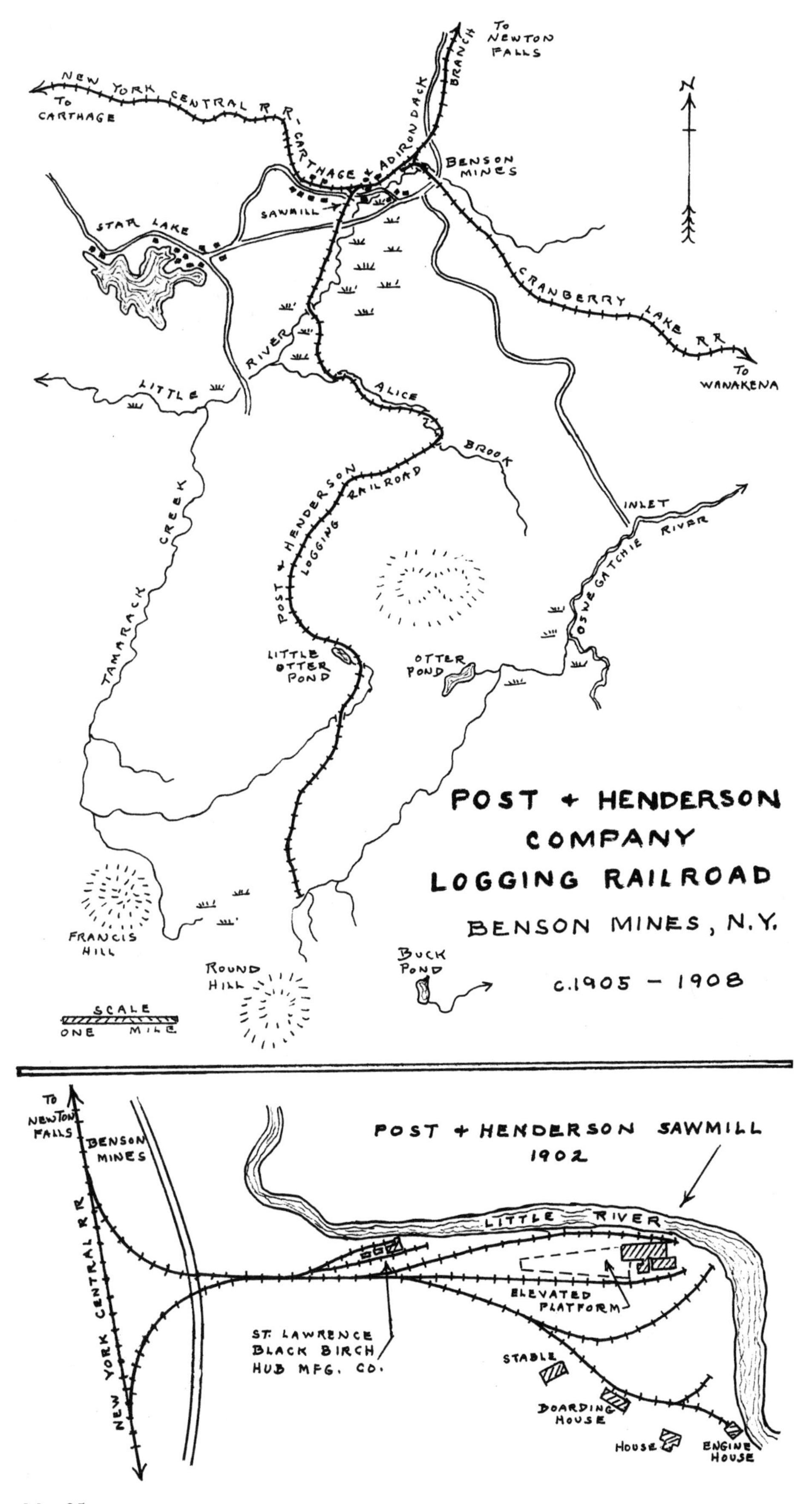
To Newton Falls
New York Central R R
To Carthage
Carthage + Adirondack Branch
Benson Mines
Sawmill
Star Lake
Cranberry Lake R R
To Wanakena
River
Little
Alice
Brook
Post + Henderson Logging Railroad
Tamarack Creek
Inlet
Oswegatchie River
Little Otter Pond
Otter Pond
POST + HENDERSON COMPANY LOGGING RAILROAD
BENSON MINES, N.Y.
c.1905 – 1908
Francis Hill
Round Hill
Buck Pond
Scale
One Mile
To Newton Falls
Benson Mines
New York Central R R
POST + HENDERSON SAWMILL 1902
Little River
Elevated Platform
St. Lawrence Black Birch Hub Mfg. Co.
Stable
Boarding House
House
Engine House

Map 35.

lowed the grade of the old Sternberg Road for a short way, the original road from Star Lake to Inlet. The end of the railroad was about a mile short of Buck Pond, reaching out a total of 8.5 miles from the mill at Benson Mines.

The railroad was powered by a single locomotive, a secondhand Baldwin 0-4-4T saddletank. Its former life had been on the Chicago and South Side elevated railway, and it was believed to have been exhibited at the 1893 Chicago World's Fair.

The operation had a short life, however. Post and Henderson sold their sixty-five hundred acres of timberland to the state in April 1908, and the abandoned roadbed became a popular access route for many years for hunters and trappers. The cut-over land became part of the present day 95,525-acre Five Ponds Wilderness Area, which contains some of the finest remote wilderness opportunities in the Adirondacks. Within the wilderness area, the Buck Pond Primitive Corridor was established, principally along the old Post and Henderson railroad bed, for access to private property at Buck Pond and vehicular access for hunters using base camps in the area.

247. The Post and Henderson locomotive was a former elevated railway engine in Chicago, an 0-4-4T Forney built by Baldwin. It was afterward sold to the Emporium Lumber Company, where it is pictured above in Galeton, Pennsylvania. The locomotive's final days were spent as a yard locomotive for the Tupper Lake Chemical Company. From the collection of Thomas T. Taber III.

CHAPTER THIRTY

Newton Falls and Northern Railroad

The Town of Clare is one of those remote Adirondack townships that has never had the resources to support a permanent village. Iron ore and timber resources each gave birth to some sizable settlements, which eventually faded out as fast as they were created. Both resources spawned railroads that now get little more than brief mention in historical accounts.

In 1866 the settlement of Clarksboro was formed on the south branch of the Grasse River to provided employees for the nearby iron mines and blast furnace of the Clifton Iron Company. The Clifton Iron Company railroad was built southeasterly from DeKalb Junction to connect the mining community with the outside world, opening in January 1868. The 23.5-mile railroad was constructed of wood rails (maple) with iron straps secured along the top. Clarksboro village had about seven hundred residents that year—but not for long. A fire in 1869 destroyed much of the equipment, operations were suspended, and the village went into a fast decline.

About thirty-five years later and four miles further up the South Branch, another temporary village was born, this one nurtured on timber. Lumberman Robert W. Higbie of Jamaica was seeking a large-scale opportunity, and an agreement was forged with the Newton Falls Paper Company.

The Newton Falls Paper Company mill had been built in 1894 on the Oswegatchie River and had acquired large timberland ownerships in that northwestern part of the Adirondacks. Timberland owners would customarily use both company logging crews and independent contractors to harvest their sawlogs and pulpwood, but Newton Falls Paper used an arrangement with Robert Higbie that was somewhat different. Involved was a large tract of company timberland in the north side of the Oswegatchie River in the towns of Clare and Clifton.

Table 32
Newton Falls and Northern Railroad

Owner	Robert W. Higbie Company
Dates	1906–1919
Product hauled	sawlogs, long length pulpwood, passengers
Length	approx. 12 miles
Gauge	standard

Motive Power

No.	Builder	Type	Builder No.	Year Built	Weight (tons)	Source	Disposition
1	Climax	2-truck		1905	38	new	unknown
2	Climax	2-truck		1907	38	new	Dexter Sulfite Pulp & Paper Co. Dexter, N.Y.

Rolling Stock
20 log cars (wood-beam skeleton)
Log loader, car mounted

In November 1905 the Newton Falls Paper Company conveyed title to 21,500 acres of timberland to Robert W. Higbie for an undisclosed price. In return, Higbie had to agree to log the property and deliver to Newton Falls Paper the softwood pulpwood and the pine and cedar sawlogs. All of the spruce and fir pulpwood timber and the hemlock not over a twelve-inch stump diameter was to be delivered in long lengths by railroad and dumped into the pond above the dam at Newton Falls. Pine and cedar sawlogs ten inches and up were also to be delivered to the pond.

Beginning in 1906 there was to be an annual delivery of five thousand cords of spruce, fir, and hemlock for which the company would pay Higbie $4.70 per cord delivered. Pine logs brought $8.66 per thousand board feet and cedar $6.66.

To harvest and manufacture the hardwood, Higbie created a village on the South Branch of the Grasse River and built a railroad. In 1906 Milo Woodcock of Edwards set up a portable sawmill at the selected village site and produced the lumber necessary to build a sizable settlement. Before long there were homes for fifty families, the inevitable boardinghouse, a general store, and a post office. There was a local doctor, and the one-room schoolhouse had thirty-five to forty students.

A large double band sawmill was built, capable of producing thirty thousand board feet of hardwood lumber daily. John Thomas was mill superintendent. Associated with the mill was a broom handle factory. The settlement was known as New Bridge because of a new bridge that had been built across the South Branch in about 1902. The principal office of the Robert W. Higbie Company remained at Newton Falls.

Higbie constructed his railroad from Newton Falls to New Bridge in 1905 and 1906. The Emporium Lumber Company from Pennsylvania was beginning to purchase Adirondack timberland at this time and had made purchases in the northern part of the Town of Clare. Higbie asked Emporium in September 1905 to assist him in building the railroad, but Emporium declined. Junctioning with the New York Central at the railroad yard next to the paper mill, Higbie's railroad crossed over the Oswegatchie River on a bridge just below the power house. Then going northerly, the line had to pass through some land of the Newton Falls Paper Company that had not been conveyed to Higbie. In

248. Leslie Spicer's logging camp at Spruce Pond. The horse barn was in the right rear. Circa 1910.

249. Cleanliness was the general rule in the camp mess halls. The agate plates and cups were neatly lined up at an operation of Robert W. Higbie Company. Circa 1910. Source: N.Y. Assembly Doc.

250. Higbie's No. 1 Climax was negotiating a river crossing with three boxcars loaded with finished lumber. Circa 1910.

these sections the paper company granted Higbie a 120-foot-wide right-of-way. According to the location survey done by Eaton and Brownell in 1906, the distance from Newton Falls village to the terminus of the line at New Bridge was 6.7 miles.

Common carrier status for the railroad known as the Newton Falls and Northern Railroad was established in June 1908, probably required because of the passenger traffic into the remote village. A railroad machine shop was maintained at New Bridge.

For motive power Higbie purchased two new forty-five-ton Climax geared locomotives. Short twenty-one-foot skeleton log cars were used, with rails on top for the big loader to move along. Rail sidings were probably built, as well as branches into the valleys, but no records remain to indicate where they might have been. Thus, actual mileage of rails laid is unknown but is reported to have been at least twelve miles.

The sawmill operated until about 1915, and the village quickly deteriorated after that. In November 1919 the Newton Falls and Northern Railroad was dissolved and the rails pulled up.

252. The rail-mounted steam log skidder of the Robert W. Higbie Company also had a boom for loading cars. The flats were lettered "Newton Falls & Northern RR." Source: N.Y. Assembly Doc.

The trustees of the Robert W. Higbie Company dissolved the corporation in 1923, and two years later the large tract of timberland once sold by Newton Falls Paper Company to Robert Higbie was sold to the Power Corporation of New York.

251. Robert W. Higbie used two Climax locomotives on his Newton Falls and Northern Railroad from Newton Falls to New Bridge. Shown in the photo is No. 1, a forty-five-ton engine built in 1907.

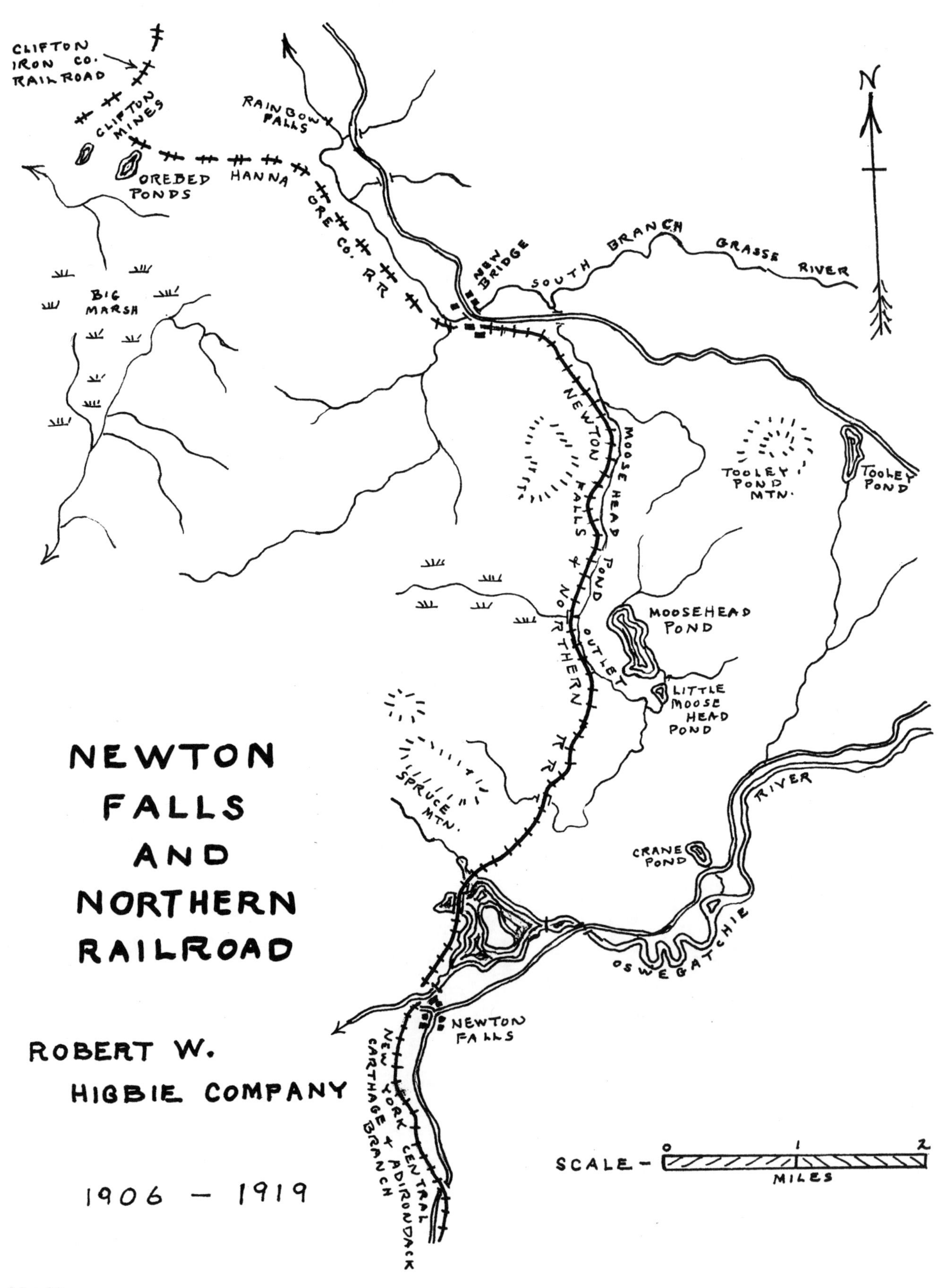
CLIFTON IRON CO. RAILROAD
CLIFTON MINES
RAINBOW FALLS
OREBED PONDS
HANNA ORE CO. RR
N
NEW BRIDGE
SOUTH BRANCH GRASSE RIVER
BIG MARSH
NEWTON FALLS & NORTHERN RR
MOOSE HEAD POND OUTLET
TOOLEY POND MTN.
TOOLEY POND
MOOSEHEAD POND
LITTLE MOOSE HEAD POND
SPRUCE MTN.
RIVER
CRANE POND
OSWEGATCHIE
NEWTON FALLS
NEW YORK CENTRAL CARTHAGE & ADIRONDACK BRANCH
NEWTON FALLS AND NORTHERN RAILROAD
ROBERT W. HIGBIE COMPANY
1906 – 1919
SCALE – 0 1 2 MILES

Map 36.

That was not quite the end of railroading over Robert Higbie's roadbed, however. During World War II the Clifton mines were opened again to obtain magnetic iron ore for the war effort. In the early 1940s the Hanna Ore Company relaid tracks from Newton Falls to the old Clifton mines. Portions of Higbie's old roadbed were used and the railroad extended three or four miles beyond New Bridge. The first train arrived at the Clifton mines in December 1941, and over a ten-year period it is said that 1,730,991 tons of ore were carried out.

CHAPTER THIRTY-ONE

Rich Lumber Company Railroad and Cranberry Lake Railroad

During the peak of Adirondack railroad logging activity, in the era of the removal of much of the remaining virgin timber, it was sometimes not enough for an operator to build just the sawmill. There were situations where an operator had to build an entire village to support the mill. None exhibited the ability and capacity to a greater degree than the Rich Lumber Company.

For thirty-three years the Rich Lumber Company created new villages in three different states and railroaded out many millions of board feet of fine old-growth timber. Four

Table 33
Rich Lumber Company Railroad and Cranberry Lake Railroad

Dates	1905–1914
Product hauled	hardwood and softwood sawlogs; passengers on Cranberry Lake RR (common carrier)
Length	38 miles, including Cranberry Lake RR
Gauge	standard

Motive Power

No.	*Builder*	*Type*	*Builder No.*	*Year Built*	*Weight (tons)*	*Cylinders Drivers*	*Source*	*Disposition*
2[a]	Lima	Shay 3-truck	464	1894	50	12" x 12" 32"	new	Jackman Lbr. Co., Jackman, Me.
4[b]	Lima	Shay 3-truck	553	1898	50	12" x 12" 32"	new	Basic Refrac., Natural Bridge, N.Y.
5	Lima	2-8-0	1015	1905	51	17" x 20" 40"	new	

Equipment
30 log cars
20 flat cars
Barnhart loader
coach
combine
speeders
rail (40 and 60 pound)

[a]Rebuilt and renumbered No. 6.
[b]Both Shays were formerly used by Rich in Pennsylvania.

253. *Letterhead.*

new villages were carved out of the wilderness, two of which still stand today. They stand as testimony to the hard drive of two cousins from Cattaraugus, New York, Herbert C. and Horace Clarence Rich.

When their partnership began in 1885, neither one had any experience in the lumber business. Herbert C. Rich, the partner soon to become the dominant figure, was the proprietor of hardware stores in the southern tier of New York State. But they became the core of a hard-working group that was to stay together for many years as they migrated from one virgin timber stand to another.

The Rich Lumber Company operations began in McKean County in northwestern Pennsylvania in that same year of 1885, with a small operation at Clermont. Success bred bigger plans and stirred ambitions. Two large operations in Pennsylvania followed, both of them creating a new village in wilderness areas of prime hemlock and hardwood timber. It was the accepted industry practice during that era to set up a large but temporary sawmill, build the necessary village facilities around it when necessary, clean the area of its virgin timber, and then move on with no intention of establishing a permanent facility or managing the timberland for sustained yield. This was the scenario that Rich Lumber Company applied in a capable manner a number of times.

Logging railroads played an essential role in each of their four major operations. In the first one, at Gardeau, Pennsylvania, the railroad was incorporated as the Keystone Railroad and operated until 1898. The second village was created under the name of Granere with a railroad incorporated as the South Branch Railroad. The company used only Shay geared locomotives, four of which were bought for the Pennsylvania sites.

The sawmill at each of the Pennsylvania villages was capable of sawing about seventy-five thousand board feet of hemlock in each eleven-hour shift. Timber was quickly exhausted; the village of Gardeau was abandoned in 1898 and the village of Granere in 1902. Nothing exists of those villages today except a few hunting camps.

The story of the Rich Lumber Company operations in Pennsylvania and in Manchester, Vermont, is told in *Rails in the North Woods: Histories of Ten Adirondack Short Lines*, published by North Country Books.

WANAKENA, NEW YORK

Cranberry Lake, located in the western portion of New York's Adirondack region, is a picturesque lake, long admired by sportsmen and camp owners. Trout fishing at the turn of the century was superb, a treasured privilege for those able to gain access to this sparkling gem in the wilderness.

When Herbert Rich came to the area in 1901 to seek out and purchase land for the next home of his lumber domain, Cranberry Lake was still a little-known lake in a primeval wilderness. Cranberry Lake village, at the foot of the lake (north end), still had only about fifteen dwellings, and the lake was just beginning to be discovered by those with the means to enjoy a summer home in the Adirondacks. The newly built Carthage and Adirondack Railroad had previously reached Oswegatchie Station and Benson Mines in 1889, and then in the early 1890s a road was cut through the woods from Newton Falls to reach Cranberry Lake village. It was the first direct link to the west. It would be the Rich Lumber Company, however, that really opened the door, with a railroad linked to the Carthage and Adirondack Railroad, allowing the wealthy to ride on iron rails right to the shore of the lake.

With the purchase of sixteen thousand acres of land covered with virgin timber on the southwest side of the lake in 1901, the Riches came back in 1902 ready to build a village, a task with which they were now experienced. The site was wisely chosen on the lake itself, at the head of a long flow in the southwest corner of the lake where the Oswegatchie River enters Cranberry Lake. The price paid for the land was reported to be $230,000, about $15 per acre. The village was given the name of Wanakena.

Actually, Rich was not the first to build at the location. Sternberg's Hotel, a sportsmen's camp, had been there since 1884, and it is said that Rich's people stayed at the small hotel when first coming into the area. This camp was situated either on or near the old Albany Road, one of two old "Military Roads" that passed near Cranberry Lake. The Albany Road had been built about 1812 but eventually deteriorated because of disuse, reverting to a hunter's trail long before Rich appeared.

Herbert Rich moved quickly; soon a steam donkey engine was set up at a point later to be the center of the village and was kept busy uprooting tree stumps. The first building erected was the company store, soon followed by the dwellings. Trainloads of hemlock lumber and personal goods were brought in from Pennsylvania for house construction, much of it salvaged from homes dismantled at Granere. Three hundred employees and family members were moved up from the Granere operation.

Cranberry Lake was still a wilderness retreat at that time. The gaunt tree stumps and driftwood in the flows created by the damming and raising of the lake in 1867 made a haven for wildlife and a home for large trout that would please anglers for many years yet to come. The rich green mantle of spruce and pine that dripped over the shoreline told the fortunate traveler that he need look no further. What are today bare rock ledges were once shorelines and islands that harbored towering evergreen trees.

The Wanakena operation was to feature an industrial complex different from the previous villages. Wanakena became a mixture of wood-manufacturing mills producing a variety of products and using most of the different species of trees available. A small stream was dammed just above the point where it entered the Oswegatchie River flow, and here the sawmill was built, soon to be followed by construction of other mills. As with previous operations, the sawmill was actually built and operated not by the company but by the Ford brothers. Royal B. and Heman R. Ford came from Rich's home town in western New York and were capable sawmillers who owned the sawmill and sawed Rich's lumber under contract.

It was probably 1903 before the sawmill got into full operation. A change in mill diet was in order, however; spruce was the principal species to be sawn. Lumber sizes and existing markets, however, had doubtless changed little from the hemlock days in Pennsylvania. Considerable white pine was also sawn at Wanakena as well as what little hemlock could be found. Annual production was about five million board feet.

Logging Railroad

As with previous operations, a railroad was again established to feed the mills, totaling about thirty-eight miles of logging

254. Loading the log train in the partially open area known as the Plains in 1907.

pike built by the company. All of the rails were laid within the township of Fine and, for the most part, south of the Wanakena site. The main branch, about ten miles in length, followed up Skate Creek and then angled over to the Oswegatchie River, which was followed up as far as Boiling Spring, a short ways below High Falls.

This branch took the trains into the area known as the Plains, a relatively level, sandy river plain with many openings and rich with large white pine. Early residents of the area traveled there to cut hay and later residents even grazed sheep there. The clearings on the Plains are of unknown origin, possibly caused by severe ground fires in the distant past.

Some of the softwood logs were driven down the small Oswegatchie River to a jackworks set up on the river's east bank at the Plains, where they could be loaded on the railroad cars. A controversy eventually arose between the Rich Lumber Company and the professional guides over the passage of sportsmen's boats up the Oswegatchie River when the river was filled with logs backed up behind a boom at the jackworks. The guides won, forcing the company to create a passageway for guide boats and canoes to pass by the boom. The guides predicted at the time that the logs in the river would mean the end of fishing, but good fishing remained long after the Rich Lumber Company was gone.

Another two-mile railroad branch worked its way southeasterly from the sawmill location to reach a jackworks at the head of Dead Creek Flow, a large backwater at the southwest extremity of Cranberry Lake. Logs were boomed down the lake to reach the sawmill via this short trip on the rails. The only railroad logging north of Wanakena village was at Crimmins Switch, an access to a rail line laid around Cathedral

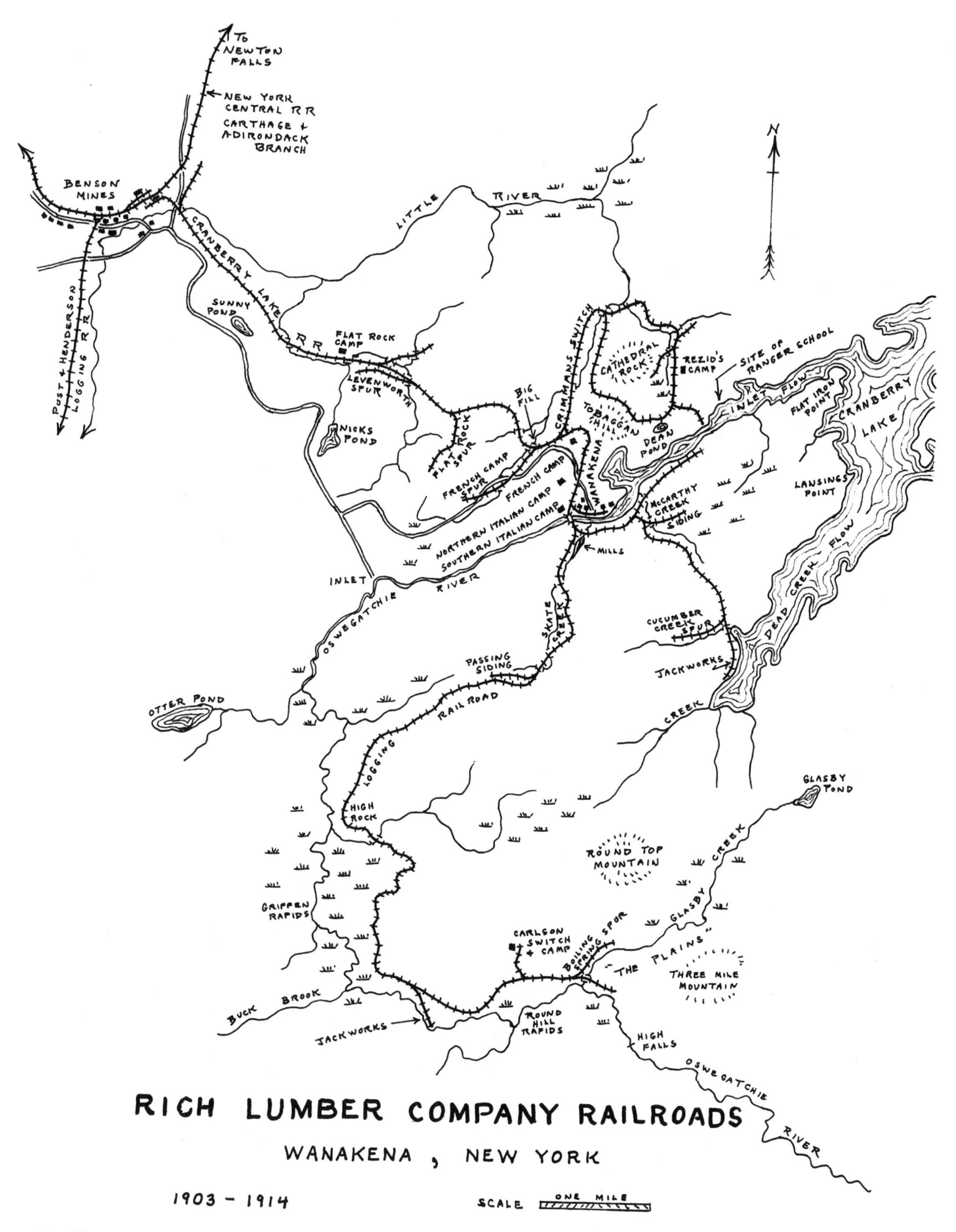

TO NEWTON FALLS
NEW YORK CENTRAL RR CARTHAGE & ADIRONDACK BRANCH
BENSON MINES
CRANBERRY LAKE RR
POST & HENDERSON LOGGING RR
SUNNY POND
LITTLE RIVER
FLAT ROCK CAMP
LEVENWORTH SPUR
NICKS POND
FLAT ROCK SPUR
BIG FILL
CRIMMANS SWITCH
CATHEDRAL ROCK
REZIO'S CAMP
SITE OF RANGER SCHOOL
INLET FLOW
FLAT IRON POINT
CRANBERRY LAKE
TOBAGGAN HILL
DEAN POND
WANAKENA
FRENCH CAMP SPUR
FRENCH CAMP
NORTHERN ITALIAN CAMP
SOUTHERN ITALIAN CAMP
McCARTHY CREEK SIDING
LANSINGS POINT
MILLS
INLET
OSWEGATCHIE RIVER
SKATE CREEK
CUCUMBER CREEK SPUR
DEAD CREEK FLOW
PASSING SIDING
JACKWORKS
OTTER POND
LOGGING RAILROAD
CREEK
HIGH ROCK
GLASBY POND
ROUND TOP MOUNTAIN
GLASBY CREEK
GRIFFEN RAPIDS
CARLSON SWITCH & CAMP
BOILING SPRING SPUR
"THE PLAINS"
THREE MILE MOUNTAIN
BUCK BROOK
JACKWORKS
ROUND HILL RAPIDS
HIGH FALLS
OSWEGATCHIE RIVER
RICH LUMBER COMPANY RAILROADS
WANAKENA, NEW YORK
1903 - 1914
SCALE ONE MILE

Map 37.

Rock, and also some logging activity along the Cranberry Lake Railroad, the common carrier line built by the Rich Lumber Company.

Original construction of the logging line is credited to John Olson for the most part, although Claude Carlson was active building rights-of-way under contract. The abundance of Swedish names shows that the Riches brought many of their Pennsylvania crew with them to Wanakena, respecting their hard work and capabilities. The logging railroad began operation in 1905.

255. Pulling a load in reverse, Shay No. 4 shows its homely side, the one opposite the three vertical cylinders. Courtesy of Bill Gleason.

Motive power on the extensive logging pike was, as per company practice, exclusively Shay locomotives. Shays Nos. 2 and 4, both fifty tons in weight, were brought up north from the Granere operation. Gene Madison was often found at the throttle of the Shays. Log cars used were the short, two-bunk, twenty-two-foot cars with rails mounted on the bunks for the loader to move along from car to car. "Shoo-fly" rails were placed between cars to allow the loader to ride over the gap between cars. Although the logging was all contracted out, Rich Lumber used their own crews to load the cars. Andy Rancier was remembered as one of the skilled loader operators who could place the loading tongs swiftly but gently into the hands of his helper.

Locomotive and car repairs were done in the large machine shop, which sat on the north side of the river at the edge of the village proper and west of the store. On a siding near the machine shop there usually were several wooden

256. Chaining four logs together for a single lift was not common practice, but the Barnhart crew had to show off a little. It appears that the hook tender became bored with it all and relaxed with a good book. Circa 1908.

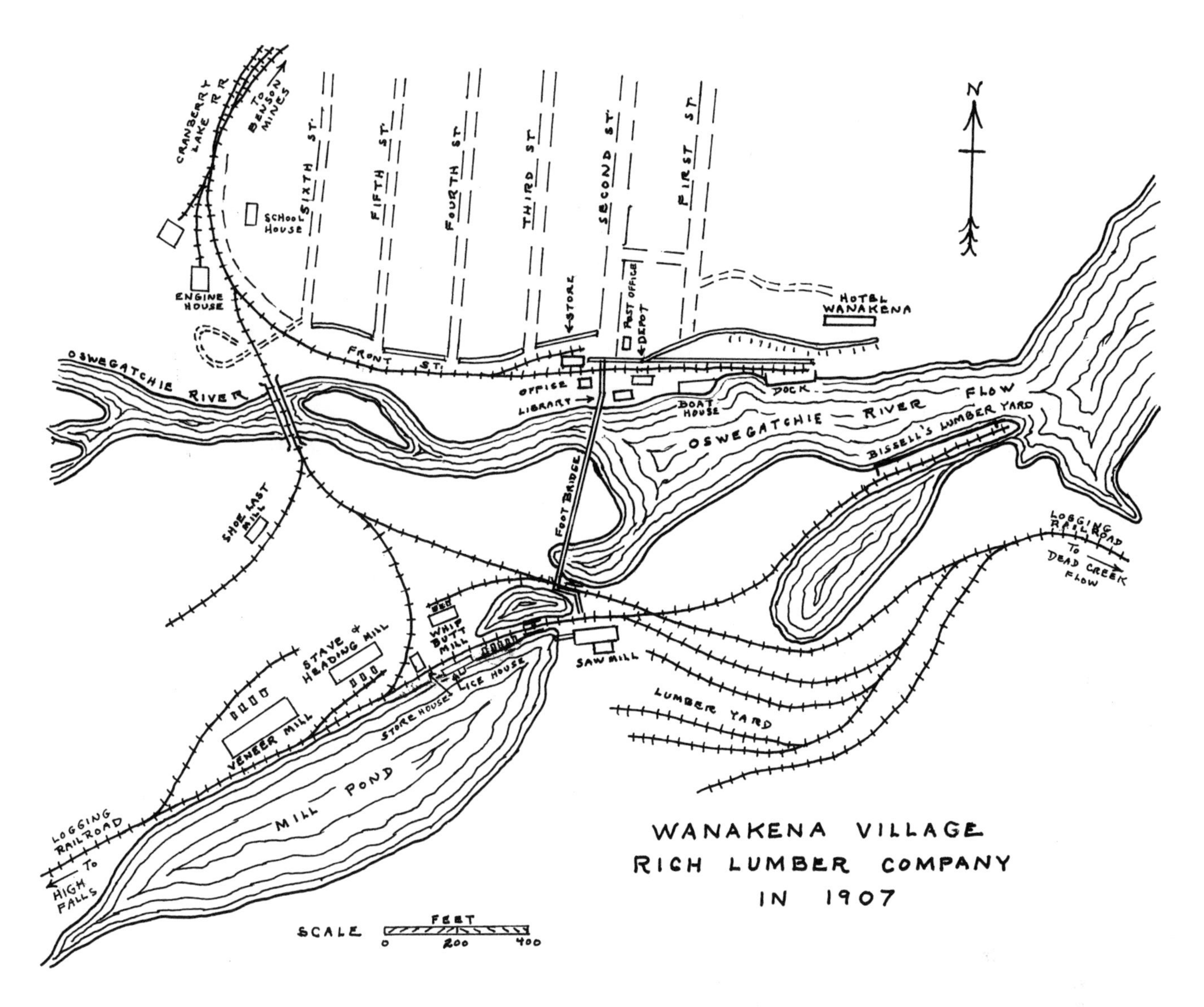

Map 38.

257. Both Shay locos were busy on a snowy day. They are shown parked in front of the engine house at Wanakena.

tank cars filled with water and loaded with several thousand feet of fire hose. Forest fires presented a constant threat around steam sawmills and railroad activity, and it was a threat that the management was not allowed to forget. The year 1908 is remembered as a year of terrible fires throughout the Adirondacks, a dry September and October creating extremely hazardous fuel in the abundant logging slash left after the customary clear cutting of pine and spruce.

Despite precautions, fires from locomotive sparks would still occur. On June 20, 1904, one of the Shays started a blaze that consumed a railside skidway of logs before being extinguished. Damage was about $5,000.

Throughout the sixteen thousand acres of timberland there were about twelve logging camps, each built and run

258. A panoramic view of the mill complex at Wanakena shows how closely the mill activities and log deliveries were intertwined. From left: veneer mill, shoe-last mill, stave-and-heading mill, storehouse (light color), whip-butt mill (in rear), icehouse (at water edge), planing mill (in rear), and sawmill.

by the individual logging contractors but coordinated by the Rich Lumber Company woods boss, Dick Hanley. Logs were usually moved to the railroad landings on double-bunk sleds, though not the large two-sleds so common in the northeast. Pine cut along the edge of Cranberry Lake during the winter was sledded directly to the mill on the ice. Very little pulpwood was cut on the Rich Lumber Company property.

INDUSTRIAL COMPLEX

The Rich Lumber Company had profited from the full-utilization operation in Pennsylvania, and at Wanakena an even greater variety of mills was established, all situated on the small mill pond south of the Oswegatchie River, on the opposite side of the river from the village. The pond was created by damming Skate Brook, and the mills were lined up along the north edge of the pond, all accessible to the logging railroad as it swung in from the southwest. In all cases the mills were constructed, owned, and operated by companies or individuals other than Rich Lumber Company. This industrial complex allowed not only full utilization of all species but also selection of the logs for best use. Overall superintendent for the Rich Lumber Company was Leonard G. Willson, a kingpin in many of the company endeavors. Willson's son was killed in 1910 when caught between cars on the company railroad.

259. The Wanakena sawmill was owned and operated by Royal and Heman Ford, sawing under contract for Rich Lumber Company. Spruce log lengths normally ran from twenty to forty-eight feet.

Sawmill

Situated on the easterly end of the mill pond was the large sawmill built by Royal and Heman Ford, sawing the lumber under contract for the Rich Lumber Company. This mill had two band saw head rigs, the longer carriage capable of sawing logs up to forty-eight feet in length. Seldom were softwood logs taken into the mill that were less than twenty feet in length. Capacity production was about seventy-five thousand board feet per day with good logs, barely exceeding the production of the Pennsylvania sawmills. Though the New York mill had two log carriages feeding the saws, the Adiron-

dack spruce was small when compared to the Pennsylvania hemlock.

The claim has been made that this sawmill was one of the first mills to heat its log pond with exhaust steam from the mill in order to keep it open all winter. No question that far colder temperatures were found here than experienced in Pennsylvania; Wanakena is located in one of the coldest parts in the Adirondacks. The sawmill pond allowed for log storage free of insect attack, for control of end checking because of drying out, for cleaner logs, and for movement of the logs to the mill entrance. The logs were also sorted in the water if necessary.

Andrew E. Race was sawmill manager for the Ford brothers, and the millwright was J. Otto Hamele, a man well respected in community affairs. Best known of the sawyers were Sid Bush on the long side and Gus Hegstrom on the smaller carriage.

260. Herbert C. Rich, general manager and one of the company founders (right), along with the company woods boss Dick Hanley (left) and J. Otto Hamele, a millwright and respected community leader (center).

On the east end of the sawmill were the usual long lumber docks, five of them, for the sorting and piling of lumber. Some of the docks were almost a half-mile long, and it meant a long journey for the lumber as it was wheeled out by hand on one-axle dollies. Charles Abramson was the foreman on the docks, a job he held at all four of the sawmill towns created by the Rich Lumber Company.

The planing shed attached to the north side of the sawmill contained three planers: a four-sider, a one-sider, and a furring planer with rip saw to make the much-in-demand one-by-twos for plaster lath. Lumber was usually air dried before planing. The foreman of this operation was a man named Burlson.

Also associated with the sawmill was a chip mill operated by Pat Ryan and George Schleider; it was a unique operation that predated by far the modern development of chippers to utilize sawmill waste. This operation received the slabs and edgings from the sawmill, salvaged what lath could be made, and then rossed off the bark and chipped the waste material for sale to a pulp mill.

Whip-Butt Mill

Next down the mill pond and behind the ice house at pond's edge was a mill to make turnings for the butt ends of buggy whips. Beech was used, apparently exclusively. The mill was claimed to be the largest of its kind in the world, but the company went out of business in 1910.

261. *Lath manufacture in the Ryan and Schleder portion of the mill was a labor-intensive task of continually ripping the lath strips from sawmill edgings. Judging from the style of pipes being smoked in the mill, many of the workers were Italian immigrants.*

262. *Bundles of spruce lath ready for shipment.*

263. *The log unloading ramp at the whip-butt mill was almost empty and the empty log cars had not been moved out as yet. The mill in the background is the planing mill.*

264. *The heading mill manufactured barrel components from various hardwood species.*

Heading Mill

Proceeding west along the pond, a large mill for the manufacture of barrel heads was located on the north side of a large storehouse. The mill could utilize any hardwood species and used a bolter saw to cut up the logs. Barrel heads were assembled and shipped out to various cooperage manufacturers. The mill was owned by Henry Venters and managed by a Mr. Buckley.

Shoe-Last Factory

Shoe lasts were turned out in the next structure, a mill owned by Charles Bates and run by Herbert Northrup. Lasts were the wooden forms once used in shoe manufacture and repair and were made exclusively from hard maple.

This mill burned on February 12, 1912, and was apparently the only mill in Wanakena ever to burn. The company's mill fire record was definitely improving, but it would be spoiled after the next move, which was to Vermont.

Veneer Mill

Setter Brothers, a concern out of the Riches' home town of Cattaraugus, New York, was persuaded by Rich to build a small veneer mill in Wanakena in 1902. This mill was the long building at the far western end of the industrial complex. Setter Brothers was an old firm established in western New York in 1885 to manufacture plywood and face stock for furniture. Ernest Horth, millwright with the company for

265. It was a cold winter day as the mill crew rolled the hard maple logs into the shoe-last mill. This mill was the last one to operate at Wanakena. It was destroyed by fire in February 1912.

266. The railroad branch from High Falls came in from the right of the photo and had been in heavy use, judging by the quantity of logs in the pond. The mill in the foreground is the veneer mill.

267. The Setter Brothers veneer mill cut high-grade yellow birch furniture veneer on a rotary lathe. The plant superintendent had persuaded the labor force to dress up a little for the photographer. Courtesy of Bill Gleason.

Cranberry Lake Railroad Co.

Wanakena, N.Y.

268. Letterhead.

fifty years, was sent to Wanakena to build the plant and remained there for the life of the mill.

The Wanakena mill operated only sporadically for nine years and used high-grade yellow birch exclusively. Possibly the better birch logs were not easy to come by at times because purchased veneer logs were reportedly brought into Wanakena by scow down the lake. Veneers cut on the rotary lathe in this relatively small mill were sent to Cattaraugus for further manufacture. The Setters went on to build expanded operations in western New York and Ontario and were eventually absorbed by United States Plywood Corporation in 1963.

From this lineup of mills, it appears that every species of log could find a home if the logging crew found the tree profitable to cut and load on the railroad. Here was a cooperative marriage of manufacturers that has seen very few equals since.

Cranberry Lake Railroad

The building of the wilderness sawmill village of Wanakena necessitated a railroad to connect with the outside world during this turn-of-the-century era. In harmony with the Rich Lumber Company's experiences in Pennsylvania, construction of the village and manufacturing mills was planned around the conception of a company-owned common carrier line that would make connection with a major railroad.

The Cranberry Lake Railroad was chartered on February 24, 1902, with the authority to lay rail from Benson Mines to Cranberry Lake. The building of the eight-mile railroad was one of the first tasks undertaken in 1902 as clearing work began at the site that was to be Wanakena. At the western terminal, the connection was made at Benson Mines with the

269. The four-hundred-foot trestle near Wanakena, known as the Big Fill, was still new when the log train was headed back from the area first logged by the company. The trestle was filled soon after with iron ore waste. Circa 1903. Courtesy of Bill Gleason.

Carthage and Adirondack Railroad, then being operated by the New York Central and Hudson River Railroad. The C&A Railroad depot was used by the Cranberry Lake line. Official opening of the Cranberry Lake Railroad came on May 18, 1903, after a hasty construction to get rail through to Wanakena.

Construction was with sixty-pound secondhand rail laid on six-inch-thick ties hewed on location from spruce, beech, birch, and maple logs. The main line had no span bridges but did have an impressive four-hundred-foot trestle constructed from round timbers. It was later filled in with iron ore sand and became known as the "big fill." On the branch across the Oswegatchie River to the mill there was an attractive three-span pony lattice bridge on concrete masonry, located where the present automobile bridge is found. Total cost of the road and equipment was almost $82,000.

Motive power for the fist two years on the Cranberry Lake Railroad was the two Shay locomotives, Nos. 2 and 4, that also served the logging line. In 1905 a new "rod" or conventional locomotive was purchased from the same Lima Locomotive Company to take over the chores on the passenger line and relieve the Shays. It was a fifty-one-ton Consolidation type (2-8-0) and cost at that time about $9,000. This able little Lima effectively handled the duties of moving the coach and combine plus the shipments of lumber and various freight items. Other rolling stock besides the two passenger cars on the Cranberry Lake Railroad were listed as twenty flat cars and about thirty log cars, all equipped with automatic couplers and air brakes. All three of the locomotives were lettered "Cranberry Lake Railroad." Small gasoline-powered cars (speeders) were used at times for small payload runs. No switch lamps were installed because there were no night trains operating.

Earnings of the railroad corporation were quite favorable, something not always true of a captive railroad serving a lumber company. After surviving a deficit in the fledgling year of 1903, net earnings fluctuated between $3,000 and $5,000 annually. The fact that between eighteen thousand and twenty-one thousand passengers were carried annually shows that it was not only residents of the little village of Wanakena that were being served; a new access had been opened up for the summer vacationers to establish camps on Cranberry Lake. Lumber and other forest products were the major freight commodities to move to Benson Mines, with a much smaller return tonnage of flour, grain, meat, and other provisions, in all totaling better than fifty thousand tons a year.

Forest products of other mill operations also found their way to the Cranberry Lake Railroad because this was really the only convenient way out except down the lower Oswegatchie River. The John McDonald Lumber Company had a rossing mill at Barber's Point on the lake to ross (debark) and slash softwood pulpwood, and they sent the wood and hemlock bark by scow to Wanakena for railroad shipment until their little plant was destroyed in the big fires of 1908. Dana and Brahm Bissell had a small sawmill at Cranberry Lake village and shipped lumber on a barge behind the boat *Merrimac* to Wanakena for loading on railroad boxcars. This, of course, predated by many years the large sawmill and railroad line from the east that Emporium Forestry Company was to give Cranberry Lake village.

The list of railroad corporation officers, slightly different from the previous South Branch Railroad, was as follows: president, Herbert C. Rich; vice president, Horace Clarence Rich; secretary, Clayton R. Rich; treasurer and superintendent, Leonard G. Willson; and auditor, E. E. Keith. There were fifteen stockholders with par value of stock listed at $80,000.

There were twenty-seven railroad employees in all, including officials. This number included a track force of eleven, which may have been kept quite busy after the 1904 visit of a New York State Railroad Commission inspector who reported that the grading was deficient and the cuts

270. The Lima-built Consolidation arrived new in 1905, pictured here at the Benson Mines depot with the company combine and coach.

271. The operator of the compact little Cranberry Lake Railroad speeder appears bored as he waited for a connection at Benson Mines depot. Circa 1908. Courtesy of Bill Gleason.

272. For selected passenger runs on the Cranberry Lake Railroad an attractive railcar was used. Circa 1908.

poorly ditched on the newly built line. Regarding the train crew, the regular engineer was Eugene Blodgett, with Chub Mattison as conductor.

About a mile before Wanakena village, there was a sharp bend in the track known as Shannons' Curve that would cause the wheel flanges of the approaching train to squeal unusually loudly. This noise became a signal to the people in the village that the afternoon train was about to arrive, and they would gather at the depot for the excitement of watching the train come in and unload and waiting for the mail to be sorted, if they had the leisure time.

Though only six miles long, the line was at times graced

273. The depot at Wanakena. A boxcar can be noted on the track that continued on down to the dock at water's edge on the Oswegatchie River.

274. Dana and Brahn Bissell had a sawmill in Cranberry Lake village, from which they barged the lumber down the lake to this shipment yard at Wanakena for loading on the Cranberry Lake Railroad. The yard crew is unloading the barge.

with dignitaries. The elegant private railroad car of New York Central president Chauncey M. DePew was occasionally seen parked on the dock spur at Wanakena.

Camps According to Nationality

At one time there were as many as fifteen hundred people getting their mail through the Wanakena post office. Not that they were all living at Wanakena village; many were in the various logging camps of different types. Noteworthy is the contrast with modern accepted standards, because at Wanakena each nationality had its own camp; there was little mingling.

The loggers' camps usually were located on the railroad and were mostly manned by Italians and Swedes. Lumberjack minister Charles Atwood found logging railroads to be most practical while traveling to the various camps he wanted to preach at in the Adirondacks, whether using a powered speeder on loan from a lumber company or peddling his track bicycle over many long miles. Camps were remote and inaccessible.

Although strong drink was not permitted in the logging camps, an exception was made in the case of the hard-

Table 34
Cranberry Lake Railroad
Local Freight and Passenger Rates

Commodities	
Apples, flour, feed, grain, potatoes, baled hay	$12 per car
Logs, sand, stovewood	$5 per car
Coal (bituminous or anthracite)	30 cents per net ton (carload)
Wood Products	
Veneer in bundles, crates or boxes	40 cents per ton
Lumber, bark, ties, lath, staves, shooks, hoop poles, shingles, chair stock, heading, baled shavings, baled sawdust, poles, piles, posts, wood ashes, sulphite chips, last blocks, mangle blocks, pulpwood	40 cents per ton
Passenger Rate (between Wanakena and Benson Mines)	30 cents one way; 50 cents round trip

275. Claude Carlson and family members at mealtime at the Flat Rock Camp. Carlson was a railroad builder and logging foreman.

working Italians. The crew of the log train was startled one day in 1906 while going out the Dead Creek Flow branch and about to pass the logging camp of jobber Joe Rezio on the east side of the track about a mile before the Flow. Joe came running out to stop the train, waving a beer bottle in each hand and jumping for joy at the birth in camp that morning of his son Daley, many later years a resident of Manchester, Vermont. This news was sufficient cause for a drink for all in the train crew.

At the large camp set up for immigrant Italian workers near Wanakena village, the company found that they had to separate the immigrants from northern and southern Italy. As bitter enemies, their arguments and fights on nonworking days could be vicious. This camp supplied workers for the various mills at Wanakena, hard-working Italians who saved their money and often returned later to Italy to finish their years. To economize they maintained a skimpy diet—dark bread and cheese.

276. The bitterly cold temperatures in the Adirondacks were enough to put frosty beards on both men and horses when sledding off the mountain.

Although beer was allowed in their camp, it was strictly forbidden for Italians to sell it. On one occasion a group from the northern Italian camp was caught selling beer for twenty-five cents per bottle, and they were ordered to appear in court at the county seat of Fine. While traveling the early morning train to the village of Fine on the day of the court appearance, the group consumed a considerable quantity of beer and in their jovial condition hatched a scheme to beat "John Law." Before the bench, only one of the Italians pleaded guilty, and all the witnesses stated they had bought the beer only from that one person. On the way home the gang all chipped in to pay the one person's fine and nobody was out very much. Needless to say, the beer continued to flow.

The French Canadian workers at the mills had their own camp, called the French Settlement, located on the railroad about two miles from Wanakena toward Benson Mines. The site is near the present Route 3 and is now covered by a pine plantation.

Wanakena Village

Most of the inhabitants of the residences in Wanakena village were those who came with the Rich Lumber Company from Pennsylvania, many of them Swedes. Fine little homes were built for them by the company, in contrast to the cruder woods camps, and many of these houses still stand today.

The village got its name back in 1902 while being built. Herbert Rich recalled the name "Wanakena," which he had seen on a street car in Buffalo, and, liking the sound of the name, he chose it for his new village. At its peak, Wanakena village had nearly five hundred inhabitants.

The boardinghouse was operated by the Conroy family and was busy housing the workers who wouldn't think of staying at the Italian or French camps. Another nephew of Herbert C. Rich, Claude Rich, was at first the manager of the express office among other duties, but was in time to take on greater responsibilities as the original Riches grew older.

To supply the village with drinking water, a pipe was laid from some large springs of cold water about two miles away.

277. A planting project was under way in front of the general store for a group of the village youth. It appears that the boys were proud of their efforts at beautification. The gasoline-powered passenger speeder was parked on the depot siding. Circa 1905.

Every house in the village was supplied by these springs. Then, after about fifty years of drinking from this source, it is said that the Albany Board of Health declared it unsafe to drink and required the installation of a chlorinator. As one old-time resident put it, "now when you take a drink, you can think of how Albany saved your life."

Wanakena became quite a busy location during the summer months with the influx of vacationers and camp residents on the lake. Steam-powered passenger boats navigated the West Inlet Flow up to the Wanakena dock except possibly during the dry summer period of late August, when the water was too low. There was a daily excursion around the lake by one of these boats, stopping at all the hotel and cottage docks along the lake for passengers, freight, and mail and reaching Wanakena by noon to connect with the train leaving for Benson Mines. Afterwards it would return to the home dock at Cranberry Lake Village. It was usually the steamboat *Helen*, but the Rich Lumber Company owned a larger gasoline-powered passenger launch, the *Wanakena*, which had a capacity of one hundred people. The company later sold the *Wanakena* and a smaller racing boat, the *Comet*, to the Holland Brothers, owners of *Helen*. It is said

278. Passengers were slowly boarding the steamboat Helen, *sharing dockage with the Rich Lumber Company gasoline-powered launch* Wanakena.

that the popular *Helen* sank at least once a year, ramming one of the numerous floating logs through her wooden hull.

Summer visitors to Wanakena included not only the sportsmen who hired guides to take them to choice fishing locations; there were many who came just to sit on the verandah of the Hotel Wanakena, which the company had built soon after arrival in 1902. The New York Central operated a sleeper for several years that would leave Grand Central Terminal every Friday night and arrive at Wanakena via Utica and Carthage. The sleeper car would be drawn to Wanakena by the Cranberry Lake Railroad and parked on the siding next to the general store until Sunday afternoon, when it would return home with a load of contented New York City guests.

THE EXODUS FROM WANAKENA

The Adirondack timber purchase lasted longer than the company's Pennsylvania operations, but within a little less than eight years the virgin timber growth was about gone. What the loggers hadn't stripped off, the terrible fires of that era had. Logs were purchased from outside sources to feed the dying gasps of the sawmill. In the summer of 1911, the Emporium Lumber Company sold Rich spruce logs for $16 per thousand board feet, towing them from the Bear Mountain Flow to Carlson's boom on the South Flow of Cranberry Lake. The sawmill shut down permanently that year, however, and the heading mill was the last to close, in 1912.

Again, the Rich Lumber Company had in mind the purchase of another tract onto which to move the entire operation, and by the time the last of the Wanakena mills shut down, a site had been selected in Vermont that was ripe for a sawmill and village. But Wanakena did not then die a sudden death as had its two previous Pennsylvania counterparts. Summer camp life on Cranberry Lake was then at its height of popularity, and the Riches' Cranberry Lake Railroad was still quite active.

The homes were gradually sold to their occupants, and camp lots along the inlet were laid out with a seventy-five-foot frontage. Price was $50 a lot on the north side of the river and $25 on the south side. The store was sold to Mrs. J. Otto Hamele, who operated it for many years.

However, Wanakena village did not retain its prominence for long as the primary link between Cranberry Lake and the outside world. The Emporium Forestry Company, another Pennsylvania concern looking for virgin territory to conquer, set up another new sawmill village at Conifer in 1911 and by 1913 had pushed the Grasse River Railroad through to Cranberry Lake village. The Grasse River line provided another link from Cranberry Lake to the outside world, and it proved to be a more practical and enduring rail access than the Rich Lumber Company's common carrier railroad. The Cranberry Lake Railroad operated one more year before ceasing operations, and the assets were sold at a receiver's sale on September 1, 1914, followed shortly thereafter by dissolution of the corporation. The rails were pulled up in the fall of 1917 and the roadbed made into an auto road.

The sixteen thousand acres of timberland were by then a desolate area of logging slash and gaunt tree skeletons left by clear cutting and forest fires, and disposal of the land was a matter of concern. An idea was born in the mind of J. Otto Hamele of the Rich Lumber Company that the cutover lands could serve a useful purpose if used as an experimental area for forestry students. From his suggestion there arose the creation of the New York State Ranger School at Wanakena, the oldest institution of its kind in North America. The Rich Lumber Company donated a 1,814-acre forest tract on the West Inlet Flow to the State University College of Forestry in Syracuse, and in the fall of 1912 the first students arrived by railroad, having as their first tasks the clearing of the site two miles east of Wanakena village and the erecting of the first buildings. The remainder of the Rich Lumber Company timberlands was sold to the State of New York in 1919 and is now part of the Five Ponds Wilderness Area. A sparse new growth of pine and spruce can be seen along the Plains now, slowly reseeding amidst the huge pine stumps still lingering

279. The original buildings for the New York State Ranger School, built 1912–13. Ice fishing was not part of the school's curriculum.

from the Rich Lumber Company days. Vigorous stands of black cherry, yellow birch, and red maple blanket the ridges where the hardwoods had been clear cut for the complex of mills at Wanakena.

The exodus of Rich Lumber Company was not the end of sawmill activity on Cranberry Lake. Emporium Lumber Company built a large band mill at Cranberry Lake village in 1917 that operated for ten years. However, since the time of Bill Streeter's sawmill in the 1930s there hasn't been a sawmill at Cranberry Lake, denying the village any opportunity to grow as the vacation trade diminished.

Wanakena still exists as a quiet little village, tucked away at river's edge on the end of a winding little road. Some of its residents work at nearby Benson Mines or the Newton Falls Paper Company. Others are retired folks, preferring picturesque Wanakena over a Florida retirement home, rejoicing in the solitude even without the primeval forest that first attracted the Rich Lumber Company.

CHAPTER THIRTY-TWO

Jerseyfield Lumber Company Railroad

The Goodyear Lumber Company was one of the largest lumber producers in Pennsylvania during the era of railroad logging in the virgin timber. The firm had its genesis at the capable hands of Frank H. and Charles W. Goodyear in 1887. Although the company had its corporate office in Buffalo, near the home town of the Goodyears, their timber operations were confined to Pennsylvania. Except, that is, for one brief excursion into the southern Adirondacks.

In 1912 the Goodyears purchased a large tract of twenty-three thousand acres in the area around Jerseyfield Lake in the north edge of the Town of Salisbury. They formed the Jerseyfield Lumber Company as the holding and operating corporation to remove the merchantable timber. The Goodyears had no plans for a company-operated mill, but a logging railroad was in the immediate plans. Rights-of-way for a roadbed were obtained from Salisbury Center north into the timber tract.

The village of Salisbury Center had received its first railroad connection in 1908, when the Dolgeville and Salisbury Railway Company opened up its new track, in reality a three-mile extension of the Little Falls and Dolgeville Railroad, which had opened in 1892. The new railroad resulted in an immediate boom for the Salisbury Steel and Iron Company, which had mines nearby and a concentration plant in Salisbury Center for separating iron from the ore. Soon after its opening, the new railroad was leased to the New York Central and Hudson River Railroad.

To create markets for their timber, the Goodyears had the task of convincing associate companies that the Jerseyfield operation was going to be a profitable one and that each should set up a mill on site. Arrangements were no doubt discussed prior to the land purchase.

Timber on the tract was primarily hardwood of varied quality. To utilize the hardwoods, the Goodyears were able

Table 35
Jerseyfield Lumber Company Railroad

Dates	1914–1926
Product hauled	hardwood sawlogs and veneer logs
Length	17 miles
Gauge	standard

Motive Power

No.	*Builder*	*Type*	*Builder No.*	*Year Built*	*Weight (tons)*	*Cylinders Drivers*	*Source*	*Disposition*
1	Lima	Shay 3-truck	2758	1914		12" x 15" 36"	new	Emporium Forestry Co., 1930
2	Lima	Shay 3-truck	974	1905	75	14" x 12" 35"	Goodyear Lumber Co.	Emporium Forestry Co., 1930
6	Lima	Shay 2-truck	874	1904	37	10" x 12" 29.5"	Kentucky & Tennessee RR no. 2	Emporium Forestry Co., 1930

Your Order No. No.

R. M. WHITNEY COMPANY
INCORPORATED
MANUFACTURERS OF BIRCH WAGON HUBS
SALISBURY CENTER, N. Y.

To 192

1% 10 Days. 30 Days Net F. O. B. Shipping Point

280. Letterhead.

to establish an industrial complex of three different manufacturers, each using a different grade and species of wood. And they all used hardwood exclusively.

The first wood-manufacturing mill established at Salisbury Center was the R. M. Whitney Company, a manufacturer of birch wagon hubs, primarily small hubs for baby and doll carriages. In December 1913 Anson C. Goodyear, president of the Jerseyfield Lumber Company, signed an agreement with Russell W. Whitney to deliver ten to fifteen million board feet of birch logs to the Whitney plant that was to be built on the Jerseyfield Lumber Company railroad line in 1914.

By far the largest market for the Jerseyfield hardwoods was the next company to agree to terms, the Brooklyn Cooperage Company. In January 1914 an agreement was inked in which Jerseyfield agreed to cut and deliver 60 percent of the merchantable hardwood on their ownership to the Brooklyn Cooperage plant, which was to be up and operating by January 1, 1915. Minimum yearly delivery was seven million board feet and maximum was ten million. The contract would extend until a total of seventy million board feet was delivered to the Brooklyn Cooperage plant, located next to the Whitney plant.

281. Birch wagon hubs, manufactured by the R. M. Whitney Company. Courtesy of the Salisbury Historical Society.

282. The Brooklyn Cooperage Company plant in Salisbury Center. The railroad car log dump is in the foreground, and the stave-drying sheds in the left rear. Circa 1916. Courtesy of the Salisbury Historical Society.

Brooklyn Cooperage did not need high-grade logs for the manufacture of barrel staves and headings, and the species of hardwood was not important, either. Therefore, the higher grades of birch and maple logs were sorted out before loading the cars for Brooklyn Cooperage; beech was delivered unsorted ("woods run"). Brooklyn Cooperage paid $14 per thousand board feet for the logs delivered.

Jerseyfield Lumber laid the tracks to connect the Brooklyn Cooperage plant and the other mills with the New York Central tracks in the village. When moving cars of finished wheel hubs to the New York Central transfer, the Jerseyfield Lumber Company billed the R. M. Whitney Company forty cents per two thousand pounds of product.

The third company established in the complex, which didn't come on board until almost two years later, was a veneer producer able to utilize the highest grade of logs. The Salamanca Panel Company, an established veneer manufacturer from western New York, built a small plant at Salisbury in 1916. The Jerseyfield railroad supplied Salamanca with birch and maple veneer logs, fifteen inches and up in diameter, for a delivered price of $26 per thousand board feet. Log delivery was limited to three million board feet per season.

The building of the Jerseyfield Lumber Company log-

283. The Jerseyfield Lumber Company used a Marion model 10 log loader and carried the logs on flat cars.

ging railroad began in November 1913; by the end of the month two miles of roadbed were graded. The following year the company put logging crews to work and began the first rail deliveries during the 1914–15 logging season. The company had its own flat cars to carry the logs, with the company name on the side. Loading was done with a Marion model 10, which moved from car to car along rails mounted on the flat car body. There may have been more than one loader. One of the loader operators was Archie Pollard, Sr.

For motive power Jerseyfield Lumber used Shay geared locomotives. Two of them were the heavy seventy- to eighty-

284. What appears to be Shay No. 1 is steamed up and ready for a cold day's trip into the Jerseyfield woods. This large eighty-ton engine was one of only six three-truck Shays used in the Adirondacks. Circa 1917.

285. This small Shay, at use in the Jerseyfield Lumber Company railroad, is of unknown identification and was well used, judging by the array of cables on board. The older gent on the left, obviously not part of the crew, is believed to be Dwight Louchs, who had one arm.

286. A bevy of smiling girls duck behind some assembled barrels made by the Brooklyn Cooperage Company. The company plant at Salisbury Center made the unassembled barrel components, the staves and the heads. Courtesy of the Salisbury Historical Society.

ton three-truck type not usually used in the Adirondacks. Ray Owens was one of the locomotive engineers. Marshall Greenly supervised the railroad operation.

The Brooklyn Cooperage Company plant made the staves and headings for the thirty-gallon dry-goods barrels. J. E. Sutter was the original superintendent but because of ill health, he was replaced in January 1920 by William Keyes. Many of the workers moved up with their families from a

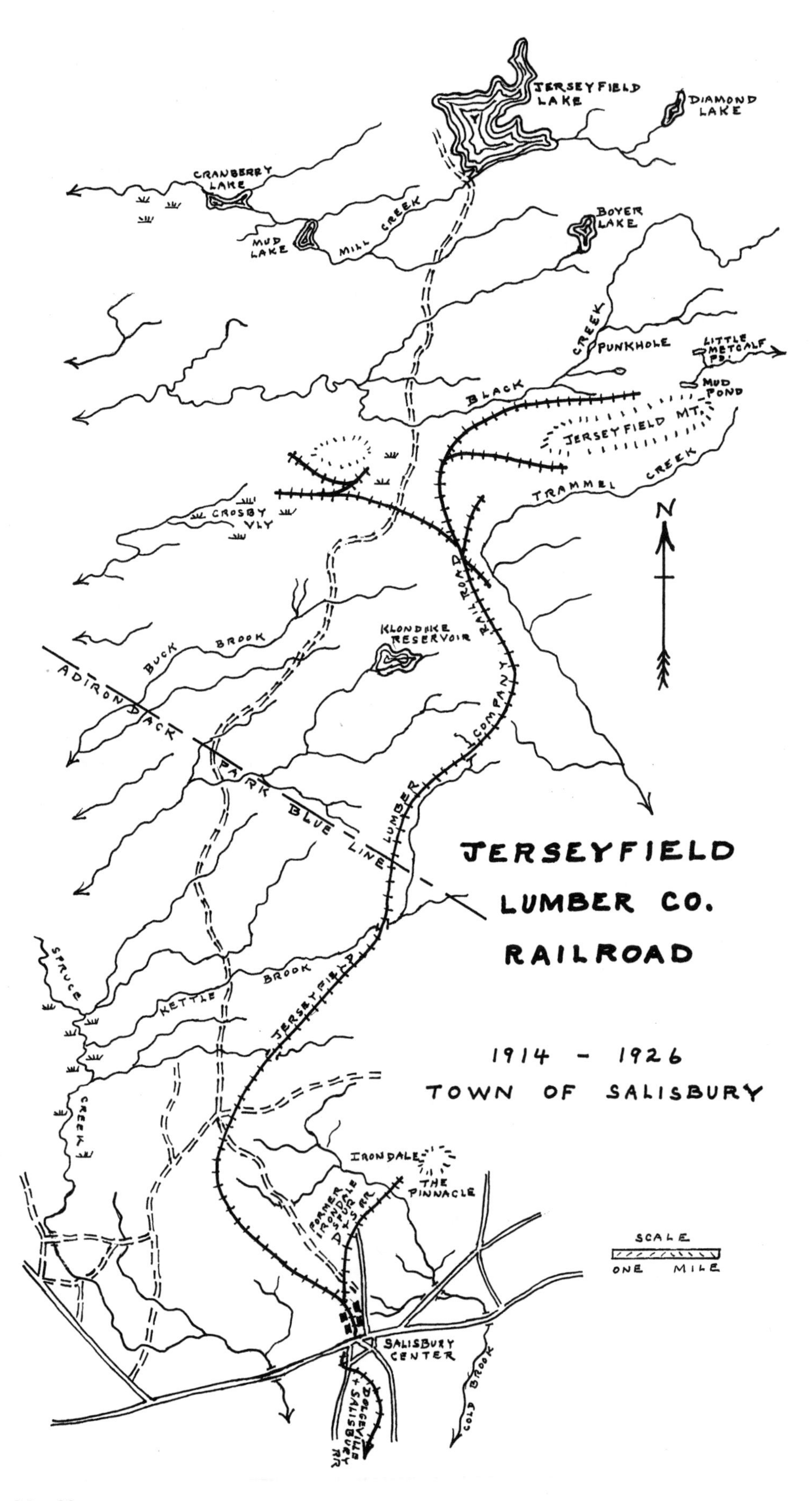
JERSEYFIELD LAKE
DIAMOND LAKE
CRANBERRY LAKE
BOYER LAKE
MUD LAKE
MILL CREEK
CREEK
PUNKHOLE
LITTLE METCALF PD.
MUD POND
BLACK
JERSEYFIELD MT.
TRAMMEL CREEK
CROSBY VLY
N
KLONDIKE RESERVOIR
BUCK BROOK
ADIRONDACK PARK BLUE LINE
LUMBER COMPANY RAILROAD
JERSEYFIELD LUMBER CO. RAILROAD
1914 - 1926
TOWN OF SALISBURY
SPRUCE
KETTLE BROOK
JERSEYFIELD
CREEK
IRONDALE
THE PINNACLE
FORMER IRONDALE SPUR D. & S. RR.
SCALE
ONE MILE
SALISBURY CENTER
COLD BROOK
DOLGEVILLE + SALISBURY RR

Map 39.

company facility in Potter County, Pennsylvania, and Brooklyn Cooperage built ten houses for employees, plus the superintendent's house, two factory buildings, and an engine house. The houses were sold to the employees for $2,000 each.

Tragedy hit the plant when one of the employees fell into a curing vat. He was pulled out by foreman John Herringshaw but died shortly afterward.

The veneer mill was managed by Lloyd Henry. These were the war years, and product demand allowed for full employment. A large portion of the work force was women, working for $14 per week.

The marriage between this small complex of industries and the Jerseyfield Lumber Company lasted at the most for the length of the contract with Brooklyn Cooperage. It appears that about ten years after the inception of activities, the timber supply became almost drained. Whether Jerseyfield Lumber Company had been unable to harvest the appraised seventy million board feet or whether other harvesting problems cropped up to bring a premature end to the arrangement, nobody now knows. There had been some rumors of discontent by the industries.

Reports have it that the veneer mill ended full-time production by 1920 and that the machinery was moved soon after to a company plant in Poland, N.Y. Their log supply agreement with Jerseyfield Lumber was not cancelled until 1928, when Jerseyfield paid for a settlement with the company.

The Brooklyn Cooperage Company shut down their doors about 1925 and began to sell off the mill property in 1926. There appears also to have been a problem with their log supply contract with Jerseyfield Lumber; the contract was assigned to another party in 1926 but not released until 1932.

287. Jake Merriman's doodle bug, rigged to travel the rails on the abandoned Jerseyfield railroad, in August 1928. Courtesy of Paul Flanders.

The Whitney hub mill was apparently the last industry to close up, reportedly remaining until 1927. They sold their property in October of that year.

Although Jerseyfield Lumber had finished their log harvest and railroad activity by 1926, the rails were left in place. In July of that year the railroad rights-of-way were conveyed to the owners of Jerseyfield Preserve, an adjoining ownership in the area of Jerseyfield Lake. The caretaker, Jake Merriman, rigged up an old car to travel in on the rails.

The majority of the timber tract of Jerseyfield Lumber Company was sold to the State of New York in 1932.

Epilogue

The Slow Bell

The year was 1948, and the trip home with a string of log cars was a slow pull. Deteriorating track conditions dictated a minimum speed, but the aging old steam pot and its forlorn attendants weren't particularly eager to hurry to the engine house. It was all coming to an end; the last run under steam with a load of logs had probably been made. A slow bell announced to the village of Conifer that the era of railroad logging in the Adirondacks was done, finished. Preparations were underway to pull all of the Emporium Forestry Company railroad track except the two miles between Conifer and Childwold Station.

Sixty-five years had passed since that first trainload of Adirondack logs had rolled into St. Regis Falls to open an era of unprecedented assault on the old-growth forests of the Adirondack Mountains. The era had been one that introduced mechanized logging techniques, had created new log markets to include about any species of tree, and had brought in the tools to reach about any remote area of merchantable

288. Well-worn rails course through the North Woods, flanked by pine and spruce logs waiting for the loading crew. Scattered pine trees tower over the next generation of spruce growth for future harvesting at Brandreth Park.

289. The small Shay once operated by the Horse Shoe Forestry Company is the only remaining relic of the Adirondack's railroad logging era. Bought new in 1904, it was later shipped to a lumber company in West Virginia and eventually to Ralph Day of Biddeford, Maine, who has placed it on display. Photograph by Alan Thomas.

timber. It was an era that introduced the chemical pulping of wood and the newfound market for smaller-sized conifer trees. It was an era that was to change the character of much of the Adirondack forests and was to introduce the sometimes-applied practice of forest management. It was a contentious era that brewed the never-ending debate over the proper forest policy for the now 2.8 million acres of state-owned forest land within the 6-million-acre Adirondack region; should this vast acreage remain forever wild and untouched or should the forest be intelligently managed for improved forest growth to benefit the industries of the state? The debate still goes on.

In the midst of those profound changes upon the timber resource during the early 1900s, steam played a major role, both on the iron rails coursing through the woods and in the boiler rooms of the mills. Steam contributed to the romance and excitement of a logging scene that could now never be duplicated. Nor could it be credibly chronicled in all of its fascination and folklore. Woods employment was physically demanding in skills, toil, and endurance; the days were long and the compensation meager, but the comradeship was a strong and rewarding bond. It was a bewitching era.

The loggers loved their locomotives, or at least most of them. The engines were captivating to watch and enchanting to listen to. They were noisy, dirty, cantankerous, and temperamental. Each one had its own peculiar characteristics, its own stubborn annoyance and quirks of operation. No wonder so many of the loggers or engineers would not let anyone else touch their locomotive. But the machines did their assigned task admirably, if not always cleanly and faithfully.

290. The former roadbed of the Mac-A-Mac railroad in Brandreth has made an excellent truck road for forest management purposes. Photograph by author.

The shrill cry of a steam logger's long whistle piercing the forest growth is a sound never forgotten, if one has been fortunate enough ever to have heard it. The vision of a well-used old steamer sitting at the log landing, waiting for the loading crew to finish their chores, breathing spurts of steam as the locomotive crew builds steam pressure for the tough pull going home, is a cherished mental picture of times gone by. The sharp staccato sound of the three cylinders on the side of the Shay locomotive straining to make a grade, the scream made by the flanges of the car wheels as a sharp downgrade curve was negotiated, are all sounds etched now only in memories. The loggers' beloved and cursed old steam pot has steamed off into oblivion, a victim of changing times.

Only one of the many locomotives used in the Adiron-

291. *Huge white pine stumps still remain in 1971 from the Rich Lumber Company operations on the Plains, lingering among the growth of new softwood trees. Photograph by author.*

293. *Where once the flooded waters of North Pond flow ground at Brandreth Park floated thousands of logs, there now flows a quiet stream, framed by a new growth of spruce. Photograph by author.*

292. *The site of Brandreth settlement along the abandoned Adirondack Division of the New York Central is barely recognizable now. Photograph by author.*

dack woods has escaped the graveyard with its accompanying cutting torch. When A. A. Low's Horse Shoe Forestry Company closed up operations in 1911, the Shay was shipped out to the Porterwood Lumber Company in West Virginia. After a succession of a half-dozen owners, the engine was acquired in 1984 by Ralph Day of Biddeford, Maine, who has the little Shay on display at Biddeford Station.

Why was there such an inglorious end for the steam locomotive in the woods? For the logger it was a matter of economics; for the landowners it was adjusting to the changing resource base. Steam locomotives were inefficient and wasteful of fuel and water. Construction, maintenance, and support facilities were costly. Along came the diesel locomotive in the 1930s, lacking the romance, warmth, and majesty of steam power but offering a much more economical and practical replacement. However, railroad logging in the Adirondacks was in its dying throes by this time, and there never was a diesel locomotive used on a log job in the north woods. Heywood-Wakefield Company had a small forty-four-ton diesel at Conifer in the 1950s, but it was only for industrial yard use.

A major factor in the decline of Adirondack railroading, as almost everywhere, was the development of high-capacity trucks, much more efficient and economical than railroading. However, the hard toil of the railroad builders was not entirely lost in history. The firmly packed and well-drained railroad beds made excellent truck roads and are still in use seventy-five years later. A little narrow maybe, but the grades are excellent.

The other factor pushing woods railroads into oblivion was the changing character of the timber resource on the private and industrial timberlands that were available for timber harvest. A large timber cut was required to justify building a new railroad, but the sizable old-growth timber stands were almost gone. Professional forestry techniques, when applied on privately owned timberlands, often dictated smaller, more frequent timber removals, ideal for trucks.

True, the railroads were a key tool in the removal of immense volumes of old-growth timber, but in fairness, it is not the fault of the loggers or the railroads that so much of the big

294. The loggers have long ago departed, and a fine mixed growth of hardwood and softwood trees once again grace the landscape. Photograph by author.

timber is gone. Timber to build the nation is what the people wanted and is what the markets demanded. The loggers responded to the demand. And it was steam railroads that were used to make a major contribution toward feeding those voracious new markets for wood.

Railroad logging in the Adirondacks unquestionably changed the landscape. Motivating the logger was the philosophy of an expanded use of the forest resource, and the public supported this view. From that bygone era when there was always considered to be plenty of trees over the next hill, there has now developed a greater public awareness of our valuable forest resource. Forest management practices are being applied to varying degrees on private and industrial ownerships, but the strong timber markets continue to exert a heavy demand for wood from the lands that remain available for timber harvest. The huge state ownership in the Adirondacks, with its large areas of mature timber, remains unavailable for forestry practices that could encourage a healthier, faster-growing, sustainable forest resource.

The great forest of the Adirondacks was harvested by the river drivers, followed by the railroad loggers and the truckers, and at the same time lamented by those with the dire predictions that the forests would be ruined. But there is a new growth of trees now, recreation is strong along the pathways and river channels of the old-time loggers, and there is still timber to harvest for the wood products industry.

The old-time logger is gone; the remains of his camp and his river dams have long since rotted into obscurity. Modern-day loggers still work hard at the dangerous occupation that supplies the public's demand for the great variety of everyday products made from the forest resource. Many are those who curse the loggers but wouldn't for one moment consider a lifestyle devoid of the essentials of life that the loggers and forest industries make available for them.

Yes, the old-time loggers are gone, along with the romance of a lifestyle peculiar to the north woods. And silent is the slow bell of the loggers' beloved steam locomotive.

GLOSSARY

BIBLIOGRAPHY

INDEX

Glossary

American locomotive. Locomotive with the old-style 2-4-0 wheel arrangement.

Banking ground. An area next to a river where logs were piled preparatory to the river drive.

Barienger brake. A mechanical brake used to hold back the descent of log sleds on a steep grade. It consisted of grooved friction wheels around which many turns of steel cable were wound and operated by two hand levers.

Board foot. A piece of lumber twelve inches long, twelve inches wide, and one inch thick, or any piece of equivalent volume. MBF refers to one thousand board feet.

Bob sled. A single set of sled runners with a cross beam or bunk on top upon which the front end of the log is placed while the rear end drags on the ground.

Boom of logs. Quantity of loose logs or pulpwood contained in the water within a large enclosure formed by a sequence of logs strung together, end to end.

Bridle chain. Chain wrapped around the sled runners to act as a brake on downgrades.

Bucked. Cutting up a fallen tree into individual logs or short pulpwood lengths.

Bullnose. A coupler box on the front of the locomotive with a number of slots that could accommodate the adjustable knuckles or link-and-pins at various heights.

Burrow pit. A depression in the ground from which sand or dirt fill has been dug out.

Chickadees. Workers that sanded and maintained sled roads.

Circular sawmill. A mill using a large circular saw for the principal breakdown of the logs into lumber.

Clear cutting. Cutting all the trees in the forest or, in some cases, cutting all of the merchantable trees.

Common carrier. A railroad chartered and authorized by the government to carry passengers.

Consist. Makeup of railroad cars behind the locomotive.

Cooperage. Barrels or tubs.

Cord. A pile of wood four feet long, eight feet wide, and four feet high, or a pile of equivalent cubic volume.

Crosscut saw. A two-man saw, generally about five and a half feet long, used to cut trees.

Crown sheet. The metal covering over the top of a steam locomotive fire box.

Dog. Metal hook having an eye in one end that was driven into log ends for loading logs with a crotch-line cable or for fastening logs together end to end when skidding.

Double a train. An operation whereby a train was separated and taken up a steep grade a section at a time.

Double head. Using two locomotives in the head end of a train.

Drawbar. A long piece of bar used to separate railroad trucks or cars, used on either link-and-pin or knuckle couplers.

Drive. The method of floating logs or pulpwood down a river or stream to a mill location.

Dumping air. A sudden release of all the air pressure in the locomotive brake system, which would result in the brakes exerting maximum stopping effort.

Extra. Another train operated in addition to the scheduled operation.

Fiddlebutts. High-quality spruce logs used for the manufacture of piano sounding boards, violins, and other musical instruments.

Fish plate. A plate used to hold the ends of two rails together, bolted to the rails from each side.

Flag stop. A location along the railroad where a train could make an unscheduled stop for passengers if the flag was put out.

Flanger. The metal scraper mounted under a railroad car, used to remove snow and ice from between the rails.

Flow ground. An area in a stream temporarily dammed and flooded for the storage or gathering of logs.

Frog (track). The point in a rail switch where the train wheels cross to another rail, having depressed channels to allow passage of the wheel flangers. It is so named because of the resemblance to a part of a horse's hoof that bears the same name.

Gandy dancers. Railroad track builders.

Gang saw. Sawing machine with a heavy frame and a sash containing several saw blades that cut while oscillating vertically.

Geared locomotive. Any engine that had the power from the cylinders transmitted to the wheels by means of gears.

Grab skipper. A short-handled hammer similar to a pick ax but with a heavier head and stouter handle. Used to detach grabs.

Grabs. Two log dogs connected by a short length of chain driven into log ends for pulling or loading logs.

Hardwood trees. Deciduous trees, those that shed their leaves each fall; in the Adirondacks especially beech, birch, maple, and ash.

Heading. The top or end piece on a barrel.

High grading (timber). Harvesting only the best quality trees and leaving a remaining stand of poor quality trees.

Hostler. The worker assigned to maintain the locomotive.

Jackworks. A powered chain conveyor (usually steam) used to convey logs out of the water onto a deck for loading onto sleds or railroad cars.

Jammer. A log loader, frequently using an A-frame for a boom.

Jim Crow. A bar used by the train brakeman to set or tighten up the hand brakes on a car.

Jobber. A logging contractor or subcontractor.

Landing. Location where logs from the woods are accumulated for rolling into a river or loading onto sleds or railroad cars.

Link-and-pin couplers. An old style of coupling railroad cars together by which a link, generally a long metal oval ring, is secured in the pockets on the ends of each car by a heavy metal pin.

Log drive. See **Drive.**

Merchantable. That portion of a timber stand or of a tree that can be logged profitably and have a market.

Mogul. A locomotive having a 2–6–0 wheel arrangement.

Narrow gauge. Any width or gauge between the train rails that is less than the standard gauge of four feet and eight and a half inches. The most common narrow gauge was three feet.

Old growth. Virgin timber that has never been logged.

Peavey. A logger's tool for rolling and manipulating logs. It consists of a long handle with a steel point on one end and a hook on a hinge near the point. One of the most useful tools in the woods.

Peeling. The removal of the bark from softwood logs that are to be used for paper manufacture.

Pickaroon. A short-handled tool with a pick-pointed head on the end for handling pulpwood sticks.

Pile-bottoms. Portable foundations on which lumber was piled for air drying.

Pulp mill. Mill that manufactures wood pulp for use in paper manufacture.

Pulpwood log. Pulpwood material that is cut and transported in long lengths and later sawn into short lengths before being processed at the pulp mill.

Rail weight. Based on the number of pounds per yard of rail.

Replacer; rerailer. Used to place derailed car wheels back on the track.

Right-of-way. The legally obtained strip of land within which the railroad track or road is built.

Road monkey. A worker who maintained sled roads.

Rod locomotive. The standard type of locomotive on which the power is transmitted directly from the cylinders to the drive wheels by means of a main rod and siderods.

Rossing. The removal of bark from logs or bolts.

Saddletank. A locomotive with the water tank draped over the boiler saddle-style, instead of being carried in a tender.

Saranac standard. An old-time standard of log size, a log twenty-two inches in diameter and thirteen feet long.

Scaler. Man who estimates the contents of logs in board feet or cords.

Scaling. Measuring the board feet or cord content of logs or trees.

Shoe last. A form used for making shoes.

Shook. A bundle of staves and headings needed to make one barrel.

Skeleton log car. Two sets of trucks held together by a center beam or frame but having no deck.

Skidder, rail-mounted. Set or built upon a flat car and used to drag logs up to railside. Steam or gasoline powered.

Skidding. Dragging logs along the ground without the use of wheels or runners.

Skid road. A cleared path over which the horses or tractor dragged out the logs.

Skidway. Elevated crib work from which logs were loaded onto sleds or railroad cars.

Slasher mill. A stationary saw set up to cut long-length pulpwood into short lengths.

Snubbing. The use of cables or ropes to regulate the descent of loaded sleds on steep slopes.

Softwood trees. Conifers with needles; in the Adirondacks especially pine, spruce, fir, and cedar.

Speeder. Gasoline- or diesel-powered small railroad car used for transporting passengers or the railroad track crew.

Sled tender. The teamster's assistant who pulls the sled around into position, loading and binding the logs and performing other duties.

Spring binder. A sapling used as a pole to tighten a sled chain. One end was placed under the chain and then the pole was bent around the other way and secured, tightening the chain.

Stumpage. Refers to standing merchantable trees.

Standard gauge. A distance of four feet eight and a half inches between the track rails.

Tires (locomotive). The removable metal rims around the drive wheels.

Tram road. A temporary railroad line used for the transportation of logs.

Trespass (timber). The unlawful cutting of trees belonging to another owner.

Truck (railroad). A framework holding the axles and flanged railroad wheels, generally two axles.

Twitching. Dragging logs along the ground without the aid of runners or wheels.

Whistle stop. A logging camp with inferior food, where, consequently, loggers didn't stay long.

Wye. A Y-shaped track siding that allows a locomotive to turn around without a turntable.

Bibliography

REPOSITORIES, COMPANY RECORDS, AND PERSONAL CORRESPONDENCE

Arfman, John. 1965. Personal correspondence, June 28.

Cardamone, Mrs. Joseph. 1979. Correspondence with the author.

Cornell University. Krock Library Rare Manuscript Collection. Ithaca, N.Y.

County Deed Registries. Franklin and St. Lawrence Counties, N.Y.

Emporium Forestry Company Records. Adirondack Museum, Blue Mountain Lake, N.Y.

Farmer, Ralph M. Personal notes on Watson Page Lumber Company.

Fernow, Bernhard. 1899–1901. Correspondence while at Axton. Carl A. Krock Library, Cornell Univ., Ithaca, N.Y.

Giblin, John. 1999. Correspondence with author.

Hull, Gerald P. 1966. Personal correspondence, July 27.

International Paper Co. Ca. 1920. Correspondence from company file at Woods Lake field office.

Kellogg, William N. Personal diary, private collection.

McDonald, Gene. 1981. Correspondence with author.

Roth, Filbert. 1899–1901. Correspondence while at Axton. Carl A. Krock Library, Cornell Univ., Ithaca, N.Y.

Smallwood Collection of newspaper clippings, Paul Smith's College Library, Paul Smiths, N.Y.

BOOKS AND ARTICLES

Aber, Ted, and Stella King. 1965. *History of Hamilton County.* Lake Pleasant, N.Y.: Great Wilderness Books.

Adams, Kramer A. *Logging Railroads of the West.* New York: Bonanza Books, 1961.

"Adirondack Enterprises." 1986. *Malone Palladium,* July 29, 1.

Alfke, Frederick H., and Charles Ten Eick. 1919. *Wood and Mill Operations in the Adirondacks in the Vicinity of Tupper Lake, New York.* New York State College of Forestry Pamphlets, no. 20-1-401. Division of Rare Manuscript Collections, Cornell Univ. Library, Ithaca, N.Y.

Allen, Richard S., William Gove, and Richard Palmer. 1973. *Rails in the North Country Woods.* Utica, N.Y.: North Country Books.

Bowen, G. Bryn. 1970. *History of Lewis County, N.Y., 1880–1965.* N.p.

"A Brief Look at Brooklyn Cooperage." 1946. *Sugar.* Jan.

Brown, Ralph Adams. 1933. *The Lumber Industry in the State of New York, 1790–1830.*

Brown, William H. 1963. *History of Warren County.*N.p.: Board of Supervisors of Warren County.

Burnett, Charles H. 1932. *Conquering the Wilderness.* [Norwood, Mass.]: Privately published.

Clark, F. Mark. 1974. "The Low Dynasty." *The Quarterly, Official Publication of the St. Lawrence County Historical Society* 19, no. 1 (Jan.), 9–15.

"A Combination Dam and Bridge." 1908. *Engineering News* 59, no. 15 (Apr. 9).

De Sormo, Maitland C. 1974. *The Heydays of the Adirondacks.* Saranac Lake, N.Y.: Adirondack Yesteryears.

Doherty, Lawrence. 1971. "Railroad History of Franklin County." *Franklin Historical Review* 8:6–22.

Donaldson, Alfred L. 1921. *History of the Adirondacks.* Vol. 2. New York: Century Co.

"Dr. Webb's Adirondack Railroad." 1973. *Franklin Historical Review,* 49–61.

Fowler, Albert V. 1968. *Cranberry Lake, 1845–1959: An Adirondack Miscellany.* Blue Mountain Lake, N.Y.: Adirondack Museum.

———. *Cranberry Lake from Wilderness to Adirondack Park.* Syracuse: Adirondack Museum and Syracuse University Press.

Fox, William F. 1976. *History of the Lumber Industry in the State of New York.* Harrison, N.Y.: Harbor Hill Books.

Graham, Frank. 1978. *The Adirondack Park: A Political History.* New York: Knopf.

Harter, Henry A. 1979. *Fairy Tale Railroad: The Mohawk and Malone.* Utica, N.Y.: North Country Books.

Herbert, Paul. *Reminiscences of Tupper Lake.* Unpublished manuscript, Paul Smith's College, Paul Smiths, N.Y.

Hochschild, Harold K. 1962. *Lumberjacks and Rivermen in the Central Adirondacks, 1850–1950.* Blue Mountain Lake, N.Y.: Adirondack Museum.

Hosmer, Ralph S., and Eugene S. Brace. 1901. *Forest Working Plan for Township 40.* Washington, D.C.: GPO.

Hough, Franklin B. 1883. *History of Lewis County.* Syracuse: D. Mason and Co.

Hull, Gerald P. 1951. "Story of Oval Wood Dish Corp." *Monitor,* Dec.

Hyde, Floy S. 1974. *Adirondack Forests, Fields, and Mines.* Utica, N.Y.: North Country Books.

Johnson, Joseph C. 1920. "Logging in the Adirondacks." *The American Sugar Family,* May 13, 12–13.

Keith, Herbert F. 1972. *Man of the Woods.* Syracuse: Syracuse Univ. Press.

Kerr, John Hartwell. 1975. *Personal Memories.* Unpublished manuscript in the possession of Oswegatchie Town Historian.

Kiernan, Betsy. 1984. "Conifer Remembered." *Ogdensburg Advance News,* Nov. 18.

Kling, Edwin M. *McKeever in the Teens.* Unpublished manuscript in the possession of the author.

Koch, Michael. 1971. *The Shay Locomotive: Titan of the Timber.* Denver: World Press.

Kudish, Michael. 1985. *Where Did the Tracks Go? Following Railroad Grades in the Adirondacks.* Saranac Lake, N.Y.: Chauncey Press.

———. 1996. *Railroads of the Adirondacks: A History.* Fleishmanns, N.Y.: Purple Mountain Press

Marleau, William R. 1986. *Big Moose Station.* Eagle Bay, N.Y.: Marleau Family Press.

McLaughlin, William G. *Lake Ozonia: An Informal History.* Privately published.

McMartin, Barbara. *The Great Adirondack Forest.* Utica, N.Y.: North Country Books.

Meigs, Ferris. 1941. *Annals of the Santa Clara Lumber Co.* Unpublished manuscript, Adirondack Museum, Blue Mountain Lake, N.Y.

New York State Forest Commission. 1891–94. Annual Reports, Albany, N.Y.

New York State Forest, Fish, and Game Commission. 1895–1907. Annual Reports, Albany, N.Y.

New York State Public Service Commission. 1953. Minutes on Hearing of July 27, 1953, re. proposal of discontinuance by NYC, Albany, N.Y.

Parker, Lucinda M. 1987. *Into Salisbury Country.* Brookfield, N.Y.: Worden Press.

Pinchot, Gifford. 1898. *The Adirondack Spruce.* New York: The Critic Co.

Recknagel, A. B. 1923. *The Forests of New York.* New York: Macmillan.

Reed, Frank A. 1965. *Lumberjack Sky Pilot.* Utica, N.Y.: North Country Books.

Rodgers, Andrew Denny III. 1951. *Bernhard Eduard Fernow: A Story of North American Forestry.* Princeton, N.J.: Princeton Univ. Press.

Seaver, Frederick J. 1918. *Historical Sketches of Franklin County.* Albany, N.Y.: J. B. Lyon Co.

Simmons, Louis. 1976. *Mostly Spruce and Hemlock.* Tupper Lake, N.Y.: Tupper Lake Free Press.

Smith, Nelda Young. 1969. "John Hurd's Railroad." *Franklin Historical Review* 6:29.

Trimm, Paula. 1989. *Looking Back Upstream.* N.p.

Valliancourt, Armand. 1979. Personal notes regarding A. A. Low. Sept. 9.

Welsh, Peter C. 1995. *Jacks, Jobbers, and Kings.* Utica, N.Y.: North Country Books.

PERIODICALS

Adirondack News. 1887–1913.

American Lumberman. 1910, 1912.

Essex County Republican. Nov. 11, 1900.

Journal of Forestry. 1917–1936.

New York Times. 1899–1913.

New York Tribune. 1899–1903.

Northeastern Logger. July 1957, May 1960.

Northwest Lumberman. 1887.

Tupper Lake Free Press. Apr. 26, 1912.

INDEX

Page number in *italic* denotes an illustration.

Abbey, Henry, 8
Abbott, Vasco P., 39
Abramson, Charles, 227
Adirondack and St. Lawrence Railroad, 66, 67, 69, 72
Adirondack Park Associates, 66
Adirondack Railroad, 7
Adirondack Scenic Railroad, 71
Adirondack Timber and Mining Company, 67
Aldrich village, 208
American Realty Company, 82
Angello, James, 191
A. Sherman Lumber Company, 189–90
Atwood, Charles, 222
Avery, Nelson, 79
Axtle, Leo D., 122
Axton, 77

Barienger brake, 45, *47*
Bates, Charles, 228
Bay Pond Inc., 184, 185–88, *185, 186, 187*
Bay Pond village, 183
Beaver River settlement, 70, 72
Benson Mines, 212, *212*
Big Mill, 158, *158*, 161
Big Moose village, 69, 70, 73
Bissell, Dana and Brahm, 230
Black River and Woodhull Railroad, 8
Blesh, Jesse, 123
Brandon village, 30, 32, 152, 183
Brandreth, Benjamin, 87
Brandreth Park, 71, 87, *87*, 91
Briggs village, 211
Brooklyn Cooperage Company, 169–70, 181; at Salisbury Center, 238–41, *238*; at Santa Clara and St. Regis Falls, 170–75, *171*; at Tupper Lake, 177–81, *178*
Brooklyn Cooperage Company Railroad: at Everton, 172; at Lake Ozonia, 173; at Meno, 172–75; at Tupper Lake, 175–80, *179*; locomotives, *172, 174, 175*
Burns, Edward M., 78
Bushey, George, 52, 83, 93

Caflisch, William S., 29, 113, 134
Canton Lumber Company, 106
Canton village, 106
Carlson, Claude, 114, 224, 233
Carlson's Camp, 88
Carter settlement, 76
Carthage and Adirondack Railroad, 203, 208, 221
Cascade Chair Company, 165
Champlain Realty Company, 83
Chateaugay Ore and Iron Company, 143
Chateaugay Railroad, 138
Childwold station, *70*, 114
Clarksboro, 215
Clearwater, 76
Cleveland tractor, 52
Clifton Iron Company, 215
Conifer village, 108–11, *112*
Cope, Henry, 131
Cotter, Thomas B., 144
Cranberry Lake Railroad, 35, 36, 229–32, *231*, 235
Cranberry Lake village, 121, 125
Creighton, Thomas, 194, 197
Cuddy, John, 186, 194
Culhea, Bill, 148, 194
Cutting, Frank A., 174

Davignon, John, 83, 133
Debar Mountain, 142
DeCamp, William, 76
Delaware and Hudson Railroad, 65–66, 143–44
DePew, Chauncey M., 232
Derrick settlement, *152*, 198
Devereaux, Henry, 8
Dix, John, 73, 75, 76–77
Dodge, Meigs and Company, 153, 157
Dolgeville and Salisbury Railway Company, 237
Douglass, Robert, 162
Dowling, Elias, 84
Downey-Snell Logging Company, 175
Draper Corporation, 134
Ducey, Patrick A., 30, 182
Durant, Thomas Clark, 76
Durant, William West, 77

Eaton, Albert, 8
Edwards, Ambrose, 111, 133
Emporium Forestry Company, 111–13, *113*; Conifer sawmill, 109–11, *110, 111*; Cranberry Lake sawmill, 122–23, *127*; logging, 127–32
Emporium Forestry Company locomotives, 107–9, 115–16, 123; Climax locomotive, *20, 140*; railcars, 125, *126, 127, 130*; rod locomotives, *17, 18, 34, 113, 119*; Shay locomotive, *16*, 17, *114, 129*
Emporium Forestry Company logging railroad, 113–18, 123–25, *129*
Emporium Lumber Company, 106
Erhart, William, 132
Everton Lumber Company, 163
Everton village, 162–64
Fernow, Bernhard E., 176–81, *177*
Flanagan, Charles W., 142
Fletcher, Charles, 82
Ford, Royal B. and Heman R., 222
Forest Land and Mill Company, 164
Forestport village, 8, 71
Fournier, Noah, 86
Fox, William F., 7
Fox, Weston and Bronson Company, 7
Franks, Bert, 97
Fuller, Clinton, 11, 118

Gardeau, Pa., 22
Gaylord, F. A., 100
Gays, Henry, 177
Goodyear Lumber Company, 237
Gould, Harry P., 94
Gould Paper Company, 51, 75, 144
Granere, Pa., 22
Grasse River Railroad, 34, 35, 118–22, *120, 124*, 135
Graves, Henry C., 99
Greco, Frank, 114
Greenly, Marshall, 237

Hamele, J. Otto, 227, *227*, 235
Hanley, Dick, *226*, 227
Hanna Ore Company, 220
Harvey, George, 70, 83
Haynes, Howard, 187
Hazelet, Irving, 122
Heltman, A. S., 108
Hemlock bark, 43, *44*
Henry, Lloyd, 241
Herkimer village, 70
Herkimer, Newport and Poland Railroad, 66
Heywood-Wakefield Company, 135–37, *136*
Higbie, Robert, 215
Holt tractor, 52, 53, 83, *83*
Horse Shoe Forestry Company Railroad, 101, 102, *102, 104*, 243
Horseshoe Lake, 101
Hotchkiss, Charles, 148, 152
Hull, William C., 191, 194, 200
Huntington, Collis P., 77, 78
Hurd, John, 7, 67, 147–61, *158*

International Paper Company, 80, 161
Iroquois Pulp and Paper Company, 75

jackworks, 25, 26, 92, 93
J. and J. Rogers Company, 50

Jerseyfield Lumber Company, 237, 241; locomotives, 239, *239*; railroad, 238–40
John McDonald Lumber Company, 230
Johnson, Fred, 48, 92
Johnston, John E., 94, 100, 129, *201*
Jones, E. J., 113

Kalurah settlement, 205
Kellogg, William N., 207, *208*
Keys, William, 239
Keystone Railroad, 221
Kildare settlement, 191, *191*, *192*, *198*
Kinsley Lumber Company, 142–44
Kivlen, John S., 71

LaFountain, George, 27, 92, 98, 186
LaFountain, Harte, 129
Lake Clear Junction station, 140, *140*
Lamora, Oliver, 183
Landry, Joe, 84
LaPlatney, A. M., 128, 133
LaVoy, Roy, *192*, 199, *201*
LeBoeuf, Albert, 150
Leighton, George, 178
Lewis, Crawford and Company, 8
Lidgerwood skidder, 198
Linn tractor, 52, *52*, 131, *133*, 187, *201*
Little Falls and Dolgeville Railroad, 237
Little Independence Lake, 86
Little Rapids Lumber Company, 94
locomotive types: Climax, 19, *20*; geared, 18; Heisler, 20, *21*; rod driven, 17; Shay, 18, *19*
logging camps, 38, *39*, *40*, *42*
log cars, 21, 22, 116, *118*
logging tractors, 51–53
log loaders: American, *24*; Barnhart, 23, *24*, 116, *119*, *133*, *224*; Marion, *239*; Raymond, 116, 194, *194*
log sleds, 44–49, *46*, *48*
log slides, 49–50, *50*, *51*, *131*, 198–200, *200*
Lothium, Gene, 122
Low, Abbot Augustus, 101, *104*

Mac-A-Mac Corporation, 89–94, 182; locomotives, *21*, 92, 93, *94*; railroad, 89–91
MacFarlane, Peter, 148
MacFarlane, Ross and Stearns, 162
Madawaska, 152, 154
Madison, Gene, 224
Malone city, 67
Marleau, Thomas, 73
Martin, Elmer B., 84, 92, 98
McAlpin, Benjamin B., 89
McCoy, George A., 108
McDonald, Eugene, 93, 186, 187
McDonald, George, 93, 95, 186
McDonald, Henry, 93, 188
McDonald, John N., 30, 89, 97
McDonald, Sandy, 92, 95
McDonald village, 27, 185
McKeever village, 72
Mecca Lumber Company, 205–6, *206*
Meigs, Ferris, 157
Meigs, Titus B., 153, 158
Merritt, Albert, 73
Mohawk and Malone Railroad, 7, 69
Moira station, 148
Moore, F. L., 211
Moose River Lumber Company, 73–75, *73*, *75*
Morrow, R. R., 84
Mountain View station, *69*
Mount Matumbla, 10, 51, 194
Moynehan, Patrick C., 72, 96

Nehasane Park, 99
New Bridge settlement, 36, 216
Newton Falls and Northern Railroad, 36, 216–18, *216*, *217*
Newton Falls Paper Company, 207, 215; locomotives, 208, *209*; railroad, 208–11, *209*, *211*
New York and Ottawa Railroad, 160
New York Central and Hudson River Railroad, 65, 160, 237
New York Central Railroad, 69
New York State Ranger School, 235, *235*
Noelk, Victor, 128, 135
Northern Adirondack Railroad, 67, 148–60, *150*
Northern New York Railroad, 160

O'Brien, William, 83
Olson, John, 224
Oncheota, 142
O'Neil, William T., 164
Onroful, Frank, 85
Oswegatchie River, 222
Ouderkirk, Firman, 72
Oval Wood Dish Company, 97, 189–91, *190*; locomotives, *28*, *193*, 194; railroad, 191–99, *193*
Owen, Arthur L., 113, 122

Page, Fairchild and Company, 8
Partlow Lake Railroad, 99
Paul Smith's Railroad, 139–41, *140*
Peryea, Joseph, 184, 188
Pinchot, Gifford, 99
Plunkett Webster Company, 94
Pollard, Archie, 116
Pope-Paine Lumber Company, 159
Post and Henderson Company, 212–14, *214*
Potter, Fred A., 96
Potter, John, 73

Race, Andrew E., 227
Raquette Lake Railway, 77–79, *77*
Raquette River Pulp Company, 205
Recknagel, Arthur, 97
Ressler Ed, 116
Rezio, Joe, 233
Rice Veneer Company, 75
Rich, Claude, 233
Rich, Herbert C., 30, 221, 227
Rich Lumber Company, 34, 220–35; locomotives, *19*, *25*, 224; mill complex, 226–29, *226*, *228*, *229*; railroad, 222–25, *222*, *224*
Riley, John P., 83
R. M. Whitney Company, 238, *238*
Rockefeller, William, 142, 164, 183
Rockefeller, William A., 184, 188
Rockefeller, William G., 184
Rome, Watertown and Ogdensbury Railroad, 65
Rondaxe Lake, 76
Ryan and Schleider chip mill, 227

Sabattis settlement, 56
Sagamore Lumber Company, 75
Salamanca Panel Company, 238
Salisbury Center, 34, 237
Salisbury Steel and Iron Company, 237
Santa Clara Lumber Company, 153–54, 157–58
Santa Clara village, 150–53, *151*, 170–71
Saranac Lake village, 68
Setters Brothers veneer mill, 228
Shaw, Luther, 209
Shephard and Morse Lumber Company, 160
Sillman, Loren, 123, 135
Sisson-White Lumber Company, 202
Smith, Paul, 138–39, 141
Snyder, Charles, 76
South Branch Railroad, 221
Spring Cove, 154
sprinkler sleds, 44, *46*
St. Regis Falls and Everton Railroad, 162–64, *163*
St. Regis Falls village, 148, *148*, 165, 171
St. Regis Paper Company, 164, 171
St. Regis River Lumber Company, 151
Stock, John W., 128
Strife, Clarence, 52, 94
Sykes, Frank P., 49, 113, 128
Sykes, George, 118, 134
Sykes, Roy, 113, 120, 125
Sykes, W. Clyde, 113
Sykes, William L., 106, 113, 125, *130*, 134
Sylvan Lake, 99

Tekene, 142
Thendara village, 68
Thomas, Paul, 111, 118, 135
Thomson, Lemon and Edward, 73
Ticonderoga Union Terminal Railroad Company, 140
Todd, John B., 52
Tupper Lake Chemical Company, 181
Tupper Lake Junction (Faust), *156*, 159
Tupper Lake village, 153, 158, 190
Twitchell Creek, 25, 69, 80

Underwood, 27

Vanderbilt, William H., 65
Van Horn, Harry C., 128
Venters, Henry, 228
Vincent, George, 83

Wanakena village, 30–31, 56, 221–22, *232*, 233–34
Watson Page Lumber Company, 165, 166–67, *167*
Wawbeck Corners, 178
Webb, William Seward, 7, 65–67, *66*, 99, 158
Weidmann Stave and Heading Company, 165, 171
Whitney, Cornelius, 96
Whitney, William C., 96, 99
Whitney Industries, 96–98, *97*
Wilbur, T. J., 53, 83
Willson, Leonard G., 226
Wood, M. O., 85
Woods Lake: locomotive, 82–83; railroad, 82–85; settlement, 70, 80–82

Zenger, Roy, 123